Dynamics and Relativity

The Manchester Physics Series

General Editors
F.K. LOEBINGER: F. MANDL: D.J. SANDIFORD:

School of Physics and Astronomy,
The University of Manchester

Properties of Matter:	B. H. Flowers and E. Mendoza
Statistical Physics: *Second Edition*	F. Mandl
Electromagnetism: *Second Edition*	I. S. Grant and W. R. Phillips
Statistics:	R. J. Barlow
Solid State Physics: *Second Edition*	J. R. Hook and H. E. Hall
Quantum Mechanics:	F. Mandl
Computing for Scientists:	R. J. Barlow and A. R. Barnett
The Physics of Stars: *Second Edition*	A. C. Phillips
Nuclear Physics:	J. S. Lilley
Introduction to Quantum Mechanics:	A. C. Phillips
Particle Physics: *Third Edition*	B. R. Martin and G. Shaw
Dynamics and Relativity:	J. R. Forshaw and A. G. Smith
Vibrations and Waves:	G.C. King

DYNAMICS AND RELATIVITY

Jeffrey R. Forshaw and A. Gavin Smith
School of Physics and Astronomy,
The University of Manchester, Manchester, U.K.

A John Wiley and Sons, Ltd., Publication

This edition first published 2009
© 2009 John Wiley & Sons Ltd

Registered office
John Wiley & Sons Ltd, The Atrium, Southern Gate, Chichester, West Sussex, PO19 8SQ, United
Kingdom

For details of our global editorial offices, for customer services and for information about how to apply
for permission to reuse the copyright material in this book please see our website at www.wiley.com.

Library of Congress Cataloging-in-Publication Data

Forshaw, J. R. (Jeffrey Robert), 1968 –
 Dynamics and relativity/Jeffrey R. Forshaw and A. Gavin Smith.
   p. cm.
 Includes bibliographical references and index.
 ISBN 978-0-470-01459-2 (cloth : alk. paper) – ISBN 978-0-470-01460-8 (pbk. : alk. paper)
 1. Special relativity (Physics) 2. Dynamics. I. Smith, A. Gavin. II. Title.
 QC173.65.F67 2009
 530.11 – dc22
                                                                            2008053366

A catalogue record for this book is available from the British Library.

ISBN 978-0-470-01459-2 (HB)
ISBN 978-0-470-01460-8 (PB)

Typeset in 10/12 Times by Laserwords Private Limited, Chennai, India

Dedicated to the memory of Howard North and Edward Swallow.

Contents

Editors' Preface to the Manchester Physics Series

The Manchester Physics Series is a series of textbooks at first degree level. It grew out of our experience at the University of Manchester, widely shared elsewhere, that many textbooks contain much more material than can be accommodated in a typical undergraduate course; and that this material is only rarely so arranged as to allow the definition of a short self-contained course. In planning these books we have had two objectives. One was to produce short books so that lecturers would find them attractive for undergraduate courses, and so that students would not be frightened off by their encyclopaedic size or price. To achieve this, we have been very selective in the choice of topics, with the emphasis on the basic physics together with some instructive, stimulating and useful applications. Our second objective was to produce books which allow courses of different lengths and difficulty to be selected with emphasis on different applications. To achieve such flexibility we have encouraged authors to use flow diagrams showing the logical connections between different chapters and to put some topics in starred sections. These cover more advanced and alternative material which is not required for the understanding of latter parts of each volume.

Although these books were conceived as a series, each of them is self-contained and can be used independently of the others. Several of them are suitable for wider use in other sciences. Each Author's Preface gives details about the level, prerequisites, etc., of that volume.

The Manchester Physics Series has been very successful since its inception 40 years ago, with total sales of more than a quarter of a million copies. We are extremely grateful to the many students and colleagues, at Manchester and elsewhere, for helpful criticisms and stimulating comments. Our particular thanks go to the authors for all the work they have done, for the many new ideas they have contributed, and for discussing patiently, and often accepting, the suggestions of the editors.

Finally we would like to thank our publishers, John Wiley & Sons, Ltd., for their enthusiastic and continued commitment to the Manchester Physics Series.

F. K. Loebinger
F. Mandl
D. J. Sandiford
August 2008

Authors' Preface

In writing this book, our goal is to help the student develop a good understanding of classical dynamics and special relativity. We have tried to start out gently: the first part of the book aims to provide the solid foundations upon which the second half builds. In the end, we are able, in the final chapter, to cover some quite advanced material for a book at this level (when we venture into the terrain of Einstein's General Theory of Relativity) and it is our hope that our pedagogical style will lead the keen student all the way to the denouement. That said, we do not assume too much prior knowledge. A little calculus, trigonometry and some exposure to vectors would help but not much more than that is needed in order to get going. We have in mind that the first half of the book covers material core to a typical first year of undergraduate studies in physics, whilst the second half covers material that might appear in more advanced first or second year courses (e.g. material such as the general rotation of rigid bodies and the role of four-vectors in special relativity).

The classical mechanics of Newton and the theory of relativity, developed by Einstein, both make assumptions as to the structure of space and time. For Newton time is an absolute, something to be agreed upon by everyone, whilst for Einstein time is more subjective and clocks tick at different rates depending upon where they are and how they are moving. Such different views lead to different physics and by presenting Newtonian mechanics alongside relativity, as we do in this book, it becomes possible to compare and contrast the two. Of course, we shall see how Newtonian physics provides a very good approximation to that of Einstein for most everday phenomena, but that it fails totally when things whizz around at speeds approaching the speed of light.

In this era of electronic communications and online resources that can be researched at the push of a button, it might seem that the need for textbooks is diminished. Perhaps not surprisingly we don't think that is the case. Quiet time spent with a textbook, some paper and a pen, reading and solving problems, is probably still the best way to do physics. Just as one cannot claim to be a pianist without playing a piano, one cannot claim to be a physicist without solving physics problems. It is a point much laboured, but it is true nonetheless. The problems that really help develop understanding are usually those that take time to crack. The painful process of failing to solve a problem is familiar to every successful physicist, as is the excitement of figuring out the way forward. Our advice when solving the problems in this book is to persevere for as long as possible before peeking at the solution, to try and enjoy the process and not to panic if you cannot see how to start a problem.

We have deliberately tried to keep the figures as simple as possible. A good drawing can often be an important step to solving a physics problem, and we encourage you to make them at every opportunity. For that reason, we have illustrated the book with the sorts of drawings that we would normally use in lectures or tutorials and have deliberately avoided the sort of embellishments that would undoubtedly make the book look prettier. Our aim is to present diagrams that are easy to reproduce.

A comment is in order on our usage of the word "classical". For us "classical" refers to physics pre-Einstein but not everyone uses that terminology. Sometimes, classical is used to refer to the laws of physics in the absence of quantum mechanics and in that sense, special relativity could be said to be a classical theory. We have nothing to say about the quantum theory in this book, except that quantum theories that are also consistent with relativity lie at the very heart of modern physics. Hopefully this book will help whet the appetite for further studies in that direction.

We should like to express our gratitude to all those who have read the manuscript and provided helpful suggestions. In particular we thank Rob Appleby, Richard Battye, Mike Birse, Brian Cox, Joe Dare, Fred Loebinger, Nicola Lumley, Franz Mandl, Edward Reeves, David Sandiford and Martin Yates.

Finally, we would like to express particular gratitude to our parents, Thomas & Sylvia Forshaw and Roy & Marion Smith, for their constant support. For their love and understanding, our heartfelt thanks go to Naomi, Isabel, Jo, Ellie, Matt and Josh.

Jeffrey R. Forshaw
A. Gavin Smith
October 2008

Part I

Introductory Dynamics

1

Space, Time and Motion

1.1 DEFINING SPACE AND TIME

If there is one part of physics that underpins all others, it is the study of motion. The accurate description of the paths of celestial objects, of planets and moons, is historically the most celebrated success of a classical mechanics underpinned by Newton's laws[1]. The range of applicability of these laws is vast, encompassing a scale that extends from the astronomical to the microscopic. We have come to understand that many phenomena not previously associated with motion are in fact linked to the movement of microscopic objects. The absorption and emission spectra of atoms and molecules arise as a result of transitions made by their constituent electrons, and the random motion of ensembles of atoms and molecules forms the basis for the modern statistical description of thermodynamics. Although atomic and subatomic objects are properly described using quantum mechanics, an understanding of the principles of classical mechanics is essential in making the conceptual leap from continuous classical systems with which we are most familiar, to the discretised quantum mechanical systems, which often behave in a manner at odds with our intuition. Indeed, the calculational techniques that are routinely used in quantum mechanics have their roots in the classical mechanics of particles and waves; a close familiarity with their use in classical systems is an asset when facing problems of an inherently quantum mechanical nature.

As we shall see in the second part of this book, when objects move at speeds approaching the speed of light classical notions about the nature of space and time fail us. As a result, the classical mechanics of Newton should be viewed as a low-velocity approximation to the more accurate relativistic theory of Einstein[2]. To look carefully at the differences between relativistic and non-relativistic theories

---

[1] After Isaac Newton (1643–1727).
[2] Albert Einstein (1879–1955).

Dynamics and Relativity   Jeffrey R. Forshaw and A. Gavin Smith
© 2009 John Wiley & Sons, Ltd

forces us to recognise that our intuitive ideas about how things move are often incorrect. At the most fundamental level, mechanics of either the classical or the quantum kind, in either the relativistic or non-relativistic limit, is a study of motion and to study motion is to ask some fundamental questions about the nature of space and time. In this book we will draw out explicitly the different underlying structures of space and time used in the approaches of Newton and Einstein.

1.1.1 Space and the classical particle

We all have strong intuitive ideas about space, time and motion and it is precisely because of this familiarity that we must take special care in our attempts to define these fundamental concepts, so as not to carry too many unrecognised assumptions along with us as we develop the physics. So let us start by picking apart what we mean by position. We can usually agree what it means for London to be further away than Inverness and we all know that in order to go to London from Inverness we must also know the direction in which to travel. It may also seem to be fairly uncontentious that an object, such as London, has a position that can be specified, i.e. it is assumed that given enough information there will be no ambiguity about where it is. Although this seems reasonable, there is immediately a problem: day-to-day objects such as tennis balls and cities have finite size; there are a number of 'positions' for a given object that describe different parts of the object. Having directions to London may not be enough to find Kings Cross station, and having directions to Kings Cross station may not be enough to find platform number nine. To unambiguously give the position of an object is therefore only possible if the object is very small – vanishingly small, in fact. This hypothetical, vanishingly small object is called a particle. It might be suggested that with the discovery of the substructure of the atom, true particles, with mass but no spatial extent, have been identified. However, at this level, the situation becomes complicated by quantum uncertainty which makes the simultaneous specification of position and momentum impossible. The classical particle is therefore an idealisation, a limit in which the size of an object tends to zero but in which we ignore quantum phenomena. Later we shall see that it is possible to define a point called the centre of mass of an extended object and that this point behaves much like a classical particle. The collection of all possible positions for a particle forms what we call space.

The mathematical object possessing the properties we require for the description of position is called the vector. A vector has both magnitude and direction and we must be careful to distinguish it from a pure number which has a magnitude, but no directional properties. The paradigm for the vector comes from the displacement of a particle from point A to point B as shown in Figure 1.1. The displacement from A to B is represented by the directed-line-segment AB. We can imagine specifying the displacement as, for example, "start at A and move 3 km to the northeast" or "start at A and go 1 parsec in the direction of Alpha Centuri". Once we have specified a displacement between the two points A and B we can imagine sliding each end of the line segment in space until it connects another two points C and D. To do this, we move each end through the same distance and in the same direction,

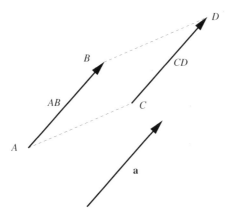

Figure 1.1   Displacement of a particle from point *A* to point *B* is illustrated by the directed line segment *AB*. Parallel transport of this line gives the displacement from point *C* to point *D*. The displacement vector **a** is not associated with any particular starting point.

an operation that is known as parallel transport. Now the displacement is denoted *CD* but its direction and magnitude are the same. It should be clear that there is an infinity of such displacements that may be obtained by parallel transport of the directed line segment. The displacement vector **a** has the magnitude and direction common to this infinite set of displacements but is not associated with a particular position in space. This is an important point which sometimes causes confusion since vectors are illustrated as directed line segments, which appear to have a well defined beginning and an end in space: A vector has magnitude and direction but not location. The position of a particle in space may be given generally by a position vector **r** only in conjunction with a fixed point of origin.

Now, all of this assumes that we understand what it means for lines to be parallel. At this point we assume that we are working in Euclidean space, which means that parallel lines remain equidistant everywhere, i.e. they never intersect. In non-Euclidean spaces, such as the two-dimensional surface of a sphere, parallel lines do intersect[3] and extra mathematics is required to specify how local geometries are transported to different locations in the space. For the moment, since we have no need of non-Euclidean geometry, we will rest our discussion of vectors firmly on the familiar Euclidean notion of parallel lines. Later, when we consider the space-time geometry associated with relativistic motion we will be forced to drop this deep-rooted assumption about the nature of space.

So far, we have been considering only vectors that are associated with displacements from one point to another. Their utility is far more wide ranging than that though: vectors are used to represent other interesting quantities in physics. For example the electric field strength in the vicinity of an electric charge is correctly represented by specifying both its magnitude and direction, i.e. it is a vector. Since it is important to maintain the distinction between vectors and ordinary numbers

---

[3] For example lines of longitude meet at the poles.

(called scalars) we identify vector quantities in this book by the use of bold font. When writing vectors by hand it is usual to either underline the vector, or to put an arrow over the top. Thus

$$\underline{a} \equiv \vec{a} \equiv \mathbf{a}.$$

Use the notation that you find most convenient, but always maintain the distinction between vector and scalar quantities. In this book both upper case (**A**) and lower case (**a**) notion will be used for vectors where **A** is in general a different vector from **a**. When a vector has zero magnitude it is impossible to define its direction; we call such a vector the null vector **0**.

1.1.2   Unit vectors

The length of a vector **a** is known as its magnitude, often denoted $|\mathbf{a}|$. To simplify the notation we shall adopt the convention that vectors are printed in bold and their magnitudes are indicated by dropping the bold font, thus $a \equiv |\mathbf{a}|$. Often we will separate the magnitude and direction of a vector, writing

$$\mathbf{a} = a\hat{\mathbf{a}},$$

where $\hat{\mathbf{a}}$ is the vector of unit magnitude with the same direction as **a**. Unit vectors, of which $\hat{\mathbf{a}}$ is an example, are often used to specify directions such as the directions of the axes of a co-ordinate system (see below).

1.1.3   Addition and subtraction of vectors

The geometrical rules for adding and subtracting vectors are illustrated in Figure 1.2. Addition of the vectors **A** and **B** involves sliding the vectors until they are "head-to-tail", so that the resultant vector connects the tail of **A** to the head of **B**. The vector $-\mathbf{A}$ is defined as a vector with the same magnitude, but opposite direction to **A**. The difference $\mathbf{B} - \mathbf{A}$ is constructed by adding **B** and $-\mathbf{A}$ as shown. Subtraction of a vector from itself gives the null vector:

$$\mathbf{A} - \mathbf{A} = \mathbf{0}.$$

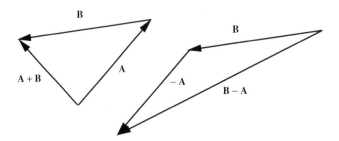

Figure 1.2   Adding and subtracting vectors.

1.1.4 Multiplication of vectors

There are two types of vector multiplication that are useful in classical physics. The scalar (or dot) product of two vectors **A** and **B** is defined to be

$$\mathbf{A} \cdot \mathbf{B} = AB \cos \theta, \tag{1.1}$$

This scalar quantity (a pure number) has a simple geometrical interpretation. It is the projection of **B** on **A**, i.e. $B \cos \theta$, multiplied by the length of **A** (see Figure 1.3). Equally, it may be thought of as the projection of **A** on **B**, i.e. $A \cos \theta$, multiplied by the length of **B**. Clearly the scalar product is insensitive to the order of the vectors and hence $\mathbf{A} \cdot \mathbf{B} = \mathbf{B} \cdot \mathbf{A}$. The scalar product takes its maximum value of AB when the two vectors are parallel, and it is zero when the vectors are mutually perpendicular.

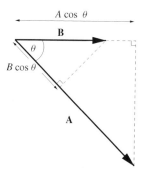

Figure 1.3 Geometry of the scalar product. $\mathbf{A} \cdot \mathbf{B}$ is the product of the length of **A**, and the projection of **B** onto **A** or alternatively the product of the length of **B**, and the projection of **A** onto **B**.

The vector (or cross) product is another method of multiplying vectors that is frequently used in physics. The cross product of vectors **A** and **B** is defined to be

$$\mathbf{A} \times \mathbf{B} = AB \sin \theta \, \hat{\mathbf{n}}, \tag{1.2}$$

where θ is the angle between **A** and **B** and $\hat{\mathbf{n}}$ is a unit vector normal to the plane containing both **A** and **B**. Whether $\hat{\mathbf{n}}$ is 'up' or 'down' is determined by convention and in our case we choose to use the right-hand screw rule; turning the fingers of the right hand from **A** to **B** causes the thumb to point in the sense of $\hat{\mathbf{n}}$ as is shown in Figure 1.4. Interchanging the order of the vectors in the product means that the fingers of the right hand curl in the opposite sense and the direction of the thumb is reversed. So we have

$$\mathbf{B} \times \mathbf{A} = -\mathbf{A} \times \mathbf{B}. \tag{1.3}$$

The magnitude of the vector product $AB \sin \theta$ also has a simple geometrical interpretation. It is the area of the parallelogram formed by the vectors **A** and **B**. Alternatively it can be viewed as the magnitude of one vector times the projection

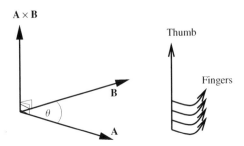

Figure 1.4   Vector product of **A** and **B**.

of the second on an axis which is perpendicular to the first and which lies in the plane of the two vectors. It is this second geometric interpretation that has most relevance in dynamics. As we shall see later, moments of force and momentum involve this type of perpendicular projection. In this book the vector product will find its principal application in the study of rotational dynamics.

The scalar and vector products are interesting to us precisely because they have a geometrical interpretation. That means they represent real things is space. In a sense, we can think of the scalar product as a machine that takes two vectors as input and returns a scalar as output. Similarly the vector product also takes two vectors as its input but instead returns a vector as its output. There are in fact no other significantly different[4] machines that are able to convert two vectors into scalar or vector quantities and as a result you will rarely see anything other than the scalar and vector products in undergraduate/college level physics. There is in fact a machine that is able to take two vectors as its input and return a new type of geometrical object that is neither scalar nor vector. We will even meet such a thing later in this book when we encounter tensors in our studies of advanced dynamics and advanced relativity.

1.1.5   Time

We are constantly exposed to natural phenomena that recur: the beat of a pulse; the setting of the Sun; the chirp of a cricket; the drip of a tap; the longest day of the year. Periodic phenomena such as these give us a profound sense of time and we measure time by counting periodic events. On the other hand, many aspects of the natural world do not appear to be periodic: living things die and decay without rising phoenix-like from their ashes to repeat their life-cycle; an egg dropped on the floor breaks and never spontaneously re-forms into its original state; a candle burns down but never up. There is a sense that disorder follows easily from order, that unstructured things are easily made from structured things but that the reverse is much more difficult to achieve. That is not to say that it is impossible to create order from disorder – you can do that by tidying your room – just that on average

---

[4] i.e. other than trivial changes such as would occur if we choose instead to define the scalar product to be $\mathbf{A} \cdot \mathbf{B} = \lambda AB \cos \theta$ where λ is a constant. We choose $\lambda = 1$ because it is most convenient but any other choice is allowed provided we take care to revise the geometrical interpretation accordingly.

things go the other way with the passing of time. This idea is central to the study of thermodynamics where the disorder in a system is a measurable quantity called entropy. The total entropy of the Universe appears always to increase with time. It is possible to decrease the entropy (increase the order) of a part of the Universe, but only at the expense of increasing the entropy of the rest of the Universe by a larger amount. This net disordering of the Universe is in accord with our perception that time has a direction. We cannot use natural processes to "wind the clock back" and put the Universe into the state it was in yesterday – yesterday is truly gone forever. That is not to say that the laws of physics forbid the possibility that a cup smashed on the floor will spontaneously re-assemble itself out of the pieces and leap onto the table from which it fell. They do not; it is simply that the likelihood of order forming spontaneously out of disorder like this is incredibly small. In fact, the laws of physics are, to a very good approximation, said to be "time-reversal invariant". The exception occurs in the field of particle physics where "CP-violation" experiments indicate that time-reversal symmetry is not respected in all fundamental interactions. This is evidence for a genuine direction to time that is independent of entropy. Entropy increase is a purely statistical effect, which occurs even when fundamental interactions obey time-reversal symmetry.

Thermodynamics gives us a direction to time and periodic events allow us to measure time intervals. A clock is a device that is constructed to count the number of times some recurring event occurs. A priori there is no guarantee that two clocks will measure the same time, but it is an experimental fact that two clocks that are engineered to be the same and which are placed next to each other, will measure, at least approximately, the same time intervals. This approximate equivalence of clocks leads us to conjecture the existence of absolute time, which is the same everywhere. A real clock is thus an imperfect means of measuring absolute time and a good clock is one that measures absolute time accurately. One problem with this idea is that absolute time is an abstraction, a theoretical idea that comes from an extrapolation of the experimental observation of the similar nature of different clocks. We can only measure absolute time with real clocks and without some notion of which clocks are better than others we have no handle on absolute time. One way to identify a reliable clock is to build lots of copies of it and treat all the copies exactly the same, i.e. put them in the same place, keep them at the same temperature and atmospheric conditions etc. If it is a reliable clock the copies will deviate little from each other over long time intervals. However, a reliable clock is not necessarily a good clock; similarly constructed clocks may run down in similar ways so that, for example, the time intervals between ticks might get longer the longer a clock runs, but in such a way that the similar clocks still read the same time. We can get around this by comparing equally reliable clocks based on different mechanisms. If enough equally-reliable clocks, based on enough different physical processes, all record the same time then we can start to feel confident that there is such a thing as absolute time. It is worth pointing out that in the 17th century reliable clocks were hard to come by and Newton certainly did not come to the idea of absolute time as a result of the observation of the constancy of clocks. Newton had an innate faith in the idea of absolute time and constructed his system of mechanics on that basis.

There is no doubt that absolute time is a useful concept; in this book we shall at first examine the motion of things under the influence of forces, treating time as though it is the same for every observer, and we will get answers accurate to a high degree. However, absolute time is a flawed concept, but flawed in such a way that the cracks only begin to appear under extreme conditions. We shall see later how clocks that are moving at very high relative velocities do *not* record the same time and that time depends on the state of motion of the observer. Einstein's Special Theory of Relativity tells us how to relate the time measured by different observers although the deviations from absolute time are only important when things start to move around at speeds approaching the speed of light. In describing the motion of things that do not approach the speed of light we can ignore relativistic effects with impunity, avoiding the conceptual and computational complications that arise from a full relativistic treatment. This will allow us to focus on concepts such as force, linear and angular momentum and energy. Once the basic concepts of classical mechanics have been established we will move on to study Special Relativity in Part II. Even then we will not completely throw out the concepts that are so successful in classical mechanics. Rather, these shall be adapted into the more general ideas of energy, momentum, space and time that are valid for all speeds.

1.1.6 Absolute space and space-time

At a fundamental level, the natural philosophy of Aristotle and the physics of Newton differ from the physics of Galileo[5] and Einstein in the way that space and time are thought to be connected. One very basic question involves whether space can be thought of as absolute. Consider the corner of the room you might be sitting in. The intersection of the two walls and the ceiling of a room certainly defines a point, but will this point be at the same place a microsecond later? We might be tempted to think so, that is, until the motion of the Earth is considered; the room is hurtling through space and so is our chosen point. Clearly the corner of the room defines a 'different' point at each instant. So would it be better to define a 'fixed' point with reference to some features of the Milky Way? This might satisfy us, at least until we discover that the Milky Way is moving relative to the other galaxies, so such a point cannot really be regarded as fixed. We find it difficult to escape completely from the idea that there is some sort of fixed background framework with respect to which we can measure all motion, but there is, crucially, no experimental evidence for this structure. Such a fixed framework is known as absolute space.

The concept of absolute space, which originates with Aristotle and his contemporaries, can be represented geometrically as shown in Figure 1.5(a). Here we have time as another Cartesian axis, tacked onto the spatial axes to produce a composite space that we call space-time. Consider two things that happen at times and positions that are measured using clocks and co-ordinate axes. We call these happenings 'events' and mark them on our space-time diagram as *A* and *B*. In the picture of absolute space, if the spatial co-ordinates of events *A* and *B* are identical we say that they represent the same point in space at different times. We can construct a path shown by the dotted line that connects the same point in space for all times. Galilean

---

[5] Galileo Galilei (1564–1642).

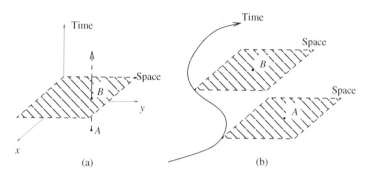

Figure 1.5 Different structures of space and time: (a) absolute space where points A and B are the same point in space; (b) a fibre-bundle structure where each moment in time has its own space.

relativity challenges this picture by rejecting the notion of absolute space and replacing it with the idea that space is defined relative to some chosen set of axes at a given instant in time. This is more like the picture in Figure 1.5(b), a structure that mathematicians call a fibre bundle. The same events A and B now lie in different spaces and the connection between them is no longer obvious. The fibre bundle is a more abstract structure to deal with than the space × time structure of (a). Imagine, for example, trying to calculate the displacement from A to B. To do this we have to assume some additional structure of space-time that allows us to compare points A and B. It is as if space is erased and redefined at each successive instant and we have no automatic rule for saying how the 'new' space relates to the 'old' one. Notice that this view still treats time as absolute; observers at different points in the $x - y$ plane agree on the common time t. In later chapters we will reconsider the geometry of space and time when we come to study the theory of Special Relativity, where universal time will be rejected in favour of a new space-time geometry in which observers at different positions each have their own local time.

1.2   VECTORS AND CO-ORDINATE SYSTEMS

As far as we can tell, space is three-dimensional, which means that three numbers are required to define a unique position. How we specify the three position-giving numbers defines what is known as the co-ordinate system. The co-ordinate system therefore introduces a sort of invisible grid or mesh that maps every point in space onto a unique ordered set of three real numbers. Figure 1.6 shows two commonly-used 3-dimensional co-ordinate systems. The Cartesian system is named after the French philosopher and mathematician René Descartes (1596–1650), who is reputed to have invented it from his bed while considering how he might specify the position of a fly that was buzzing around his room. This co-ordinate system consists of three mutually perpendicular axes, labelled x, y and z, that intersect at the point O, called the origin. The position of a particle at P may be specified by giving the set of three distances (x, y, z). Another frequently used co-ordinate system, the spherical-polar system, is obtained when the position of the particle is given instead by the distance from the origin r and two angles: the polar angle

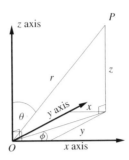

Figure 1.6   Two 3-dimensional co-ordinate systems covering the same space. The Cartesian co-ordinates consist of the set (x, y, z). The spherical polar co-ordinates consist of the set (r, θ, ϕ).

θ and the azimuthal angle ϕ. The Cartesian and the spherical polar systems are just two possible ways of mapping the same space, and it should be clear that for any given physical problem there will be an infinite number of equally-valid co-ordinate systems. The decision as to which one to use is based on the nature of the problem, and the ease or difficulty of the calculation that results from the choice.

Choosing a co-ordinate system immediately gives us a way to represent vectors. Associated with any co-ordinate system are a set of unit vectors known as basis vectors. Each co-ordinate has an associated basis vector that points in the direction in which that co-ordinate is increasing. For example, in the 3D Cartesian system $\mathbf{i}$ points in the direction of increasing x, i.e. along the x-axis, while $\mathbf{j}$ and $\mathbf{k}$ point along the $y-$ and $z-$axes, respectively. Suppose that the position of a particle relative to the origin is given by the vector $\mathbf{r}$, known as the 'position vector' of the particle. Then $\mathbf{r}$ can be written in terms of the Cartesian basis vectors as

$$\mathbf{r} = x\mathbf{i} + y\mathbf{j} + z\mathbf{k}, \tag{1.4}$$

where the numbers (x, y, z) are the Cartesian co-ordinates of the particle. The magnitude of the position vector, which is the distance between the particle and the origin, can be calculated by Pythagoras' Theorem and is

$$r = \sqrt{\mathbf{r} \cdot \mathbf{r}} = \sqrt{x^2 + y^2 + z^2}. \tag{1.5}$$

We have focussed upon a position vector in the Cartesian basis but we could have talked about a force, or an acceleration or a magnetic field etc. Any vector $\mathbf{A}$ can be expressed in terms of its components (A_x, A_y, A_z) according to

$$\mathbf{A} = A_x\mathbf{i} + A_y\mathbf{j} + A_z\mathbf{k}. \tag{1.6}$$

It is not our aim here to present a full discussion of the algebraic properties of vectors. Some key results, which will prove useful later are listed in Table 1.1.

Very often, the motion of an object may be constrained to a known plane, such as in the case of a ball on a pool table, or a planet in orbit around the Sun. In such situations the full 3D co-ordinate system is not required and a two-dimensional

TABLE 1.1 Vector operations in the Cartesian basis. **A** and **B** are vectors, λ is a scalar.

Operation	Notation	Resultant
Negation	$-\mathbf{A}$	$(-A_x)\mathbf{i} + (-A_y)\mathbf{j} + (-A_z)\mathbf{k}$
Addition	$\mathbf{A} + \mathbf{B}$	$(A_x + B_x)\mathbf{i} + (A_y + B_y)\mathbf{j} + (A_z + B_z)\mathbf{k}$
Subtraction	$\mathbf{A} - \mathbf{B}$	$(A_x - B_x)\mathbf{i} + (A_y - B_y)\mathbf{j} + (A_z - B_z)\mathbf{k}$
Scalar (Dot) Product	$\mathbf{A} \cdot \mathbf{B}$	$A_x B_x + A_y B_y + A_z B_z$
Vector (Cross) Product	$\mathbf{A} \times \mathbf{B}$	$(A_y B_z - A_z B_y)\mathbf{i} + (A_z B_x - A_x B_z)\mathbf{j} + (A_x B_y - A_y B_x)\mathbf{k}$
Scalar Multiplication	$\lambda \mathbf{A}$	$\lambda A_x \mathbf{i} + \lambda A_y \mathbf{j} + \lambda A_z \mathbf{k}$

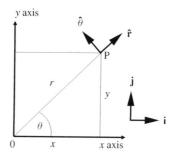

Figure 1.7 2D co-ordinate systems. The Cartesian co-ordinates consist of the set (x, y). The plane polar co-ordinates consist of the set (r, θ).

system may be used. Two of these are shown in Figure 1.7. The Cartesian 2D co-ordinate system has basis vectors **i** and **j** and co-ordinates (x, y). The plane-polar co-ordinates are (r, θ)[6] where

$$r = \sqrt{x^2 + y^2} \quad \text{and} \quad \theta = \tan^{-1} \frac{y}{x}. \tag{1.7}$$

The plane-polar system has basis vectors $\hat{\mathbf{r}}$ and $\hat{\boldsymbol{\theta}}$. These may be expressed in terms of **i** and **j** as

$$\hat{\mathbf{r}} = \mathbf{i} \cos\theta + \mathbf{j} \sin\theta,$$

$$\hat{\boldsymbol{\theta}} = -\mathbf{i} \sin\theta + \mathbf{j} \cos\theta. \tag{1.8}$$

The general position vector in the plane may therefore be written as

$$\mathbf{r} = r\hat{\mathbf{r}} = r(\mathbf{i} \cos\theta + \mathbf{j} \sin\theta) = x\mathbf{i} + y\mathbf{j}. \tag{1.9}$$

Some care is required when using polar co-ordinates to describe the motion of a particle since the basis vectors depend on the co-ordinate θ, which may itself depend on time. This means that as the particle moves, the basis vectors change direction.

---

[6] Note the conventional use of θ for the angle to the x axis rather than ϕ, which is used for the corresponding angle in the spherical (3D) polar system.

This will lead to more complicated expressions for velocity and acceleration in polar co-ordinates than are obtained for Cartesian co-ordinates, as will be seen in the next section.

1.3 VELOCITY AND ACCELERATION

A particle is in motion when its position vector depends on time. The Ancient Greek philosophers had problems accepting the idea of a body being both in motion, and being 'at a point in space' at the same time. Zeno, in presenting his 'runner's paradox', divided up the interval between the start and finish of a race to produce an infinite sum for the total distance covered. He argued that before the runner completes the full distance (l) he must get half-way, and before he gets to the end of the second half he must get to half of that length and so on. The total distance covered can therefore be written as the infinite series

$$l\left[\frac{1}{2} + \frac{1}{4} + \frac{1}{8} + \cdots\right].$$

Zeno argued that it would be impossible for the runner to cover all of the sub-stretches in a finite time, and would therefore never get to the finish line. This contradiction forced him to decide that motion is impossible and that what we perceive as motion must be an illusion. We now know that the resolution of this paradox lies in an understanding of calculus. As the series continues, the steps get shorter and shorter, as do the time intervals taken for the runner to cover each step and we tend to a situation in which a vanishingly short distance is covered in a vanishingly small time.

Assuming that the position is a smooth function of time, we define the velocity as

$$\mathbf{v}(t) = \frac{d\mathbf{r}(t)}{dt} = \lim_{\Delta t \to 0}\left(\frac{\mathbf{r}(t + \Delta t) - \mathbf{r}(t)}{\Delta t}\right). \tag{1.10}$$

Notice that it involves a difference in the position vector at time $t + \Delta t$ and at time t. This difference, divided by the time interval Δt, only becomes the velocity *in the limit* that Δt goes to zero. Thus the velocity is defined in terms of an infinitesimally small displacement divided by an infinitesimally small time interval. Notice that the vector nature of $\mathbf{v}$ follows directly from the vector nature of $\mathbf{r}(t + \Delta t) - \mathbf{r}(t)$, which differs from $\mathbf{v}$ only by division by the scalar Δt. Often it is useful to refer to the magnitude of the velocity; this is known as the speed v, i.e.

$$v = |\mathbf{v}|.$$

With the notion that the ratio of two infinitesimally small quantities can be a finite number, we return to the Runner's Paradox. Zeno's argument does not rely on the particular choice of infinite series stated above. So we can simplify things by instead using a series made of equal-length steps. First we divide l up into

N equal lengths $\Delta x = \frac{l}{N}$. Assuming that the runner has a constant speed v in a straight line, we can write

$$l = \sum_{i=1}^{N} \Delta x = \sum_{i=1}^{N} \frac{\Delta x}{\Delta t} \Delta t, \qquad (1.11)$$

where Δt is the time taken for the runner to cover the distance Δx. If we now let $N \to \infty$, then $\Delta t \to 0$ and $\frac{\Delta x}{\Delta t} \to v$, so that

$$l = v \sum_{i=1}^{\infty} \Delta t.$$

The time taken to run the whole race is therefore

$$t = \sum_{i=1}^{\infty} \Delta t = \frac{l}{v}.$$

Thus, provided that we are happy that the limit Eq. (1.10) exists, and that v is a non-zero number, then we can explain why the runner finishes the race in a finite time: there is no paradox. We may have laboured the point rather, the bottom line is of course that the distance travelled involves both integration and differentation:

$$l = \int \frac{dx}{dt} \, dt, \qquad (1.12)$$

which works even if the speed is varying from point to point.

Just as velocity captures the rate at which a displacement changes so we introduce the acceleration, in order to quantify the rate of change of velocity:

$$\mathbf{a}(t) = \frac{d\mathbf{v}(t)}{dt} = \frac{d^2\mathbf{r}(t)}{dt^2} = \lim_{\Delta t \to 0} \left(\frac{\mathbf{v}(t + \Delta t) - \mathbf{v}(t)}{\Delta t} \right). \qquad (1.13)$$

Again, $\mathbf{a}$ is a vector since it is defined as a vector divided by a scalar. In the Cartesian system, the velocity and acceleration take on a particularly simple form, since the basis vectors $\mathbf{i}$, $\mathbf{j}$ and $\mathbf{k}$ do not depend on time. Thus, if

$$\mathbf{r}(t) = x(t)\mathbf{i} + y(t)\mathbf{j} + z(t)\mathbf{k},$$

use of the definitions Eq. (1.10) and Eq. (1.13) leads to

$$\mathbf{v}(t) = \frac{dx}{dt}\mathbf{i} + \frac{dy}{dt}\mathbf{j} + \frac{dz}{dt}\mathbf{k},$$

$$\mathbf{a}(t) = \frac{d^2x}{dt^2}\mathbf{i} + \frac{d^2y}{dt^2}\mathbf{j} + \frac{d^2z}{dt}\mathbf{k}. \qquad (1.14)$$

Example 1.3.1 *Calculate the acceleration of a particle with the time-dependent position vector given by*

$$\mathbf{r}(t) = A\sin(\omega t)\mathbf{i} + \frac{1}{2}at^2\mathbf{j}.$$

Solution 1.3.1 *Differentiation once gives*

$$\mathbf{v}(t) = A\omega\cos(\omega t)\mathbf{i} + at\mathbf{j}$$

and again to obtain

$$\mathbf{a}(t) = -A\omega^2\sin(\omega t)\mathbf{i} + a\mathbf{j}.$$

1.3.1   Frames of reference

To describe the motion of a particle we need a position vector and a point of origin. We have seen that a position vector may be represented by components in a given co-ordinate system. The question immediately arises as to how one chooses a co-ordinate system to best suit a given physical situation.

For example, consider a cabin attendant who pushes a trolley along the aisle of an aircraft in flight. For a passenger on the aircraft a natural co-ordinate system to use would be one fixed to the aircraft, perhaps a Cartesian system with one axis pointed along the aisle. On the other hand, an observer on the ground might prefer a co-ordinate system fixed to the Earth. The reason why the observers tend to choose different co-ordinate systems is that each observer is surrounded by a different collection of objects that appear to be stationary. The passenger on the aircraft regards the structure of the aircraft as fixed whereas the observer on the ground regards objects on the Earth as stationary. We say that the passenger and the observer on the ground have different frames of reference.

A frame of reference is an abstraction of a rigid structure. We might think of a collection of particles whose relative positions do not change with time. However, it is not necessary for the particles to actually exist in order to define a frame of reference, we simply understand that the particles *could* exist in some sort of static arrangement that defines the frame of reference. Within a particular frame of reference there is always an infinite choice of co-ordinate systems. For example, if the observer on the ground chooses Cartesian co-ordinates, there are an infinite number of ways in which the axes may be oriented. Alternatively, latitude, longitude and distance from the centre of the Earth may be chosen as the three co-ordinates, with an arbitrary choice of where the meridian lines lie. The choice of co-ordinate system implies a particular frame of reference, but we can discuss frames of reference without commitment to a particular co-ordinate system.

1.3.2   Relative motion

In describing the motion of two particles it is often advantageous to use relative position and velocity vectors. The relative position vector $\mathbf{R}_{ab}(t) = \mathbf{r}_b(t) - \mathbf{r}_a(t)$ is the displacement from the position of particle a to that of particle b and it is, in general, a function of time. We differentiate with respect to time to obtain the relative velocity, $\mathbf{V}_{ab}(t)$,

$$\mathbf{V}_{ab}(t) = \frac{d\mathbf{R}_{ab}(t)}{dt} = \mathbf{v}_b(t) - \mathbf{v}_a(t), \tag{1.15}$$

where $\mathbf{v}_a(t)$ and $\mathbf{v}_b(t)$ are the velocities of particles a and b.

Example 1.3.2 *Consider an air-traffic controller tracking the positions of two aircraft. The controller knows the positions and velocities of the aircraft at some instant in time ($t = 0$). Assuming that the aircraft maintain their velocities, show that the relative velocity can be used to decide whether there is a danger of a collision at some later time.*

Solution 1.3.2 *The relative position vector at $t = 0$ is*

$$\mathbf{R}_0 = \mathbf{R}_{ab}(0) = \mathbf{r}_b(0) - \mathbf{r}_a(0).$$

The relative velocity is computed from the velocities of the aircraft:

$$\mathbf{V}_0 = \mathbf{V}_{ab}(0) = \mathbf{v}_b(0) - \mathbf{v}_a(0).$$

Since the aircraft have constant velocities the relative velocity is also constant and it can be integrated with respect to time to obtain

$$\mathbf{R}_{ab}(t) = \mathbf{R}_0 + \mathbf{V}_0 t.$$

The aircraft will collide if at some time t, $\mathbf{R}_{ab}(t) = \mathbf{0}$, i.e. when $\mathbf{R}_0 = -\mathbf{V}_0 t$. This is a vector equation and it can only be satisfied if both the directions and magnitudes of both sides of the equation are the same. Clearly we can only obtain a solution for $t > 0$ if $\mathbf{R}_0$ and $\mathbf{V}_0$ are anti-parallel i.e. if $\mathbf{R}_0 = R_0 \hat{\mathbf{n}}$ and $\mathbf{V}_0 = -V_0 \hat{\mathbf{n}}$, where R_0 and V_0 are positive magnitudes and $\hat{\mathbf{n}}$ is a unit vector. If the vectors are anti-parallel, the collision time is R_0/V_0.

In the previous example, we worked entirely in the frame of reference in which the air traffic controller is at rest. It is tempting to identify the relative velocity $\mathbf{V}_{ab}$ also as the velocity of the aircraft b relative to the pilot of aircraft a. Strictly speaking we have not proved this: $\mathbf{V}_{ab}$ is the velocity of b relative to a as determined by the air traffic controller, not by the pilot of aircraft a. In classical mechanics, where time is universal, the two are equivalent and specifying the relative velocity between two bodies does not need us to further specify who is doing the observing. That the assumption of universal time enters into this matter can be seen by exploring the expression $\mathbf{V}_{ab}(t) = \frac{d\mathbf{R}_{ab}(t)}{dt}$. Whose time is represented by t? That of the air-traffic controller or that of the pilot in aircraft a? If we accept the concept of absolute time then it doesn't matter and both record the same relative velocity. But we really ought to recognise that the assumption of universal time is just that: an assumption. This is not an irrelevant matter for, as we shall see in Part II, the universality of time breaks down, becoming most apparent when relative velocities start to approach the speed of light.

1.3.3   Uniform acceleration

In many physical situations the acceleration does not change with time. Integration of Eq. (1.13) then gives

$$\mathbf{v} = \int \mathbf{a}\,dt = \mathbf{v}_0 + \mathbf{a}\,t, \tag{1.16}$$

where $\mathbf{v}_0$ is the velocity at time $t = 0$. Since $\mathbf{v}_0$ is a constant vector, integration again yields

$$\mathbf{r} = \mathbf{r}_0 + \mathbf{v}_0 t + \frac{1}{2}\mathbf{a}\,t^2. \tag{1.17}$$

In general the vectors $\mathbf{r}_0$, $\mathbf{v}_0$ and $\mathbf{a}$ will have different directions and each of the vector equations, Eq. (1.16) and Eq. (1.17), is shorthand for three different scalar equations, one for each of the three spatial components. An important simplification occurs in situations where the velocity, acceleration and displacement are all collinear (i.e. all in the same direction). Then we need only consider the components of the vectors along the direction of motion, i.e.

$$v = v_0 + at \tag{1.18}$$

and

$$r = r_0 + v_0 t + \frac{1}{2}at^2. \tag{1.19}$$

Squaring Eq. (1.18) and substituting using Eq. (1.19) yields a third equation that is often useful in solving problems that don't deal explicitly with time:

$$v^2 = v_0^2 + 2a(r - r_0). \tag{1.20}$$

Even if $\mathbf{r}$, $\mathbf{v}$ and $\mathbf{a}$ are not collinear then Eq. (1.18), Eq. (1.19) and Eq. (1.20) can still be applied to each of the Cartesian components of the vectors since the basis vectors $\mathbf{i}$, $\mathbf{j}$ and $\mathbf{k}$ are independent of time.

As an example, let us consider the problem of projectile motion in a uniform gravitational field. Close to the Earth's surface any object accelerates towards the centre of the Earth. This acceleration has magnitude

$$g \approx 9.81 \text{ ms}^{-2}$$

although the exact value depends on where you are on the surface of the Earth. The fact that all objects fall at the same rate is rather amazing, but we will defer a discussion of that until the next chapter. Here we only want to use the result that the acceleration is uniform, which is true so long as we stick to low altitudes and ignore the effects of air resistance.

Example 1.3.3 *Determine the path of a projectile fired with speed u at an angle θ to the horizontal. Neglect air resistance. Use the path to determine the range of the projectile.*

Solution 1.3.3 *The choice the co-ordinate system is up to us. Since we want to separate the description of the motion into Cartesian components, we choose the y-axis to be upwards and the x-axis to be horizontal and in the same plane as the initial velocity. We then have*

$$\mathbf{a} = -g\mathbf{j}, \quad \text{and}$$

$$\mathbf{u} = u_x\mathbf{i} + u_y\mathbf{j} = u\cos\theta\mathbf{i} + u\sin\theta\mathbf{j}.$$

We write the position of the projectile as

$$\mathbf{r} = x\mathbf{i} + y\mathbf{j},$$

where x and y depend on time. For convenience we let $\mathbf{r} = \mathbf{0}$ *at* $t = 0$. *Our choice of co-ordinate system means that there is no acceleration in the x-direction. Thus we have,*

$$x = u_x t = ut\cos\theta.$$

In the y-direction we have

$$y = u_y t - \frac{1}{2}gt^2 = ut\sin\theta - \frac{1}{2}gt^2.$$

These are parametric equations for x and y (with time as the parameter). To obtain the path of the projectile we eliminate t to get y as a function of x:

$$y = \frac{u_y}{u_x}x - \frac{g}{2u_x^2}x^2 = x\tan\theta - \frac{g}{2u^2\cos^2\theta}x^2.$$

This is the equation of a parabola (see Figure 1.8). To obtain the range of the projectile we need to find values of x such that $y = 0$. *These are* $x = 0$, *the launch position, and* $x = (2u^2\sin\theta\cos\theta)/g = (u^2\sin 2\theta)/g$, *the horizontal range of the projectile. Notice that the range is maximal for* $\theta = 45°$.

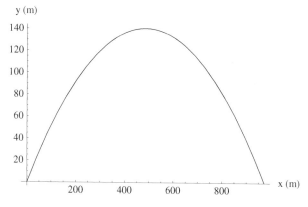

Figure 1.8   Parabolic path of a projectile fired at 30° to the horizontal with an initial speed of 105 ms^{-1}. Note that the distance scales are different on the horizontal and vertical axes.

1.3.4   Velocity and acceleration in plane-polar co-ordinates: uniform circular motion

Circular motion arises frequently in physics. Examples may be as simple as a mass whirled on a string, but also include the orbits of satellites around the Earth, and the motion of charged particles in a magnetic field. Where circular motion is concerned, problems are often most easily solved in polar co-ordinates. In this section we determine the equations for the velocity and acceleration in polar co-ordinates.

The position of a particle moving in a plane is

$$\mathbf{r} = r\hat{\mathbf{r}}. \tag{1.21}$$

As the particle moves, both r and $\hat{\mathbf{r}}$ may change, i.e. they are both implicitly time-dependent. The velocity of the particle is calculated by differentiation of the product $r\hat{\mathbf{r}}$.

$$\mathbf{v} = \frac{d}{dt}(r\hat{\mathbf{r}}) = \frac{dr}{dt}\hat{\mathbf{r}} + r\frac{d\hat{\mathbf{r}}}{dt}. \tag{1.22}$$

Since $\hat{\mathbf{r}} = \cos\theta\,\mathbf{i} + \sin\theta\,\mathbf{j}$,

$$\frac{d\hat{\mathbf{r}}}{dt} = -\sin\theta\frac{d\theta}{dt}\mathbf{i} + \cos\theta\frac{d\theta}{dt}\mathbf{j} = \frac{d\theta}{dt}\hat{\boldsymbol{\theta}}, \tag{1.23}$$

where we have used the definition Eq. (1.8) for $\hat{\boldsymbol{\theta}}$. Thus,

$$\mathbf{v} = \frac{dr}{dt}\hat{\mathbf{r}} + r\frac{d\theta}{dt}\hat{\boldsymbol{\theta}}. \tag{1.24}$$

The tangential contribution $r\frac{d\theta}{dt}\hat{\boldsymbol{\theta}}$ is zero if the particle moves radially (constant θ) whereas the radial velocity $\frac{dr}{dt}\hat{\mathbf{r}}$ is zero for motion in a circle (constant r). We introduce the angular speed $\omega = d\theta/dt$, to simplify the notation. The velocity is then

$$\mathbf{v} = \frac{dr}{dt}\hat{\mathbf{r}} + r\omega\hat{\boldsymbol{\theta}}. \tag{1.25}$$

The general expression for acceleration can be obtained by differentiation of Eq. (1.24) and further application of Eq. (1.8). However, at this point we will concern ourselves with the case of uniform circular motion, i.e. $\frac{dr}{dt} = 0$ and ω constant. In which case, we only need worry about the tangential term in (1.24) and

$$\mathbf{a} = \frac{d}{dt}(r\omega\hat{\boldsymbol{\theta}}) = \frac{dr}{dt}\omega\hat{\boldsymbol{\theta}} + r\frac{d\omega}{dt}\hat{\boldsymbol{\theta}} - r\omega^2\hat{\mathbf{r}}, \tag{1.26}$$

where we have used

$$\frac{d\hat{\boldsymbol{\theta}}}{dt} = -\omega\hat{\mathbf{r}}.$$

The first two terms in Eq. (1.26) are zero for uniform circular motion, so we obtain

$$\mathbf{a}(\text{uniform circular motion}) = -r\omega^2 \hat{\mathbf{r}}. \tag{1.27}$$

Notice that the acceleration here is not a result of a change in the magnitude of $\mathbf{v}$; this is constant. Rather, the direction of $\hat{\boldsymbol{\theta}}$ (and hence that of $\mathbf{v}$) is constantly changing and this gives rise to the acceleration in the radial direction. Notice also that the acceleration in Eq. (1.27) points towards the centre of the circular orbit, i.e. in the direction of $-\hat{\mathbf{r}}$.

We have derived Eq. (1.27) using the formal differentiation of the time-dependent position vector $r\hat{\mathbf{r}}$. We can also understand the result geometrically. We begin by sketching the important vectors in Figure 1.9. We show the position of the particle at times t and $t + \Delta t$ (points A and B respectively) as well as the corresponding velocity vectors. The velocity vectors are tangential to the path of the particle and have equal magnitudes ($v = r\omega$). Let's construct the velocity difference $\Delta \mathbf{v} = \mathbf{v}(t + \Delta t) - \mathbf{v}(t)$: you should be able to see from the diagram that $\Delta \mathbf{v}$ points approximately towards the centre of the circle from the midpoint of the circular arc between A and B. In the triangle of velocity vectors formed by $\mathbf{v}(t + \Delta t)$, $\mathbf{v}(t)$ and $\Delta \mathbf{v}$ we can approximate the magnitude of $\Delta \mathbf{v}$ by a circular arc, and write $\Delta v \approx v\Delta\theta = v\omega\Delta t = r\omega^2\Delta t$. In the limit $\Delta t \to 0$ the approximation becomes exact, $a = \Delta v/\Delta t \to r\omega^2$, and the acceleration points exactly in the direction $-\hat{\mathbf{r}}$. We are therefore led to Eq. (1.27).

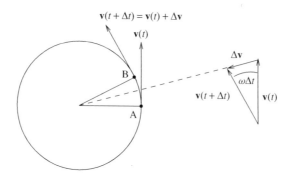

Figure 1.9   Uniform circular motion. Notice that the changing direction of the velocity vector results in a vector $\Delta \mathbf{v}$ that points approximately towards the centre of the circle. In the limit of vanishingly-small Δt this vector corresponds to the acceleration and points exactly towards the centre.

1.4   STANDARDS AND UNITS

In this chapter we have introduced the concepts of space and time without saying too much about measurement. Measurement of a physical quantity consists of making a comparison of that quantity, either directly or indirectly, with a standard. A standard is something on which we must all be able to agree and which defines the unit in which the measurement will be expressed. We will illustrate the idea by considering the legendary origins of the yard as a unit of length. Legend has it that the yard was originally defined to be the distance from tip of King Henry I of England's nose to the end of his thumb. Clearly the direct use of this standard of measurement would have been a little inconvenient; you can be sure that pretty soon a rod would have been cut to the correct length and used as a substitute for the King's own person. The use of this rod for measurement is an example of indirect measurement, though still using the same standard yard it doesn't require the King to be present. Desirable characteristics of a standard are reproducibility and precision. Reproducibility means that the standard can be used over and over again to give a consistent definition of the unit, one which doesn't vary with time. If the English people had reason to suspect that the King had grown or shrunk (perhaps by later comparisons with the rod) then they might have faced a dilemma: reject the King as the means to define the standard yard (in favour of the rod) or keep the definition using the King and face the problems associated with their not choosing a reproducible standard of length. Furthermore, the distance from the tip of the King's nose to the end of his thumb is not a terribly precise standard. Just consider the question of how he should hold his head while the measurement is taking place. The yard defined in this way clearly can only be expected to be accurate at the level of a few percent. It is easy to think of standards for length that are both more reproducible and more precise than this legendary definition.

Units are either fundamental, as is the case with the second (s), the kilogram (kg) and the metre (m), or they are derived units, such as the unit of velocity (m s^{-1}). For each of the fundamental units, there must be a precise and reproducible laboratory standard. In the case of the S.I. unit of mass, the kilogram, the standard is a lump of platinum-iridium alloy kept at the International Bureau of Weights and Measures (BIPM), at Sèvres in France. The SI unit of time, the second, was originally 1/86,400 of the mean solar day, and then later defined as a fraction of the mean tropical year. Neither of these standards could approach the accuracy of those based on the frequency of radiation emitted by certain atoms and in 1967 the second was redefined as exactly 9,192,631,770 cycles of the transition between two hyperfine levels in ^{133}Cs. In practice this standard uses a cavity filled with an ionised vapour of ^{133}Cs. Standing electromagnetic waves are created in the cavity using a radio-frequency oscillator circuit. When the frequency of the oscillator matches that of the atomic transition a resonance is observed. At resonance the oscillator circuit will then, by definition, make precisely 9,192,631,770 cycles in one second. Clocks based on sophisticated versions of this technique, such as those at the National Institute of Standards and Technology in the USA, are capable of measuring time to an accuracy of better than one nanosecond in a day.

The standard unit of length, the metre was once defined as one ten-millionth of the distance on the Meridian through Paris from the pole to the equator. This standard was replaced in 1874 and 1889 by standards based on the length, at zero degrees centigrade, of a prototype platinum-iridium bar. In 1984, standards based on prototype bars were superseded by the current standard distance that light travels in vacuum during a time interval of exactly 1/299,792,458 of a second. The effect of this definition is to fix the speed of light in vacuum at exactly 299,792,458 ms^{-1}. The justification for this choice of standard relies on our belief in the constancy of the speed of light in vacuum, a phenomenon that will be discussed in later chapters.

PROBLEMS 1

1.1 For the vectors $\mathbf{a} = \mathbf{i} + \mathbf{j} - 2\mathbf{k}$ and $\mathbf{b} = 3\mathbf{i} - \mathbf{j} + \mathbf{k}$, find:

   (a) the vectors $\mathbf{c} = \mathbf{a} + \mathbf{b}$ and $\mathbf{d} = \mathbf{a} - \mathbf{b}$;
   (b) the magnitudes of $\mathbf{a}$, $\mathbf{b}$ and $\mathbf{c}$;
   (c) a unit vector in the same direction as $\mathbf{a}$.

1.2 A bird leaves its nest and flies 100 m NE (i.e. at a bearing of 45°) then 150 m at a bearing of 150° and finally 50 m due W, where it lands in a tree. How far is the tree from the nest? In what direction is the tree from the nest?

1.3 The vertices of a triangle, A, B and C, have position vectors $\mathbf{r}_A$, $\mathbf{r}_B$ and $\mathbf{r}_C$. Write down $\overrightarrow{AB}$, the position of B relative to A, in terms of $\mathbf{r}_A$ and $\mathbf{r}_B$. Hence show that the vector sum of the successive sides of the triangle ($\overrightarrow{AB} + \overrightarrow{BC} + \overrightarrow{CA}$) is zero. Draw a sketch to demonstrate this result geometrically.

1.4 Hubble found that distant galaxies are receding from us with a speed proportional to their distance from the Earth. The velocity of the i-th galaxy is given by

$$\mathbf{v}_i = H_0 \mathbf{r}_i,$$

where $\mathbf{r}_i$ is the position vector of that galaxy with respect to us at the origin, and H_0 is a constant (known as Hubble's constant). Show that this recession of the galaxies does *not* imply that we are at the centre of the Universe.

1.5 For the vectors

$$\mathbf{A} = \mathbf{i} + \mathbf{j} + \mathbf{k},$$
$$\mathbf{B} = 2\mathbf{i} - 2\mathbf{j} - 2\mathbf{k},$$
$$\mathbf{C} = 4\mathbf{i} - \mathbf{j} - 3\mathbf{k},$$
$$\mathbf{D} = -\mathbf{i} + \mathbf{j} + \mathbf{k},$$

find their magnitudes and the scalar products $\mathbf{A} \cdot \mathbf{B}$, $\mathbf{A} \cdot \mathbf{C}$, $\mathbf{A} \cdot \mathbf{D}$ and $\mathbf{B} \cdot \mathbf{D}$. Hence find the angles between $\mathbf{A}$ and each of $\mathbf{B}$, $\mathbf{C}$ and $\mathbf{D}$ and that between $\mathbf{B}$ and $\mathbf{D}$. Evaluate the vector products $\mathbf{A} \times \mathbf{B}$, $\mathbf{A} \times \mathbf{D}$ and $\mathbf{B} \times \mathbf{D}$. Check that the magnitudes of these agree with the corresponding geometrical expressions ($|\mathbf{A} \times \mathbf{B}| = |\mathbf{A}||\mathbf{B}| \sin \theta$ etc.).

1.6 A charged particle is accelerated uniformly from rest in an electric field. If after 1.0 nanoseconds the particle has travelled 10 μm, work out its acceleration.

1.7 A coin is dropped from the top of a tall building. If an observer on the ground measures the speed of the coin immediately before impact to be 65.0 ms^{-1}, how tall is the building? For how long was the coin falling? Neglect effects due to air resistance.

1.8 A missile malfunctions in flight and has a subsequent trajectory described by the position vector (s) at time (t), given by,

$$\mathbf{s} = 0.3t\,\mathbf{i} + 0.5t\,\mathbf{j} - 0.005t^2\,\mathbf{k},$$

where t is measured in seconds and the magnitude of **s** is measured in km.

(a) What is the speed of the missile at $t = 0$? In which plane is the velocity at this time?

(b) What is the speed of the missile at $t = 30$ s? What is the angle between the velocity vector and the (positive) z axis at this time?

1.9 An airport travelator of length 50.0 m moves at a speed of $1.0\,\text{ms}^{-1}$. An athlete capable of running at a speed of $10.0\,\text{ms}^{-1}$ bets a friend that he can run to the end of the travelator and back again in exactly 10.0 s, as long as the time to change direction and restart is not included. The athlete loses the bet. What mistake has the athlete made? (Assume that at the start of each leg, the athlete is already running at full speed.)

1.10 A ferryman crosses a fast-flowing river. The ferryman knows that her boat travels at a speed v in still water, and that with the engine off the boat will drift at a speed u in a direction parallel to the bank, where u is less than v. If the line joining the two ferry stations makes a right-angle with the bank, and the stations are separated by a distance d, derive an expression for the time taken to cross the river. What happens if u is greater than v?

2

Force, Momentum and Newton's Laws

A force is something that pushes or pulls. The push or pull of a force may set an object in motion, as is the case when we throw a ball, push a book along the surface of a table or drop something from a height. Alternatively, forces may be used to stop objects already in motion; the friction between the brake-pads and the wheels of a car and between the tyres of the car and the road surface can quickly arrest the car's motion. In these dynamical situations the direction of the applied force is crucial in achieving the desired effect: a stationary object starts to move in the direction of an applied force; to stop a moving object we apply a force in the opposite direction to the motion. This directional property suggests that forces may be represented mathematically by vectors. Forces are often found in static, rather than dynamic situations. Medieval cathedrals are impressive examples of how gravitational forces can be balanced by the electrostatic forces between atoms to create structures that are stable for many hundreds of years. To start this chapter we shall seek to place these intuitive ideas on a firmer footing and establish a useful definition of force. We shall do this first by looking at forces in static situations and only once we have a definition of force will we strive to link forces to the motion of things.

2.1 FORCE AND STATIC EQUILIBRIUM

How are we to define force? We wish to do so in a way that relies on as few other concepts as possible and to this end we remove the complication of motion and look at a case in which nothing is moving. Figure 2.1 shows a situation in which a mass is held stationary on a very smooth horizontal surface between three stretched springs. The springs each pull on the mass and we say that each spring exerts a

---

Dynamics and Relativity   Jeffrey R. Forshaw and A. Gavin Smith
© 2009 John Wiley & Sons, Ltd

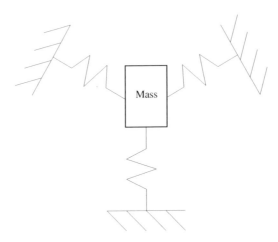

Figure 2.1   Identical, equally-stretched springs with a mass in static equilibrium in the horizontal plane. The mass is supported vertically on a low-friction surface (such as an air-hockey table). Equally you could imagine the experiment as being performed in outer space.

force on the mass. Let us further say that the springs are carefully constructed to be as similar as possible; for the sake of this argument they can be considered to be identical. To further make sure that the springs behave identically we change the angles that the springs make with each other and look for a situation in which each spring is stretched by the same amount. When we perform this experiment in the lab we observe that the mass is stationary if and only if the angle between any pair of equally-stretched adjacent springs is 120°. Our definition of force must take the result of this type of experiment into account. Since we know from experience that pushing or pulling can produce motion we assert that our experiment with three springs, in which the mass doesn't move, corresponds to a total force of zero. In this way we are led to the idea that force must be a vector quantity, which sums to give zero in our experiment. That the vector should also point along the axis of the spring can be deduced from a similar experiment constructed with two collinear springs: there is then no special direction other than the axis of the springs and any physical property of the system should not break this symmetry, so the force must point along the length of the spring. Force is thus to be regarded as a vector quantity representing a push or pull, which: (a) points along the axis of a stretched spring; (b) is additive when several springs are involved; (c) results in no motion when that sum is zero. We also know from experience that there are things other than springs which may push or pull, so we state as part of our definition that any thing that can potentially replace one of the springs in the above experiment also exerts a force on the mass.

This sounds pretty close to a good definition of force, albeit in a fairly specific scenario, but there is a weakness that you may have already noticed. We stated that the force is zero when the mass is stationary. That is quite reasonable for experiments performed at rest on the Earth but what happens if we do the experiment in outer space? How do we agree on what frame of reference to use for our force experiment and for our definition of force? An observer moving relative to us will claim that the mass in Figure 2.1 is in motion whilst we assert that it is stationary. We need, as a matter of some urgency, to encorporate the frame of reference into our definition of force. To help us choose a good frame of reference we shall consider a situation in which no forces are present.

For a force on a particle to exist there must be something else somewhere in the Universe that is responsible, i.e. a spring or something that can counter a spring. A particle, completely alone in the Universe would experience no forces. This hypothetical object, of point size and subject to no forces, is referred to as an isolated particle. What sort of motion do we expect for such an isolated particle? Similar problems troubled ancient thinkers who concluded that force was necessary to maintain the motion of a body[1]. This is a conclusion very close to our everyday experience. If you push a book along a table it may move, but when you decide to stop pushing, the book stops moving: the force is needed to maintain the motion. If however, one looks at a rolling ball, then the behaviour is noticeably different. A hard sphere set in motion on a flat, hard, horizontal surface travels a long way before stopping. So the rule that force is needed to maintain motion appears suspect. It was Galileo, studying the motion of rolling spheres on inclined planes who proposed that a moving body continues moving, i.e. we might say that the body has "inertia". This means that the behaviour of the rolling ball is closer to that of the isolated particle than is that of the book on the table. Galileo's genius was to realise that the motion of everyday objects is complicated by friction and that to see the raw, unhindered, motion of an isolated particle we need to devise careful experiments that are insensitive to friction.

Newton's First Law, as written in *Principia* is is restatement of Galileo's Principle of Inertia: "every body preserves its state of rest, or of uniform motion in a right line, unless compelled to change that state by forces acting upon it." In other words, the isolated particle will have a constant velocity vector, and this velocity may be zero. Forces are responsible for changes in the velocity of a body. On the surface this sounds very clear; a watertight rule for the motion of bodies in the absence of forces. There is however, an important weakness in the First Law as stated above. Specifically, there is no statement as to what frame of reference should be used, and this is crucial for the complete description of the state of motion of the particle. Consider the situation illustrated by the cartoon in Figure 2.2. Two observers, called A and B are measuring the motion of an isolated particle. B observes that the particle is stationary and, according to the First Law, the particle will remain

Figure 2.2   Two observers and the motion of an isolated particle.

---

[1] A common view among the Ancient Greek philosophers was that the ability to cause motion was a sign of life. The apparent motion of the heavenly bodies was taken as a sign of their divine nature.

in that state. Observer A is moving, relative to a frame of reference in which B is at rest, with velocity **v** and acceleration **a**. A will therefore measure the particle to be moving with velocity −**v** and acceleration −**a**. So what does A conclude? The particle is accelerating, so there must be a force acting on it (according to the First Law). However, as the particle is isolated, by definition it cannot be subject to any forces. This contradiction is a direct result of observing the particle from an accelerating frame of reference. If **a** = **0** then A observes the particle with velocity −**v**, but with zero acceleration. Since this is uniform motion, the First Law still holds for the isolated particle. Thus there are two classes of frames of reference, those in which **a** = **0**, called inertial frames of reference, and those for which **a** ≠ **0**, called non-inertial frames of reference. The First Law thus becomes essentially a statement upon the existence of inertial frames of reference:

> There exist inertial frames of reference, with respect to which an isolated particle moves in a straight line of constant velocity (including zero).

This reformulation of the first law supposes that we can find an isolated particle. Clearly there is no real object so alone in the Universe that it is devoid of all forces; the very act of observing something involves an interaction at some level, even if it is only the force involved in reflecting light. So how do we ever find, in practice, a good inertial frame? From a practical point of view we must find ways of isolating a particle other than by removing it to a remote region of the Universe. This involves using our knowledge of forces to arrange things in such a way that there is no *net* force on a body. An air-hockey table is just such a construction: air is blown through tiny holes to create a force on the puck that cancels the effect of the Earth's gravity. In addition, supporting the puck on a layer of air means that frictional forces are greatly reduced for most laboratory experiments. With the table adjusted properly, a puck will glide at nearly constant velocity across the table, with only a small change in speed. So does the air-hockey table define an inertial frame of reference? Approximately, yes, but at some level of precision the effects of the Earth's rotation will become apparent. As was shown in Section 1.3.4, an object moving in a circle at constant angular speed is accelerating towards the centre of the circle. Thus any laboratory fixed to the surface of the Earth is accelerating and therefore constitutes a non-inertial frame of reference. Similarly the rotation of our neighbouring stars about the galactic centre means that even the "fixed" stars cannot be counted upon to define a perfect inertial frame. We cannot take the principle of inertia as a statement that can be verified experimentally in isolation of the rest of mechanics. By itself, the First Law may be thought of as a statement of faith in the existence of inertial frames of reference. In practice it matters little that we cannot find perfect inertial frames. Approximate ones are good enough for the development of classical mechanics and experiment confirms the results to a sufficiently high degree of accuracy under a wide range of conditions.

Now that we have hammered out the definition of an inertial frame we are in a position to clarify what we mean by 'no motion' in our force experiment of Figure 2.1. We define a particle to be in static equilibrium when it is acted on by forces and yet is at rest in some inertial frame. This is equivalent to saying that the particle must be moving with constant velocity when measured in any inertial frame.

In other words, a particle in static equilibrium behaves like an isolated particle. Thus observers in all inertial frames agree upon the static nature of our force experiment.

For the moment, let us not delve too deeply into the complexities that arise when forces act upon extended bodies (ensembles of particles). We will work on this in detail in the next section when we show that the centre of mass of a system of particles behaves very much like a classical particle. For a particle in static equilibrium the vector sum of all forces acting on it is the null vector. For example, if forces $\mathbf{F}_1$, $\mathbf{F}_2$ and $\mathbf{F}_3$ act on a particle, the condition for static equilibrium is

$$\mathbf{F}_1 + \mathbf{F}_2 + \mathbf{F}_3 = \mathbf{0}. \qquad (2.1)$$

This condition may be interpreted geometrically (Figure 2.3) as the three vectors forming a triangle when placed head-to-tail. For larger numbers of forces, $\mathbf{F}_i$, $i = 1, 2, 3 \ldots N$, static equilibrium occurs when $\sum_{i=1}^{N} \mathbf{F}_i = \mathbf{0}$.

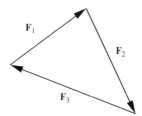

Figure 2.3   Forces in static equilibrium.

To measure the magnitude of forces we need a force meter. Let's figure out how we might make one. Imagine a situation of static equilibrium whereby one extended spring is balanced by N others. Figure 2.4 shows the setup for $N = 3$. For ideal springs, the extension of the single spring will be N times that of the springs on the other side, i.e. $x_1 = N x_N$. Using this result we can obtain an expression for the force exerted by a spring as a function of distance. Since we have static equilibrium and all of the springs are collinear we can write

$$F(x_1) = NF(x_N) = NF(x_1/N), \qquad (2.2)$$

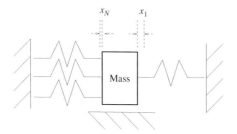

Figure 2.4   Static equilibrium with several identical springs. The extensions of the springs x_1 and x_N are measured relative to the length of an unstretched spring.

where $F(x)$ is the magnitude of $\mathbf{F}(x)$. Putting $x = x_1/N$ allows us to write, for fixed x_1,

$$\frac{F(x)}{x} = \frac{F(x_1)}{x_1} = k. \qquad (2.3)$$

Since Eq. (2.3) must hold for any N the function $g(x) = F(x)/x$ must take on the same value at an infinite number of different points, i.e. $g(x_1) = g(x_1/2) = g(x_1/3) = \cdots$ for $N = 1, 2$ and 3 etc. Assuming that $F(x)$ is smooth allows us then to conclude that $F(x)/x = k$ for all values of x. Thus we deduce that the force produced by a stretched spring is proportional to its extension x, a result known as Hooke's Law:

$$F = -kx. \qquad (2.4)$$

The choice of sign fixes the direction of the force and k is some constant characteristic of the spring, usually called the "spring constant".

It may seem rather restrictive that we should be using mechanical springs to define the magnitude of a force. But remember, once we have defined our standard force meter we can in principle use it to measure the magnitude of any other force. Also, and as we shall see in Section 3.2.3, very many systems actually behave just like springs, in that for small deviations from equilibrium they experience restoring forces that satisfy Hooke's Law.

Now we have established a force meter we can begin to look at other forces. The condition of static equilibrium allows us to put forces that arise from different sources on an equal footing. An example of this is illustrated schematically in Figure 2.5 in which a mass is in static equilibrium. There are two forces acting on the mass: the elastic force of the stretched spring is given by Hooke's Law, and is upwards with a magnitude kx; and the gravitational force or weight. Experiments like this on different masses show that the gravitational force is proportional to the amount of matter in the block, i.e. for identically constituted blocks, a doubling of the volume doubles the gravitational force (etc.) and so $F \propto m'$, where m' characterizes the amount of matter within the body and

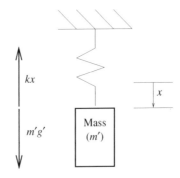

Figure 2.5   Static equilibrium of a mass under the influence of the force of gravity and that of a stretched spring.

is called the gravitational mass. Introducing a constant of proportionality g' we write

$$F = m'g'. \qquad (2.5)$$

As we shall soon see we can choose g' to be the acceleration of *all* bodies in free-fall. Since the mass is in static equilibrium, then the two forces that act upon it must be equal in magnitude and opposite in direction, and we can write $kx = m'g'$. Thus the elastic extension of the spring is related to the gravitational mass of the object, the principle behind some types of weighing scales. We can do similar experiments with other forces; Millikan's oil-drop experiment uses the the static equilibrium between the electrostatic, viscous and gravitational forces on charged oil drops to determine the electrical charge of the electron.

By using static equilibrium in various experiments we can investigate the properties of different forces, putting them on a common scale. While this procedure is practicable for many physical systems, it is not always convenient to observe forces in static equilibrium. Forces more generally affect the motion of things, as is the case with the gravitational attraction between the Moon and the Earth, or the force on an electron moving in a magnetic field. It is now time for us to study the way in which forces affect motion.

2.2   FORCE AND MOTION

We return to the experiment with springs depicted in Figure 2.4. Suppose we were to suddenly sever the three springs on the left-hand side of the mass, what would we observe? Such an experiment can be easily performed in the lab, and is depicted in Figure 2.6. Now the mass moves so we write the extension of the spring as a function of time $x(t)$. From experiment the observed motion is oscillatory and can be expressed as[2]

$$x(t) = A \cos \omega t, \qquad (2.6)$$

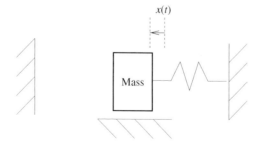

Figure 2.6   Mass on a spring in motion. The mass moves horizontally over a very smooth surface.

---

[2] Strictly speaking the motion will have some degree of damping and the amplitude of the oscillation will not be constant. We will assume that the damping is small enough to be ignored.

where ω is the angular frequency of the oscillation. It is related to the time taken to complete one cycle (i.e. the period) T by $\omega = \frac{2\pi}{T}$. What can we deduce from this result? Differentiating with respect to time gives us

$$v(t) = \frac{dx}{dt} = -A\omega \sin \omega t \qquad (2.7)$$

and

$$a(t) = \frac{d^2 x}{dt^2} = -A\omega^2 \cos \omega t. \qquad (2.8)$$

So we can write

$$a(t) = -\omega^2 x. \qquad (2.9)$$

We can connect the force with the motion by simply substituting for x using Hooke's Law, i.e.

$$F = \frac{k}{\omega^2} a. \qquad (2.10)$$

Here k is constant and ω is measured in the experiment. We can reasonably ask the question "on what does ω depend?". To answer this we need to do some more experiments. We can imagine using different calibrated springs and different masses on the end. We could for example replace the single block on the end of the spring with N identical blocks of the same material all glued together, i.e. we increase the amount of 'stuff' on the end of the spring by N times. Doing this experiment results in a decrease in ω by a factor $\sqrt{N}$. We repeat this process with springs of different k and we start to observe a remarkable pattern in the data. The coefficient $\frac{k}{\omega^2}$ is proportional to the number of blocks no matter which spring we choose, i.e.

$$\frac{k}{\omega^2} = Nm_0, \qquad (2.11)$$

where m_0 is independent of k. What this means is that our experiments have revealed a property of the motion that has nothing to do with the particular spring we use but which depends on whatever is on the end of the spring. Not only that, but the property depends linearly on the amount of 'stuff' in motion. We define $Nm_0 = m$ to be the inertial mass of what is on the end of the spring. In which case we can rewrite Eq. (2.10) as

$$F = ma. \qquad (2.12)$$

We are now in a position where we can try to write down a law of motion. But first we shall introduce the momentum, $\mathbf{p}$, of a classical particle:

$$\mathbf{p} = m\mathbf{v}, \qquad (2.13)$$

where $\mathbf{v}$ is the velocity and m is the inertial mass of the particle. We postulate that the equation of motion is

$$\mathbf{F} = \frac{d\mathbf{p}}{dt}. \tag{2.14}$$

For a particle of fixed mass this equation reduces to

$$\mathbf{F} = \frac{d(m\mathbf{v})}{dt} = m\mathbf{a} \tag{2.15}$$

and this is in accord with Eq. (2.12). Of course Eq. (2.14) looks like a more general equation for it has the capacity to describe systems of variable mass. That is indeed the case, as we shall explore in the following example. Eq. (2.14) is a very important equation: it is Newton's Second Law.

Example 2.2.1 *Flour falls from a hopper onto a railroad truck at a rate of 30 kg s^{-1}. What horizontal force is required to pull the truck at a constant speed of 5 ms^{-1}.*

Solution 2.2.1 *The speed of the truck is constant, but the mass of the truck, and therefore the momentum, changes continuously. Using Eq. (2.14), we obtain*

$$\mathbf{F} = \frac{d\mathbf{p}}{dt}$$

$$= \frac{d(m\mathbf{v})}{dt}$$

$$= \mathbf{v}\frac{dm}{dt}.$$

The force needed has magnitude $F = v\frac{dm}{dt} = 30 \times 5 \, \text{kg m s}^{-2} = 150 \, \text{N}$. Note that in this problem we have considered the horizontal forces only; the vertical force that stops the sand is provided by the normal reaction of the truck.

Momentum lies at the heart of Newton's Second Law. For a particle we have defined it to be the product of mass and velocity and through Newton's Second Law we see that a change in momentum can be induced by applying a force. The bigger the force, the more one can change the momentum. Thus we understand that the momentum of something expresses how hard it is to stop or deflect it.

We can also use Eq. (2.14) to fix the scale of inertial mass. Two bodies, subject to the same force, will experience accelerations a_1 and a_2 in the direction of the applied force. The ratio of the masses of the two bodies is then

$$\frac{m_1}{m_2} = \frac{a_2}{a_1}. \tag{2.16}$$

Choosing a standard mass (a lump of platinum-iridium alloy in the case of the S.I. system of units) fixes the mass scale. We can then fix the unit of force, using

Eq. (2.15). One newton of force (N) causes a mass of one kilogram to accelerate at one $m\,s^{-2}$, i.e. $1\,N \equiv kg\,m\,s^{-2}$.

The attentive reader may well have noticed that we have introduced two different types of mass. There is the mass m' that appeared in our definition of the force due to gravity and then there is the mass m that is the constant of proportionality in Newton's Second Law. A priori they are two different quantities and we were quite right to keep them distinct. But careful experiments reveal something remarkable: all bodies fall with the same acceleration in the vicinity of the surface of the Earth. We shall use the symbol g to denote that special acceleration. Now we know the gravitational force acting on any body close to the Earth (Eq. (2.5)) and that can be inserted into Newton's Second Law to give

$$m'g' = mg. \tag{2.17}$$

The fact that g is a constant leads (since g' is also a constant) to the conclusion that $m' \propto m$. Since we haven't yet defined the scale for gravitational masses we are perfectly at liberty to fix the constant of proportionality to unity (i.e. to choose $g' = g$) and henceforth $m' = m$ and we need not distinguish between the two different types of mass, although we ought to be impressed that Nature has arranged for their equivalence. This all may seem like pedantry but it is not. Einstein took very seriously the equivalence of gravitational and inertial mass and it played a crucial role in his development of the General Theory of Relativity. The General Theory is our modern theory of gravity and we shall introduce it in Section 14.2. It is characterized by the fact that it offers an explanation for the equivalence of inertial and gravitational mass – in fact gravitational mass never appears in Einstein's theory.

So far we have only been talking about the effect of gravity on objects close to the surface of the Earth using $F = mg$. Actually, this is a special case of the more general result discovered by Newton, building on the earlier studies of Johannes Kepler following observations of the planets within our Solar System. The more general result states that the gravitational force acts between any two massive bodies according to

$$\mathbf{F} = -G\frac{Mm}{r^2}\mathbf{e}_r, \tag{2.18}$$

where M and m are the two masses, r is the separation of their centres and $\mathbf{e}_r$ is a unit vector pointing from the centre[3] of the mass M to the centre of the mass m. G is a constant of proportionality to be fixed by the data (it is usually called Newton's gravitational constant). The force $\mathbf{F}$ is then the force acting upon mass m (the force on the mass M is equal in magnitude but opposite in direction). We have taken care to completely specify the force, making an appropriate use of vectors, but the maths should not obscure the simple fact that this is an inverse square law of attraction (i.e. the forces act to pull the bodies towards each other), which grows

---

[3] Gravitation will be studied in much more detail in Chapter 9 but for now it suffices to consider only the gravitational forces between spherical bodies.

in size in proportion to each of the masses. Notice that if the distance r is fixed so that it is equal to the radius of the Earth, R_E, then Eq. (2.18) simplifies to

$$\mathbf{F} = -mg\mathbf{e}_r, \qquad (2.19)$$

where

$$g = \frac{GM}{R_E^2}$$

and M is the mass of the Earth. In this way we can view our earlier expression, $F = mg$ as a special case and we have the bonus of relating g to the mass and radius of the Earth once we know G.

2.2.1 Newton's Third Law

So far we have not worried too much about just how forces act on extended bodies. What about all of the forces internal to the body? They certainly do not appear to play a role in the motion of the body as a whole so it seems they must cancel each other out somehow. Similarly, when we speak of the acceleration of an extended body, it is not immediately clear whether we are speaking about all parts of the body or perhaps one special point within it. The example of a spinning ball thrown through the air illustrates the point because different parts of the ball clearly accelerate differently (remember that rotation is associated with acceleration). In this section we shall make progress towards resolving these matters by considering the behaviour of extended bodies, although we shall have to wait until Chapter 10 before we finally solve the problem of a spinning object thrown through the air. As a bonus, we shall also solve another problem that we have left hanging in the air – just how do we define an inertial frame? That is a serious problem because we have shown that non-inertial frames are characterized by the fact that isolated particles accelerate. But there is a nasty loophole since we can presumably never be sure that the acceleration has not arisen as the result of a force and that the particle is not actually isolated.

Progress in addressing these matters can be made once we have a grasp of Newton's Third Law, which expresses the empirical fact that real forces are found in pairs. Applied to particles it asserts that:

If particle B exerts a force on particle A given by $\mathbf{F}_{AB}$, then A will exert a force on B ($\mathbf{F}_{BA}$) such that $\mathbf{F}_{BA} = -\mathbf{F}_{AB}$.

In other words the force exerted on particle A by particle B is equal in magnitude, but opposite in direction, to the force exerted on particle B by particle A.

Immediately we see that the Third Law gives us a mechanism for distinguishing between acceleration caused by a real force, and that which is the result of choosing a non-inertial frame of reference. Let us consider particle A. If this particle is observed to be accelerating, then either there is a force acting on it, or the observer is using a non-inertial frame. The observer may then look for another particle that is responsible for the force. If that particle (B) can be identified, it must be subject to a force equal in magnitude but opposite in direction to that acting on A. If such a mutual interaction cannot be identified, the only conclusion that can be reached (other than we did not look hard enough) is that the force is fictitious; a result of starting from a non-inertial frame.

The Third Law also allows us to generalize from the mechanics of particles to the mechanics of extended bodies. To do this we will consider a body as being composed of N classical particles. These particles may interact with each other as well as with other particles that are not part of the body. We consider two particles within the body i and j (see Figure 2.7). The mutual interaction between theses particles consists of two forces: $\mathbf{F}_{ij}$ acting on particle i and $\mathbf{F}_{ji}$ acting on particle j. The Third Law states that these forces must be equal in magnitude but opposite in direction. The net external force acting on particle i is $\mathbf{F}_i^{(e)}$ and for particle j it is $\mathbf{F}_j^{(e)}$. Since the remote particles responsible for these forces, and the nature of the interactions, are unspecified, we cannot deduce any relationship between $\mathbf{F}_i^{(e)}$ and $\mathbf{F}_j^{(e)}$ in the general case.

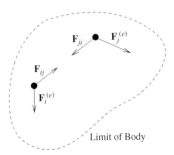

Figure 2.7   Internal and external forces on particles i and j in an extended body.

Using this separation into internal and external forces, the net force on particle i may be written

$$\mathbf{F}_i = \mathbf{F}_i^{(e)} + \sum_{j=1}^{N} \mathbf{F}_{ij}, \tag{2.20}$$

where the sum over j does not include a contribution from $j = i$ (i.e. particles do not act upon themselves). We now consider the total force acting on the body. This is the sum

$$\mathbf{F} = \sum_{i=1}^{N} \mathbf{F}_i = \sum_i \mathbf{F}_i^{(e)} + \sum_i \sum_j \mathbf{F}_{ij}. \tag{2.21}$$

The second term is a summation of all the forces that arise from mutual interactions between particles within the body. We can group terms corresponding to the same pair of particles and rewrite this sum as

$$\sum_i \sum_j \mathbf{F}_{ij} = \sum_i \sum_{j<i} (\mathbf{F}_{ij} + \mathbf{F}_{ji}).$$ (2.22)

The Third Law requires that each term $(\mathbf{F}_{ij} + \mathbf{F}_{ji})$ is null, therefore the sum is null and

$$\mathbf{F} = \sum_{i=1}^{N} \mathbf{F}_i = \sum_i \mathbf{F}_i^{(e)},$$ (2.23)

i.e. the net force acting on any body is the sum of all the forces of external origin. This allows us to determine a version of the Second Law that is valid for an extended body. Applying the Second Law to each particle

$$\mathbf{F} = \sum_{i=1}^{N} \mathbf{F}_i = \sum_{i=1}^{N} m_i \mathbf{a}_i = \sum_i \mathbf{F}_i^{(e)}$$ (2.24)

now

$$\sum_{i=1}^{N} m_i \mathbf{a}_i = \sum_{i=1}^{N} m_i \frac{\mathrm{d}^2}{\mathrm{d}t^2} \mathbf{r}_i$$

$$= \frac{\mathrm{d}^2}{\mathrm{d}t^2} \sum_{i=1}^{N} m_i \mathbf{r}_i$$ (2.25)

since the mass of each particle is independent of time. Furthermore, if we divide by the total mass $M = \sum_i m_i$, we obtain

$$\frac{\mathbf{F}}{M} = \frac{\mathrm{d}^2}{\mathrm{d}t^2} \left[\frac{1}{M} \sum_{i=1}^{N} m_i \mathbf{r}_i \right].$$ (2.26)

The term on the right hand side is the second time derivative of the weighted-mean position of all the particles in the body, where the "weights" are the masses of the particles. This special position is known as the centre of mass of the body, it is located at

$$\mathbf{R} = \frac{1}{M} \sum_{i=1}^{N} m_i \mathbf{r}_i.$$ (2.27)

We are thus led to a version of the Second Law valid for any extended body or collection of particles:

$$\mathbf{F} = M \frac{\mathrm{d}^2 \mathbf{R}}{\mathrm{d}t^2} = M\mathbf{A},$$ (2.28)

where $\mathbf{A}$ is the acceleration of the centre of mass. Eq. (2.28) represents an extremely important simplification of the dynamics of complex systems. In a macroscopic object there are of the order of 10^{23} particles. If these particles all interact with each other, there will be of the order of 10^{46} forces to consider; clearly, solving the motion for so many particles is an impossible task. The beauty of Eq. (2.28) is that irrespective of the details of the internal forces, the motion of the centre of mass is governed only by external forces.

Example 2.2.2 *Show that the centre of mass of an extended body falls with a uniform acceleration g near the Earth's surface. Neglect air resistance.*

Solution 2.2.2 *Each particle within the body experiences an external force $m_i g$. The total force on the body is thus*

$$F = \sum m_i g = g \sum m_i = Mg.$$

The acceleration of the centre of mass is therefore $A = F/M = g$, as required.

The Third Law applies equally to extended bodies as it does to particles. If we consider two bodies, A and B, then because the internal forces sum to zero in both bodies, the total force that A exerts on B is the sum of the forces that the particles in A exert on the particles in B. This sum is equal in magnitude but opposite in direction to the force that the particles in B exert on the particles in A. So the Third Law can be stated for extended bodies:

If body B exerts a force on body A given by $\mathbf{F}_{AB}$, then A will exert a force on B ($\mathbf{F}_{BA}$) such that $\mathbf{F}_{BA} = -\mathbf{F}_{AB}$.

It is important to be clear that these two forces act on *different* bodies. Figure 2.8 illustrates this with two bodies connected by a massless spring. The spring in the figure is a symbolic representation of any real force.

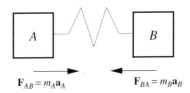

Figure 2.8   Equal magnitudes but opposite directions for forces acting on mutually interacting bodies A and B.

In many practical situations it is impossible to consider explicitly all the particles that make up a system to determine the position vector of the centre of mass. Instead, a macroscopic body can often be approximated as a continuous distribution of matter with a spatially-dependent density function. The calculation of the centre of mass position then becomes an integral rather than a discrete sum. Depending on the situation this integral may be either over a line, a surface or a volume.

Example 2.2.3 *Calculate the position of the centre of mass of a uniform thin rod of length l and linear mass density $\rho(x) = \frac{dm}{dx} = \kappa x$ where x is the distance from one end of the rod.*

Solution 2.2.3 *We consider the rod as being made up of many tiny pieces, each of length dx and mass $dm = \rho(x)dx$ (see Figure 2.9). The position of the centre of mass is given by*

$$\mathbf{R} = \frac{1}{M} \sum_i m_i x_i \, \mathbf{i} = \frac{1}{M} \sum_i \rho(x_i) x_i \, dx \, \mathbf{i}$$

and in the limit that $dx \to 0$ the sum can be replaced by an integral, i.e.

$$\mathbf{R} = \frac{1}{M} \int_0^l x \rho(x) \, dx \, \mathbf{i} = \frac{1}{M} \int_0^l \kappa x^2 \, dx \, \mathbf{i},$$

where $M = \int_0^l \kappa x \, dx = \frac{1}{2}\kappa l^2$ and $\mathbf{i}$ is the unit vector in the direction of increasing x. Thus

$$\mathbf{R} = \frac{2}{3} l \, \mathbf{i},$$

i.e. the centre of mass lies two-thirds of the way along the rod, on the high-density side of the geometric centre.

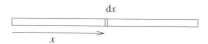

Figure 2.9   Slicing a rod into pieces in order to compute the centre of mass of a thin rod of non-uniform density.

2.2.2   Newton's bucket and Mach's principle

Earlier in this chapter we discussed the idea that an isolated particle viewed from an inertial frame of reference moves with constant velocity. This is the essential content of the First Law in Newtonian mechanics; it provides a way of selecting inertial frames from non-inertial ones. Within an inertial frame, accelerations are the result of pairs of forces operating between particles. We have shown that all inertial frames move at constant relative velocities and are thus led to the idea of an infinite set of inertial frames, which are all equally valid for doing physics. The concept of the inertial frame thus underpins classical mechanics, but there is no deeper explanation given as to *why* these particular frames of reference are so special in our Universe. In this section we will try to probe a little deeper.

In classical mechanics there are no absolute velocities and only relative velocities have any meaning. The same thing cannot be said for accelerations. Acceleration

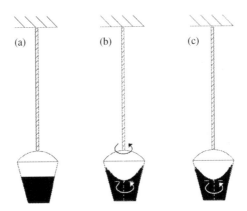

Figure 2.10   A variant of Newton's rotating bucket.

relative to the infinite set of inertial frames is absolute. Newton used the following experiment to demonstrate this point. A bucket filled with water is hung from a ceiling on a long rope. The bucket is slowly rotated many times so that the rope twists. Once the rope is sufficiently twisted, and any motion of the water has died away the surface of the water is horizontal as shown in Figure 2.10(a). The bucket is then released and starts to rotate with increasing angular speed about the axis of the rope. The water is dragged along by the inner wall of the bucket until the situation is as shown in Figure 2.10(b), where both the water and the bucket are rotating in the same sense. Once this state is established we grab hold of the bucket to stop the rotation and we observe that the water continues to rotate even though the bucket is stationary as shown in Figure 2.10(c). What can we conclude? The fact that the surface of the water makes a concave shape in (b) and (c) seems to imply that the water is accelerating. We might therefore venture to propose that the water's surface can be used to identify the existence of accelerations. But we must be careful because the water does not rotate relative to the bucket in (b) yet it is still pushed up the sides of the bucket. This is because the rotating bucket in (b) constitutes a non-inertial frame, so what we really mean is that curvature of the water's surface defines a non-zero acceleration relative to any inertial frame.

If you feel uneasy about the specialness of inertial frames then you are in good company. The physicist Ernst Mach (1838–1916) attempted to eliminate the distinction between inertial and non-inertial frames by attributing the curvature of the water's surface in Newton's bucket to an interaction between the water and the rest of the Universe (Mach's Principle). This interaction is constrained by the fact that we should not be able to tell the difference between a situation in which the bucket rotates and the rest of the Universe is fixed, and one in which the bucket is stationary and the Universe rotates. Inertial frames are then special only in so far as they are the set of frames in which the force of interaction with the rest of the Universe just happens to vanish. The obvious question to ask is whether such a force actually exists. Since it is both additive and operates over large distances, the prime candidate is the force of gravity. There have been several attempts to investigate the gravitational forces produced by distant rotating shells of matter and they do show that it is possible to obtain gravitational accelerations similar to the centrifugal and Coriolis accelerations that appear in frames of reference that rotate relative to the distant stars. That said, the agreement is not exact and depends on the assumed distribution of matter in the shell. To a large extent, the specialness of inertial frames is diminished in Einstein's General Theory of Relativity (see in particular the discussion following Eq. (14.35)) and Mach's concerns become much less pressing.

2.3 APPLICATIONS OF NEWTON'S LAWS

We have discussed at some length the theoretical content of Newton's laws of motion and are now in a position to apply them to dynamical problems. While this is in some cases a straightforward exercise, the versatility of classical mechanics ensures that there exist a huge range of different types of problems that can be posed. Different problems often require different approaches. While many people can happily follow the solution to a given problem, it is often the case that when facing a fresh problem on their own they cannot see how to start. Problem solving in dynamics is therefore a skill that needs to be learnt and which improves greatly with practice. In this section we will look at a technique for solving dynamical problems based on "free-body" diagrams. We will also show how some problems can be more easily solved using the principle of momentum conservation. Friction and viscous forces play an important role in the behaviour of macroscopic systems and they will also be discussed in the present section.

2.3.1 Free Body Diagrams

Solving problems in dynamics usually involves a sequence of several steps. While the following programme is not always the most efficient or elegant way to proceed (the use of conserved quantities often works better) it represents a direct approach to finding the solution and is a good fall-back position when you cannot spot a clever trick to use.

The problem-solving recipe is generally as follows:

- Identify the forces acting on the system
- Choose a system of co-ordinates (and associated basis vectors) appropriate to the geometry.
- Apply the Second Law to the components of the system to get second order differential equations of motion.
- Solve the equations of motion together with any constraints or boundary conditions.

To aid in the first step of this procedure it is almost always crucial to draw a good diagram. We will try to make this diagram as uncluttered as possible, while including all forces. To this end we draw each body as "free" in the sense that there are no supports drawn on the diagram (although the forces exerted by them should certainly be included). In addition we shall represent the position of a body by the position of its centre of mass. We indicate the forces as arrows and draw any acceleration vectors to the side of the diagram. Figure 2.11 shows a free-body diagram for a book in static equilibrium on a table. The weight of the book and the normal force from the table are in opposite directions. The condition of static equilibrium implies that the magnitudes of these two forces are equal and that the acceleration of the centre of mass is zero.

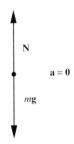

Figure 2.11   Free-body diagram for a book in static equilibrium on a table.

2.3.2   Three worked examples

In order to illustrate the methodology that we presented in the last section we will work through three specific examples. Like so often in physics, it is useful first to work through problems that are selected because they really help to develop an ability to apply the key ideas and principles that we have spent so much time developing. The problems are not chosen because they represent particularly exciting phenomena. The study of exciting things comes later, once one has a grasp of the key ideas. In this section we shall take a look at the motion of a pendulum, two spaceships connected by a cable and two masses hung over a pulley.

Example 2.3.1 *When the bob of a pendulum is made to describe a circular orbit in the horizontal plane, rather than executing the usual oscillatory motion in a vertical plane, it is known as a conical pendulum. Determine the period of revolution of a conical pendulum of mass m and length l which makes an angle α to the vertical (see Figure 2.12). Ignore the mass of the pendulum string.*

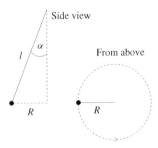

Figure 2.12   Geometric diagram of a conical pendulum.

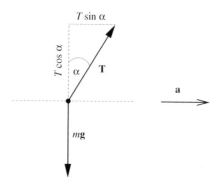

Figure 2.13   A free body diagram for the bob of a conical pendulum.

Solution 2.3.1 *The forces acting on the pendulum bob are its weight* mg *and the tension in the string* **T**. *We construct a free-body diagram as shown in Figure 2.13 where the force vectors are resolved into vertical and horizontal components. Since the bob describes a circular orbit in the horizontal plane, the acceleration is horizontal and points towards the centre of the orbit. There is no acceleration in the vertical direction so*

$$mg = T \cos \alpha.$$

Applying the Second Law in the horizontal plane we have

$$T \sin \alpha = ma = mR\omega^2,$$

where $R = l \sin \alpha$ *is the radius,* $\omega = 2\pi/\tau$ *the angular frequency of the orbit and* τ *is the period. Substitution for* R *gives*

$$T = ml\omega^2,$$

which can be used to eliminate the unknown tension in the vertical equation to give

$$\omega = \sqrt{\frac{g}{l \cos \alpha}}.$$

Often we are interested in the motion of several parts of a system, as the following example illustrates. We divide the system into discrete parts, each with its own free-body diagram and in so doing we must be careful to identify which body each force acts upon.

Example 2.3.2 *Spacecraft A and B with masses* M_A *and* M_B *are adrift in outer space and connected by a cable (see Figure 2.14). Winches on both craft are used to wind up the cable to reduce their separation. The winch on A is capable of producing a force* F_A *on the cable and that of B a force* F_B. *Determine the acceleration of both A and B in the limit that the mass of the cable is negligible compared to the masses of the spaceships. Ignore gravitational forces.*

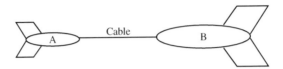

Figure 2.14   Spacecraft connected by a cable.

Solution 2.3.2 *We construct three free body diagrams for the problem; one for each spacecraft and one for the cable as shown in Figure 2.15. The Third Law can be applied to give $F_A = F'_A$ and $F_B = F'_B$. The equation of motion for the cable is*

$$F_B - F_A = M_C a_C,$$

where M_C is the mass and a_C is the acceleration of the centre of mass of the cable. If we take the limit where we can ignore the mass of the cable then $F_A = F_B$, i.e. irrespective of the relative capabilities of the two winches, each spaceship experiences a force of the same magnitude. Ship A therefore experiences an acceleration of magnitude

$$a_A = \frac{F_A}{M_A}$$

and B experiences an acceleration

$$a_B = \frac{F_B}{M_B} = \frac{F_A}{M_B}$$

with the directions of the acceleration vectors as shown in Figure 2.15.

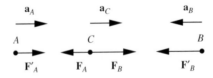

Figure 2.15   Free body diagram for two spacecraft connected by a cable.

Example 2.3.3 *Atwood's machine consists of two masses m_1 and m_2 connected by an inextensible rope of length l which is slung over a frictionless pulley of negligible mass (Figure 2.16). Determine the acceleration of the masses and the tension in the rope (you may assume that the tension is constant throughout the rope[4]).*

Solution 2.3.3 *The masses are linked by a rope of constant length, which couples their motion such that $y_1 + y_2$ is a constant. Differentiation of this equation*

---

[4] After reading Chapter 4 you might like to convince yourself that this is a good approximation if the mass of the pulley is small compared to m_1 and m_2.

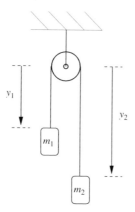

Figure 2.16 Two masses attached by a string which passes over a pulley. The device is known as Atwood's machine.

twice leads to an equation of constraint between the accelerations: $a_1 = d^2 y_1/dt^2 = -a_2 = a$. Since the pulley is massless, it cannot alter the tension, hence each mass experiences a force T due to the rope (see the free-body diagram in Figure 2.17). Applying the Second Law to each mass in turn leads to the following equations of motion:

$$m_1 g - T = m_1 a,$$

$$m_2 g - T = -m_2 a.$$

Subtraction of the equations eliminates T and yields

$$(m_1 - m_2)g = (m_1 + m_2)a$$

or

$$a = \frac{m_1 - m_2}{m_1 + m_2} g.$$

If the masses are only slightly different a may be small and hence easily measured, leading to a simple method for the determination of g.

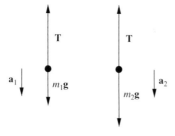

Figure 2.17 Free body diagrams for the masses of Atwood's machine.

T may be determined by the elimination of a from the equations of motion,

$$(m_1 + m_2)g - 2T = (m_1 - m_2)a$$

then

$$T = \frac{1}{2}(m_1(g - a) + m_2(g + a)) = \frac{2m_1 m_2 g}{m_1 + m_2}.$$

Notice that if $m_1 = m_2 = m$ then $a = 0$ and $T = mg$.

2.3.3 Normal forces and friction

Although there are only four known fundamental forces in Nature (the electro-magnetic, gravitational, strong and weak nuclear forces) it often seems like there are many more. We speak of the force of the wind and the sea or the tension in a rope. Each of these has their origins in the electromagnetic force, but it seldom helps in everday life to think in such terms. In this section we turn our attention to another essentially electromagnetic force that often plays a very important role when it comes to understanding the dynamics of everyday things: friction.

When solid surfaces are brought into contact the interaction between them is primarily electrostatic and depends on the structure of the two surfaces. No surfaces are perfectly smooth, zooming in on them would reveal microscopic ridges and valleys whose prominence and depth are determined by the material and the way it was prepared, see Figure 2.18. We can quite generally express the force that results when two surfaces are put into contact as the sum of components parallel and perpendicular to the surface. The perpendicular component is known as the normal force. It is the result of a microscopic compression of the layers of atoms within the surface. It does not take much displacement of the layers of atoms to support everyday objects on solid surfaces, and for the most part the compression giving rise to the normal force goes unnoticed, i.e. things are solid. Experiments with reflected laser beams from polished metal surfaces are able to measure this compression. If we imagine pushing the two surfaces together, the amount of compression increases. Usually, the contact area is a tiny fraction of the total

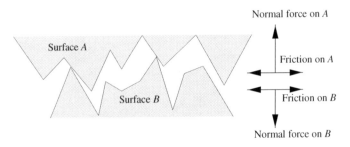

Figure 2.18   Illustration of two surfaces touching. The contact area is a tiny fraction of the area of either surface.

surface area and increasing the compression causes this fraction to increase by deformation of the points of contact. The parallel component of the force is called friction; it clearly acts to resist any motion if we attempt to drag one surface across the other. Already we can guess that as the normal force increases so will the friction because of the more intimate contact between the two surfaces.

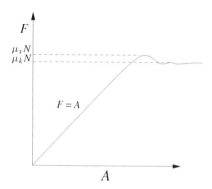

Figure 2.19   The magnitude of friction F plotted against the applied force A.

Now consider a simple experiment. A wooden block rests on a table. A force **A** is applied to the block in some direction parallel to the plane of the table. As the magnitude of **A** is increased from zero the following is observed. For small values of A the block remains stationary. The fact that the block does not accelerate as A is increased implies that friction must be equal in magnitude but opposite in direction to the applied force. This is represented by the linear portion of the graph of F against A in Figure 2.19. At some value of A the block begins to move indicating that there must be a maximum value of friction, F_{max}, acting between the block and the table. It is found experimentally that F_{max} is proportional to the normal force. This is hardly surprising since the two forces have a similar origin, and both are dependent on the actual contact area between the surfaces. This connection is expressed via

$$F_{max} = \mu_s N, \qquad (2.29)$$

where the constant μ_s is known as the coefficient of static friction and is dependent on the nature of the surfaces. μ_s is typically found to lie in the range $0.1 - 0.9$ for everyday objects, although it is possible to manufacture materials that have μ_s considerably greater than unity. When the applied force exceeds F_{max} the object will start to move and the magnitude of friction will decrease slightly to a roughly constant value of $\mu_k N$ where μ_k is the coefficient of kinetic friction. Although μ_k does have a weak and complicated dependence on velocity, for simplicity of calculation we will treat it as a constant, dependent only on the nature of the surfaces in contact. Certain fluids can have a dramatic effect on the friction between two surfaces. When oil is used to coat surfaces it acts as a buffer between the ridges and drastically reduces the value of μ_s. Note that Eq. (2.29) is *not* a vector equation. It expresses the relationship between the magnitudes of two vectors $\mathbf{F}_{max}$ and $\mathbf{N}$ that are perpendicular to each other; Eq. (2.29) written with F_{max} and N as vectors would incorrectly imply that these vectors were parallel.

Example 2.3.4 *A wooden block is at rest on a horizontal wooden plank. One end of the plank is slowly raised. Determine the angle that the plank makes with the horizontal when the block begins to slide. The coefficient of static friction between the block and the plank is 0.3.*

Solution 2.3.4 *Let the mass of the block be m. The weight of the block has components $mg \sin\theta$ parallel, and $mg \cos\theta$ perpendicular, to the surface of the plank. Static friction opposes the force down the slope and has a maximum value $\mu_s mg \cos\theta$. When the component of the weight down the slope reaches this maximum value, the block begins to slide. This occurs at an angle θ_m where*

$$mg \sin\theta_m = \mu_s mg \cos\theta_m.$$

Hence

$$\theta_m = \tan^{-1} \mu_s.$$

Using $\mu_s = 0.3$ this gives $\theta_m = 16.7°$.

Electrostatic forces between surfaces are of crucial importance to everyday life. That you are able to stand on the floor is only possible because a condition of static equilibrium exists with the normal force counteracting your weight. You are able to write only if the maximum value of static friction between your fingers and the pen exceeds the force that the writing surface exerts on the pen in a direction along its length. In some circumstances we try to reduce friction to facilitate motion - the application of oil to a rusty lock, for example. In other situations friction is essential to motion: when we walk, it is the friction between the floor and our feet that causes us to accelerate horizontally. The maximum horizontal force that we can apply to the floor is equal in magnitude but opposite in direction to the maximum horizontal force that the floor can apply to our feet and this force is limited by the maximum static friction. Therefore, running shoes and car tires are both designed to produce large static friction so that the runner or car engine may exert large forces on the ground without slipping.

Example 2.3.5 *Calculate the maximum acceleration that a four-wheel drive car may achieve going (a) straight uphill or (b) straight downhill, on a slope that makes an angle θ with the horizontal.*

Solution 2.3.5 *As with the block in the previous example, the weight will have components $mg \sin\theta$ and $mg \cos\theta$ parallel and perpendicular to the slope. Since there is no component of acceleration perpendicular to the slope, the normal force must have magnitude $mg \cos\theta$ and the maximum value of static friction will be $\mu_s mg \cos\theta$. This is the maximum force that the engine can exert without causing the wheels to spin. Applying the Second Law to components parallel to the slope, and taking acceleration uphill as positive gives*

$$ma = mg(\mu_s \cos\theta - \sin\theta)$$

or

$$a = g(\mu_s \cos\theta - \sin\theta)$$

in case (a) where the car is going uphill. Under normal circumstances a will be positive, but we can well imagine the situation where the road is icy; μ_s is then very small and we may easily get $\tan\theta > \mu_s$. Then the acceleration becomes negative and the car slides downhill. Under these circumstances, the wheels will be spinning and we no longer have a condition of static friction between the tyres and the road and μ_k replaces μ_s to give

$$a = g(\mu_k \cos\theta - \sin\theta).$$

For case (b) where the car is heading downhill we simply reverse the sign of the friction to give

$$a = g(-\mu_s \cos\theta - \sin\theta).$$

A couple of comments are perhaps in order following this example. Firstly, it may seem strange that static friction is used to describe the behaviour of a wheel, which after all is designed to roll. A little thought should enable you to visualise what is happening. At any instant, there is some area of the wheel that makes contact with the road and over this contact area the surfaces of the wheel and the road may look something like those shown in Figure 2.18: that part of wheel that is in contact with the road is instantaneously at rest. The wheel pushes against the road but does not slip, rather the forward motion creates a new region of contact between the tyre and the road, and again we have instantaneous static friction over this new area. The continual making and breaking of these contact regions does have a cost, however, producing what is known as rolling friction, characterised by the coefficient of rolling friction μ_r. Rolling friction accounts for the slowing down of a wheel rolled on a surface in vacuum. In many dynamical problems μ_r can be ignored, since it is typically an order of magnitude smaller than μ_s. Secondly, in the solution to the previous example we did not explicitly consider the fact that there were four wheels in contact with the road. This simplification is partially justified because the final result is independent of the mass. Thus is doesn't matter how the weight is distributed over the four wheels, the maximum value of a is the same. In practice, uneven forces on the wheels might lead to other problems, such as the rotation of the car about its centre of mass, but we shall not concern ourselves with such matters.

2.3.4 Momentum conservation

Consider a system of particles and assume that the ith particle experiences a force $\mathbf{F}_i$ and has momentum $\mathbf{p}_i$. Since each particle obeys Newton's Second Law (Eq. (2.14)) we can write

$$\mathbf{F} = \frac{d\mathbf{P}}{dt}, \tag{2.30}$$

where $\mathbf{F} = \sum_i \mathbf{F}_i$ is the net force acting on, and $\mathbf{P} = \sum_i \mathbf{p}_i$ is the total momentum of, the system of particles. If the net force is zero then it follows that the total momentum of the system is constant irrespective of the details of the internal forces of the system. This very important result is known as the law of conservation of momentum. For us, momentum conservation follows immediately as a consequence of Newton's Second Law. However, that is to undermine the significance of what is now understood to be a fundamental law of physics: momentum conservation applies even in circumstances where Newton's Second Law does not[5]. Momentum conservation will also play a very important role in Chapter 11. From the perspective of this chapter, it will provide a very powerful tool to help us solve problems in dynamics. The beauty is that it can be used in circumstances where unknown forces act within a system, for example when two objects collide elastically the forces that are present during the impact are not generally known but since they are internal to the system as a whole the total momentum of the system is unchanged. It means that we can go ahead and compute the momentum of the system before the action of the forces and then again afterwards, and the two must be the same.

Example 2.3.6 *A cannon fires a cannonball at an angle θ to the horizontal and at a speed v_0 relative to its muzzle. The cannon is constrained so that it recoils along horizontal rails. Use momentum conservation to calculate the recoil velocity of the cannon. Assume that the rails are frictionless.*

Solution 2.3.6 *Let the mass of the cannon be M, the mass of the cannonball m and the recoil velocity be horizontal and of magnitude v. We will take the "system" to be the cannon and cannonball. Since the rails are frictionless there are no external forces acting in the horizontal direction and we can therefore use momentum conservation in this direction. Naturally enough, we work in a frame of reference in which the cannon and cannonball are both initially at rest. Momentum conservation in the horizontal direction then gives*

$$0 = m(v_0 \cos\theta - v) - Mv,$$

where the recoil speed of the cannon has been subtracted from the horizontal component of the cannonball's velocity relative to the muzzle. This can be rearranged to give

$$0 = mv_0 \cos\theta - v(M + m),$$

$$i.e. \quad v = \frac{mv_0 \cos\theta}{m + M}.$$

In this last example, the importance of the rails being frictionless should be emphasised. Friction is to be viewed as an external force, for if it is not negligible then we would need to take account of the fact that some of the horizontal momentum is transferred to the Earth. Momentum conservation would still apply but now only if we widen the definition of our system to include the Earth. Just how much recoil momentum is transferred, via friction, to the Earth requires an

---

[5] e.g. the study of quantum mechanical systems.

understanding of the frictional forces that act. In such circumstances, the law of momentum conservation is not so useful.

Momentum conservation will recur frequently throughout the rest of this book, not least when we come to discuss collisions in Section 3.3.

2.3.5   Impulse

When forces act over a short time, as in the collision of a ball with the ground, the detailed time-dependence of the force is usually unknown. We can in some cases instead work with the time integral of the force, which is equivalent to the change in momentum. Suppose a force $\mathbf{F}(t)$ acts on a particle for a short time, from t_1 to t_2. The impulse is defined to be

$$\int_{t_1}^{t_2} \mathbf{F}(t)\, dt = \int_{t_1}^{t_2} \frac{d\mathbf{P}}{dt}\, dt = \mathbf{P}(t_2) - \mathbf{P}(t_1) = \Delta\mathbf{P}. \qquad (2.31)$$

Example 2.3.7 *Estimate the force involved in serving a tennis ball at a speed of 120 km hr^{-1}.*

Solution 2.3.7 *The mass of a tennis ball is roughly 60 g. The ball is hit at the highest point of the toss where it is stationary, so the change in momentum is equal to the momentum of the ball after serving. The collision takes typically about 2 milliseconds.*

$$\Delta P = 60 \times 120 \times \frac{10^{-3} \times 10^3 \,\mathrm{kg\,m}}{60 \times 60\,\mathrm{s}} = 2.0\,\mathrm{kg\,m\,s}^{-1}.$$

This leads to a mean value of the force:

$$\mathbf{F}_{av} = \frac{\Delta\mathbf{P}}{\Delta t} = \frac{2}{2 \times 10^{-3}}\,\mathrm{N} = 10^3\,\mathrm{N}.$$

This is roughly the weight of a 100 kg mass and that may seem surprising given that the mass of a typical tennis player is less than 75 kg. To determine what effect this force has on the tennis player, we can calculate their speed of recoil assuming that they are off the ground when contact is made:

$$v = \frac{\Delta P}{M} \sim \frac{2.0\,\mathrm{kg\,m\,s}^{-1}}{75\,\mathrm{kg}} = 0.027\,\mathrm{m\,s}^{-1}$$

or about 0.1 km hr^{-1}. Thus, despite the large force involved in propelling the ball, the duration is short enough, and the mass of the player large enough, such that the player experiences only a very small recoil velocity.

2.3.6   Motion in fluids

When a solid body moves through a fluid, some of the fluid is dragged along with the body and the fluid acquires momentum in the direction of the object's motion. This change in momentum of the fluid is associated with a force that the body exerts on the fluid and, according to Newton's Third Law, the fluid must exert an equal and opposite force upon the body. In this way the body experiences a resistive "viscous" force. The details depend upon the nature of the fluid and on the efficiency with which the body interacts with it. The latter is in turn dependent on the geometry of the body and on the nature of the surfaces that drag the fluid, as well as the velocity of the object relative to the fluid. Empirical evidence supports the following expression for the viscous force:

$$F_v = -Cv - Dv^2. \tag{2.32}$$

For motion in liquids $C \gg D$ and $F_v \approx -Cv$, at least for low speeds. For motion in air the quadratic term dominates and $F_v \approx -Dv^2$.

Example 2.3.8 *A ball bearing is released at rest in a tall cylinder of glycerol and falls under the influence of gravity. Show that the speed tends to a limiting value.*

Solution 2.3.8 *The forces acting on the ball bearing are gravity and the viscous force. The Second Law gives*

$$ma = m\frac{dv}{dt} = mg - Cv.$$

This is a first order differential equation in v. Rearranging gives

$$\int \frac{dv}{1 - \dfrac{C}{mg}v} = \int g\,dt$$

which can be integrated (substitute for $1 - \frac{C}{mg}v$) to yield

$$\ln\left(1 - \frac{C}{mg}v\right) = -\frac{C}{m}t + B,$$

where B is the constant of integration. Rearrangement, and fixing the constant of integration by the requirement that $v = 0$ at $t = 0$, gives

$$v = \frac{mg}{C}\left(1 - e^{-\frac{C}{m}t}\right).$$

The above equation contains the essential features of an object falling in a fluid. There is a limiting (or terminal) velocity $v_t = \frac{mg}{C}$ that the speed tends towards for large t. The time taken to reach a speed of $(e-1)/e$ of v_t is $\frac{m}{C}$, which is shorter the more viscous the medium. Note that objects of different mass but the same value of C have different terminal speeds. This is in contrast to the behaviour of falling objects in vacuum, where the acceleration g is constant and independent of the mass. For a medium of vanishing small viscosity, or for very small t, a Taylor expansion of the exponential leads to

$$v \approx \frac{mg}{C}\left(1 - 1 + \frac{C}{m}t + \ldots\right) \approx gt,$$

which is as expected.

PROBLEMS 2

2.1 (a) Two masses m_1 and m_2 have position vectors $\mathbf{r}_1$ and $\mathbf{r}_2$ respectively. Write down the position vector of the centre of mass of this system. Find the vector equation for the straight line through the two masses and hence show that the centre of mass lies on this line.

(b) Write down the position vector of the centre of mass for three equal masses with position vectors $\mathbf{a}$, $\mathbf{b}$ and $\mathbf{c}$. Show that the centre of mass lies at the intersection of the medians of the triangle defined by the positions of the masses. [A median of a triangle is a line running from one of its corners to the midpoint of the opposite side.]

2.2 A system is composed of three isolated particles with masses $m_1 = 10\,\mathrm{g}$, $m_2 = 30\,\mathrm{g}$ and $m_3 = 40\,\mathrm{g}$ at positions $\mathbf{r}_1 = 2\mathbf{i} - 2\mathbf{j}$, $\mathbf{r}_2 = -3\mathbf{i} + \mathbf{j}$ and $\mathbf{r}_3 = 5\mathbf{i} + 6\mathbf{j}$ respectively (all distances are measured in metres). Calculate the position vector of the centre of mass. If a force $F = 3\mathbf{i}$ (newtons) is applied to one of the particles, work out the acceleration of the centre of mass. Does it matter to which particle the force is applied?

2.3 A hose sends a stream of water at a rate of $2\,\mathrm{kgs}^{-1}$ and a velocity of $10\,\mathrm{ms}^{-1}$ against a wall. If all the water runs down the wall, what is the force on the wall?

2.4 A prisoner of mass $80\,\mathrm{kg}$ plans an escape using a rope made of strips of bed sheets. If his window is $20\,\mathrm{m}$ from the ground, and the makeshift rope will not support a tension of greater than $600\,\mathrm{N}$ without breaking, find the minimum speed at which the prisoner hits the ground assuming he descends by sliding vertically down the rope without making contact with the wall.

2.5 Ancient Egyptians push a 5.00 tonne block of stone on rollers up a slope with an acceleration of $0.30\,\mathrm{ms}^{-2}$. The slope is inclined at $20.0°$ to the horizontal. Assuming that the rollers produce a frictionless surface, and that the mass of the rollers can be ignored, calculate the force applied by the Egyptians to the block. Calculate the normal force acting on the block.

2.6 A skier of mass $75.0\,\mathrm{kg}$ skis over a hemispherical mound of snow of radius $10.0\,\mathrm{m}$. At the top of the mound the skier's velocity vector is horizontal with a magnitude of $30.0\,\mathrm{km\,hr}^{-1}$. Assuming the snow to be frictionless, calculate the magnitude and direction of the force exerted by the skier on the snow at the top of the mound.

2.7 Two blocks of mass $m_1 = 3.0\,\mathrm{kg}$ and $m_2 = 1.0\,\mathrm{kg}$ rest in contact on a frictionless horizontal surface. If a force of $3.0\,\mathrm{N}$ is applied to m_1 such that both blocks accelerate, deduce the contact force between the blocks.

2.8 A $2\,\mathrm{kg}$ block rests on a $4\,\mathrm{kg}$ block that rests on a frictionless surface. The coefficients of friction between the blocks are $\mu_s = 0.3$ and $\mu_k = 0.2$.

(a) What is the maximum horizontal force F that can be applied to the $4\,\mathrm{kg}$ block if the $2\,\mathrm{kg}$ block is not to slip?

(b) If F has half the value found in (a), find the acceleration of each block and the force of friction acting on each block.

(c) If F has twice the value found in (a), find the acceleration of each block.

2.9 Rocket propulsion is achieved by the emission of exhaust material at high speed. A rocket of mass m travelling with velocity $\mathbf{v}$ emits exhaust gases at constant velocity $\mathbf{u}$ relative to the rocket. Show that the net external force on the rocket $\mathbf{F}$ satisfies

$$\mathbf{F} = m\frac{d\mathbf{v}}{dt} - \mathbf{u}\frac{dm}{dt}.$$

Prove that a free rocket accelerating from rest attains a final velocity

$$\mathbf{v}_f = -\mathbf{u}\ln\frac{m_i}{m_f},$$

where m_i and m_f are the initial and final masses of the rocket.
If the rocket accelerates from rest, for a time t, in a uniform gravitational field $\mathbf{g}$ show that

$$\mathbf{v}_f = -\mathbf{u}\ln\frac{m_i}{m_f} + \mathbf{g}t.$$

Use the above expression to explain why space rockets are designed to burn their fuel quickly.

3

Energy

Newton's laws of motion lay down the foundations for a complete understanding of dynamical systems. Starting from Newton's laws we showed, in Section 2.3.4, that momentum is conserved for isolated systems and that it turns out to be very useful when tackling problems. We also remarked that momentum conservation really ought to be thought of as possessing a fundamental significance in its own right. In particular, it is not right to think of momentum conservation as being only a consequence of Newton's Second Law, rather we should think that Newton's laws had better be consistent with the law of momentum conservation. Why do we say that? Why is momentum conservation so fundamental that even Newton's laws are destined to respect it? The answer is easy to state but less easy to prove: momentum is conserved for the same reason that an experiment performed in Manchester should deliver the same result as the same experiment performed in New York. Of course that is true provided the experiment does not depend upon local differences, such as the difference in temperature etc. The key point is that moving experiments around should not, in itself, change the outcome. It is a pity that the proof is a little too sophisticated for us to include it here and that we are reduced to stating the result without any proof. Now it is clear why Newton's laws are duty bound to satisfy momentum conservation. The idea that it does not matter where an experiment is performed, all other things equal, is an example of a symmetry of Nature and the link between symmetries of Nature and conservation laws is not unique to momentum. It turns out that for every symmetry in Nature there is a corresponding conserved quantity. Perhaps chief amongst the other conserved quantities is the one associated with the fact that it does not matter *when* we perform an experiment: an experiment performed today should give the same result as the same experiment performed tomorrow (all other things being equal). That quantity is called "energy"[1].

[1] The conservation of angular momentum is a result of the fact that experiments can be turned around (i.e. rotated) without affecting their result.

Dynamics and Relativity   Jeffrey R. Forshaw and A. Gavin Smith
© 2009 John Wiley & Sons, Ltd

Unlike momentum, we will see that energy can appear in many different forms. Kinetic energy is associated with the motion of things. In a macroscopic body made of very many atoms it is usual to distinguish between the kinetic energy of the body as a whole, which arises as a result of the coherent motion of all of the atoms, and the kinetic energy possessed by the atoms as they jiggle randomly around within the body. The latter is commonly called the thermal energy of the body: hotter bodies have more thermal energy than colder ones. Potential energy is the energy that is stored up within a system. It might be the energy stored up as a result of the specific chemical arrangements of molecules in a mouthful of food or a drop of petrol. Or it could be the energy stored up by gravity at the start of a roller-coaster ride, or in a collapsing star. What is important is that energy can be converted from one type to another and yet, provided we account for all forms of energy, the total is a conserved quantity. This is hugely significant. Provided we do the book keeping correctly, and add up the numerical values of all the forms of energy of an isolated system, the total will always be the same, irrespective of the details. A roller coaster starting its decent can be described in terms of a transformation of gravitational potential energy into kinetic energy; a tennis player may use the chemical energy stored in a banana to help her complete a match; a rocket converts chemical energy into gravitational potential energy and kinetic energy following its launch.

Just as the conservation of momentum can be derived using Newton's laws so we will see that energy conservation is also already encoded within them. We shall see this soon when we encounter the Work-Energy Theorem. But to pave the way we first need to introduce the idea of work.

3.1 WORK, POWER AND KINETIC ENERGY

The work done dW by a force $\mathbf{F}$ acting on a particle as it moves through an infinitesimal displacement $d\mathbf{r}$ is defined to be

$$dW = \mathbf{F} \cdot d\mathbf{r}. \tag{3.1}$$

The SI unit of work is the joule[2] ($1.0\,\text{J} \equiv 1.0\,\text{N}\,\text{m}$).

The utility of this definition will become apparent very soon, for now we shall explore some of its properties. Consider the situation depicted in Figure 3.1. A force $\mathbf{F}$ is pushing a block against a wall. We can resolve $\mathbf{F}$ into components parallel and perpendicular to the wall but the displacement is constrained always to be parallel to the wall. Notice that it is only the component of force parallel to the displacement that contributes to the work; the perpendicular component of the force does not contribute to the scalar product in Eq. (3.1) and hence does no work.

So how do we calculate the work done when a particle travels between two points along an arbitrary path? It may not be immediately obvious how the definition Eq. (3.1) for the infinitesimal work dW is to be turned into an expression for the work done along a path of finite length. Some insight can be obtained by

---

[2] After James Prescott Joule (1818–89).

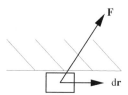

Figure 3.1   The forces involved when pushing a block against a wall.

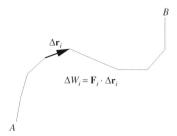

Figure 3.2   The path between A and B is made up of a series of discrete steps.

considering a given smooth path to be approximated by a series of N small but finite displacements $(\Delta \mathbf{r}_i)$ each corresponding to an interval of time Δt. Figure 3.2 illustrates this for a particle making the journey from point A to point B. You might like to consider the vertices on the path as successive positions of the particle when illuminated by a stroboscope where the interval between the flashes is Δt. The displacement vector that takes us from A to B can then be written as

$$\mathbf{r}_{AB} = \mathbf{r}_B - \mathbf{r}_A = \sum_{i=1}^{N} \Delta \mathbf{r}_i, \qquad (3.2)$$

where $\mathbf{r}_A$ and $\mathbf{r}_B$ are the position vectors of A and B, respectively. For each of the steps we can calculate the work done:

$$\Delta W_i = \mathbf{F}_i \cdot \Delta \mathbf{r}_i, \qquad (3.3)$$

where $\mathbf{F}_i$ is the force vector acting on the particle at the ith step. We can then approximate the total work done in going from A to B as

$$W_{AB} \approx \sum_{i=1}^{N} \Delta W_i. \qquad (3.4)$$

This approximation becomes more accurate as Δt gets smaller (i.e. N gets larger) and the displacement in each step gets smaller. In the limit that $N \to \infty$ the displacement $\Delta \mathbf{r}_i \to d\mathbf{r}$ and it becomes parallel to a tangent vector to the curve.

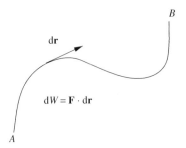

Figure 3.3   The path taken by a particle as it moves from A to B. The work done is the integral of $\mathbf{F} \cdot d\mathbf{r}$ along the path. The direction of $d\mathbf{r}$ is indicated.

Simultaneously the ratio $\frac{\Delta \mathbf{r}_i}{\Delta t} \rightarrow \mathbf{v}(t)$, the velocity at time t. This limit is shown in Figure 3.3. Now the work done in going from A to B is expressed exactly as the integral

$$W_{AB} = \int_A^B \mathbf{F}(\mathbf{r}) \cdot d\mathbf{r}, \tag{3.5}$$

where the force $\mathbf{F}(\mathbf{r})$ is written explicitly as a function of position.

Example 3.1.1 *Calculate the work done by gravity on a projectile of mass m that is fired to an altitude h.*

Solution 3.1.1 *We use a co-ordinate system where* $\mathbf{i}$ *is horizontal and* $\mathbf{j}$ *is vertical. Then (see Figure 3.4)*

$$d\mathbf{r} = dx\,\mathbf{i} + dy\,\mathbf{j}.$$

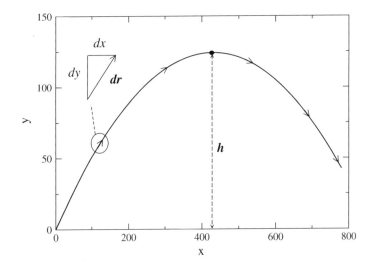

Figure 3.4   Parabolic path of a projectile launched from the origin.

The force of gravity is represented (for low altitudes) as $\mathbf{F} = -mg\,\mathbf{j}$. *So*

$$\mathrm{d}W = -mg\,\mathbf{j} \cdot (\mathrm{d}x\,\mathbf{i} + \mathrm{d}y\,\mathbf{j}) = -mg\,\mathrm{d}y,$$

i.e. only the vertical component of the path contributes to the work. We integrate the above expression to obtain

$$W = \int \mathrm{d}W = -mgh.$$

So although the path is parabolic we need only know the maximum height reached in order to determine the work done.

We are now ready to establish why work is such a useful concept. To do this we shall use Newton's Second Law in order to evaluate the work done on a particle. Consider

$$W_{AB} = \int_A^B \mathbf{F}(\mathbf{r}) \cdot \mathrm{d}\mathbf{r} = \int_A^B \frac{\mathrm{d}(m\mathbf{v})}{\mathrm{d}t} \cdot \mathrm{d}\mathbf{r}. \tag{3.6}$$

We can use $\mathrm{d}\mathbf{r} = \mathbf{v}\,\mathrm{d}t$, allowing us to write

$$W_{AB} = \int_A^B \frac{\mathrm{d}(m\mathbf{v})}{\mathrm{d}t} \cdot \mathbf{v}\,\mathrm{d}t = \int_A^B m\frac{\mathrm{d}(\mathbf{v})}{\mathrm{d}t} \cdot \mathbf{v}\,\mathrm{d}t. \tag{3.7}$$

This integral is easily evaluated once we recognise that

$$\frac{\mathrm{d}(v^2)}{\mathrm{d}t} = \frac{\mathrm{d}(\mathbf{v} \cdot \mathbf{v})}{\mathrm{d}t} = 2\mathbf{v} \cdot \frac{\mathrm{d}\mathbf{v}}{\mathrm{d}t},$$

where we have used the product rule for differentiation of the scalar product. We are therefore able to rewrite Eq. (3.7) as

$$W_{AB} = \int_A^B \frac{\mathrm{d}\left(\frac{1}{2}mv^2\right)}{\mathrm{d}t}\mathrm{d}t = \frac{1}{2}mv_B^2 - \frac{1}{2}mv_A^2 = T_B - T_A, \tag{3.8}$$

where v_A and v_B are the speeds of the particle at positions A and B respectively. The quantity

$$T = \frac{1}{2}mv^2 \tag{3.9}$$

is the kinetic energy and the effect of the force is to alter the kinetic energy by doing work. Eq. (3.8) is called the Work-Energy Theorem.

Example 3.1.2 *Let us consider again the example of the projectile. We can now use Eq. (3.8) to calculate the speed of the projectile at an altitude y having been launched from $y = 0$.*

Solution 3.1.2 *Recall that the initial velocity is* **u**. *The Work-Energy Theorem allows us to compute the answer straight away:*

$$-mgy = \frac{1}{2}m(v^2 - u^2)$$

so

$$v = \sqrt{u^2 - 2gy}.$$

Note that we could have just as easily derived this result using Eq. (1.20).

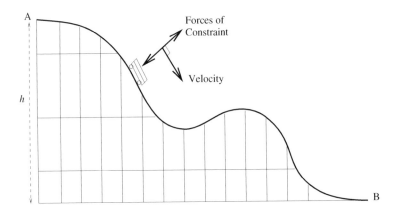

Figure 3.5   A roller-coaster travels along a track. The forces of constraint do no work.

It often happens that a body is forced to follow a well determined path in a field of force as a result of some constraint. For example, the roller coaster shown in Figure 3.5 is constrained to follow the tracks. The car begins its journey at A then descends along the track, rises again before falling to B. If we ignore friction and air-resistance then the only forces acting on the car are gravity and the normal forces between the rails and the wheels that serve to keep the roller-coaster on the track. Because the wheels of a roller-coaster form an interlocking structure with the track, the normal forces may act either inwardly or outwardly depending on whether the tendency is for the car to 'push into the track' or 'fly off the rails'. At some point on the path between A and B we determine the infinitesimal work to be

$$dW = \mathbf{F} \cdot d\mathbf{r} = (-m\mathbf{g} + \mathbf{N}) \cdot \mathbf{v}\, dt, \qquad (3.10)$$

where $\mathbf{N}$ is the normal force. Since $\mathbf{N}$ and $\mathbf{v}$ are orthogonal this gives

$$dW = -m\mathbf{g} \cdot \mathbf{v}\, dt = -mg\, dy. \qquad (3.11)$$

The work done in going from A to B is then obtained by integration:

$$W_{AB} = \int_A^B dW = -\int_{y_A}^{y_B} mg\, dy = mg(y_A - y_B) = mgh. \qquad (3.12)$$

This shows that while work is done by gravity, no work is done by the forces of constraint. This is a specific example of a more general result, which states that, for many systems, the work done by the forces of constraint is zero. In the roller-coaster example the work done depends on the change in height, as it did with the free projectile. This is an important simplification, which allows us to calculate the kinetic energy at any point on the path knowing only the height, despite the fact that we know neither the detailed equation of the path, nor any details about the forces keeping the roller coaster on the rails.

Power is defined as the rate at which work is done. If a force $\mathbf{F}$ acts on a body, which undergoes a displacement $d\mathbf{r}$, then the infinitesimal work done is $dW = \mathbf{F} \cdot d\mathbf{r}$. The instantaneous power P is simply this work divided by the time interval dt in which the displacement occurs

$$P = \frac{dW}{dt} = \mathbf{F} \cdot \frac{d\mathbf{r}}{dt} = \mathbf{F} \cdot \mathbf{v}. \qquad (3.13)$$

In the SI system the unit of power is the watt[3] (W):

$$1\,\text{W} = 1\,\text{J}\,\text{s}^{-1}.$$

3.2 POTENTIAL ENERGY

In the previous section we defined work in terms of the integral over a path between two points. We also proved that for a uniform gravitational field, the work done is proportional to the difference in height between the two points, but does not depend on the path taken between them. Such dependence on the initial and final positions but not on the path taken is the defining feature of a conservative field of force.

> A field of force $\mathbf{F}(\mathbf{r})$ is conservative if the work done in going between positions $\mathbf{r}_A$ and $\mathbf{r}_B$ is independent of the path taken. A field of force that is not conservative is known as non-conservative or dissipative.

A corollary to the above definition is that if a path ends at the starting point to form a closed loop the work done is always zero for a conservative force.

Gravity and electrostatic forces provide us with two examples of conservative forces whilst friction and air-resistance are both non-conservative. It is easy to see that the work done by friction must always be path-dependent. A longer path will result in more work being done since the frictional force always acts in opposition to the motion. For such a situation we obtain

$$W = \int_A^B \mathbf{F} \cdot d\mathbf{r} = -\mu_k N \int_A^B dr = -\mu_k N l, \qquad (3.14)$$

where l is the length of the path and μ_k is the coefficient of kinetic friction. Notice that, even for a closed loop like that illustrated in Figure 3.6, the work done by friction is still negative (which simply means that the kinetic energy reduces).

---

[3] James Watt (1736–1819).

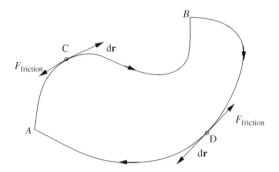

Figure 3.6   The work done against friction is negative, even for a closed loop.

What if all of the forces acting on a particle are conservative? In that case we are free to construct a special function of position, which we call the potential energy, $U(\mathbf{r})$. It is defined by

$$U(\mathbf{r}_B) - U(\mathbf{r}_A) = -W_{AB},\qquad(3.15)$$

which is the negative of the work done by the net force in going from A to B. Note that no such special function can be found for non-conservative forces because the work done in going from A to B depends on more than just the position of the end-points. The motivation for taking the negative of the work done in the definition is because, when combined with Eq. (3.8), we get

$$U(\mathbf{r}_B) - U(\mathbf{r}_A) = T_A - T_B$$

and therefore

$$T_A + U(\mathbf{r}_A) = T_B + U(\mathbf{r}_B).\qquad(3.16)$$

Thus for motion of a particle under the influence of conservative forces only we see that the sum of the kinetic and potential energy is a conserved quantity. We call this sum the mechanical energy, E:

$$E = \frac{1}{2}mv^2 + U(\mathbf{r}).\qquad(3.17)$$

Note that the opposite sign in Eq. (3.15) would be equally valid, we would just have to flip the sign of the potential energy in Eq. (3.16) and it would be the difference rather than the sum of the kinetic and potential energies that would be conserved. Also note that Eq. (3.15) does not uniquely define the function $U(\mathbf{r})$, since it is possible to add any constant to it and yet still maintain the same difference in potential energy between any two given points. The mechanical energy of a particle is therefore not an absolute quantity but rather it is always defined relative to an arbitrary position of zero potential energy.

Example 3.2.1 *Obtain an expression for the potential energy of a particle in the gravitational field of a planet.*[4]

Solution 3.2.1 *The force is given by*

$$\mathbf{F}_G = -\frac{GMm}{r^2}\hat{\mathbf{r}},$$

where M is the mass of the planet, m is the mass of the particle, r is the distance and $\hat{\mathbf{r}}$ is a unit vector directed from the centre of the planet to the particle. To calculate the potential energy we will determine the work done in bringing the particle in from a point at infinity along a path anti-parallel to the unit vector $\hat{\mathbf{r}}$. Starting the particle out at infinity is our choice, any other point would do equally well. In this case $d\mathbf{r} = dr\,\hat{\mathbf{r}}$ and

$$W_{\infty B} = \int_{\infty}^{r_B} -\frac{GMm}{r^2}\hat{\mathbf{r}} \cdot \hat{\mathbf{r}}\,dr = -\int_{\infty}^{r_B} \frac{GMm}{r^2}dr.$$

This integral is evaluated to give

$$W_{\infty B} = \frac{GMm}{r_B}.$$

Thus the work done in bringing the particle from an infinite separation to a distance r_B from the centre of the planet is a positive quantity. This is what we expect, since together with the Work-Energy Theorem this implies an increase in kinetic energy, consistent with the effect of an attractive force. To obtain the potential energy we need to choose where it should be zero. The natural choice is to pick the potential to be zero at infinity, in which case

$$U(r) = -\frac{GMm}{r}. \tag{3.18}$$

Example 3.2.2 *Calculate the escape speed of a projectile launched from the surface of the Earth.*

Solution 3.2.2 *If we ignore the effects of air-resistance, which is a dissipative force, then the mechanical energy is conserved. We want to consider a projectile launched with an initial velocity v_0 from the Earth's surface ($r = r_E$) that is able to reach a separation $r \to \infty$. Since this is, by definition, the separation at which the potential energy is zero in Eq. (3.18) we get*

$$\frac{1}{2}mv_0^2 - \frac{GMm}{r_E} = 0$$

since the final kinetic will be identically zero if the projectile has just enough energy

---

[4] The proof that the gravitational force is conservative can be found in Section 9.1.

to escape. Rearranging gives

$$v_0 = \sqrt{\frac{2GM}{r_\mathrm{E}}}.$$

3.2.1 The stability of mechanical systems

In Chapter 2 we defined static equilibrium to hold when the net force on a particle is zero. Suppose we have a system in which all the forces acting on the particle are conservative, and which the particle is free to move in one dimension only (we go beyond one dimension in Chapter 9). As we have seen, we can then define a potential energy $U(x)$ where x represents the position of the particle. Using Eq. (3.1), $\mathrm{d}W = -\mathrm{d}U = F(x)\,\mathrm{d}x$ and so

$$F(x) = -\frac{\mathrm{d}U(x)}{\mathrm{d}x}. \qquad (3.19)$$

The condition of static equilibrium can thus be written

$$\frac{\mathrm{d}U}{\mathrm{d}x} = 0, \qquad (3.20)$$

i.e. the potential energy has a stationary point where the force is zero. This stationary point can be either a maximum or a minimum, as illustrated in Figure 3.7. Figure (a) illustrates the potential energy in the case that there is a maximum at $x = x_0$. The particle is in static equilibrium at this point, but if we consider even tiny departures from $x = x_0$ we see that the equilibrium is unstable. For example, moving the particle to $x < x_0$ results in a potential energy surface with positive gradient. Thus $F = -\frac{\mathrm{d}U}{\mathrm{d}x}$ is negative and the force drives the particle away from

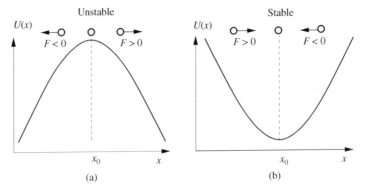

Figure 3.7   Potential energy as a function of distance for two one-dimensional systems. In (a) a maximum in U results in unstable equilibrium: for $x \neq x_0$ the force acts to move the particle away from x_0. In (b) for $x \neq x_0$ the force acts to move the particle towards x_0 and the equilibrium is stable.

$x = x_0$. If we move the particle to $x > x_0$ then $\frac{dU}{dx}$ is negative, the force positive and the particle is driven to higher values of x. Thus for system (a), any small departure from $x = x_0$ results in a force that works in the same direction as the displacement. The equilibrium is therefore unstable to small perturbations away from $x = x_0$. In Figure 3.7(b) a minimum in the potential energy is illustrated. Now the behaviour is reversed and small displacements from equilibrium result in forces that act in opposition to the displacement. These restoring forces ensure that small perturbations do not produce large effects; the equilibrium in (b) is stable.

In practice, positions of unstable equilibrium can never be achieved. While we can imagine setting up a system in unstable static equilibrium by ensuring that the condition $x = x_0$ is perfectly observed, this is impossible to achieve in practice. However, in a situation of stable equilibrium, small perturbations produce small effects that tend to return the system back to its starting point.

The relationship between potential energy and stability is of course not limited to systems involving one degree of freedom. A good example of a potential energy surface in two dimensions is obtained upon considering a marble on the surface of a hemispherical bowl. Since the gravitational potential energy is proportional to the height of the marble it can be written as a function of the marble's position in the horizontal plane using the equation for the surface of the bowl. The contact forces between the marble and the bowl act as forces of constraint and do not contribute to the potential energy. Consider first the bowl placed with its rim on a flat horizontal surface and the opening downwards. If the marble is placed on the highest point of the outer surface of the upturned bowl then this is a position of unstable equilibrium, similar to Figure 3.7(a) and is impossible to achieve. If, however, the bowl sits so that the opening is upwards and the marble is positioned at the lowest point of the inner surface, this represents stable equilibrium. In this example, the potential energy can be expressed as a function of the two position co-ordinates in the horizontal plane, x and y. To achieve static equilibrium, the potential energy must be a stationary point with respect to variations in both x and y. In other words, both the x and y components of the force in the horizontal plane must be zero in order to have static equilibrium. Notice that since the potential energy close to the surface of the Earth is proportional to height, it follows that the surface of the bowl just happens to provide a visual map of the potential energy surface appropriate to the marble's motion.

3.2.2 The harmonic oscillator

Using the idea of potential energy, we can explore what is perhaps the single most important physical system in the whole of physics, both classical and quantum: the simple harmonic oscillator. In so doing we will learn, in this section and the next, just why an object so ordinary as a stretched spring should provide the prototype for the behaviour of a vast range of physical systems close to equilibrium. We start by considering a one-dimensional system for which the only force F acting on a particle of mass m satisfies Hooke's Law, $F = -kx$ (think of a stretched spring if you like). Such a system is known as a simple harmonic oscillator.

The Second Law specifies the equation of motion:

$$m\frac{d^2x}{dt^2} = -kx \qquad (3.21)$$

for which the general solution is

$$x(t) = A\sin(\omega t) + B\cos(\omega t), \qquad (3.22)$$

where A and B are arbitrary constants and $\omega = \sqrt{\frac{k}{m}}$ is the angular frequency of the oscillation. If you cannot already see it, then you should check that Eq. (3.22) is a solution to Eq. (3.21) and that you agree with the expression for ω. Thus a simple harmonic oscillator is characterised by an oscillation of constant frequency. We are at liberty to define $x(0) = 0$, that is we put the particle at the origin at time zero. In which case, we have $B = 0$ and

$$x(t) = A\sin(\omega t). \qquad (3.23)$$

We can deduce an expression for the potential energy of a harmonic oscillator by computing the work done in going from $x = x_A$ to $x = x_B$

$$W_{AB} = -\int_{x_A}^{x_B} kx\, dx = \frac{1}{2}k\left(x_A^2 - x_B^2\right). \qquad (3.24)$$

As usual we are free to select the point where the potential energy is zero, and we choose the origin $x = 0$. So, setting $x_A = 0$ and defining $U(0) = 0$ we obtain

$$W_{AB} = U(x_A) - U(x_B) = -U(x_B) = -\frac{1}{2}kx_B^2. \qquad (3.25)$$

Thus, the potential energy of a harmonic oscillator is given by

$$U(x) = \frac{1}{2}kx^2. \qquad (3.26)$$

The harmonic oscillator constitutes a conservative system and we can write the mechanical energy:

$$E = \frac{1}{2}mv^2 + \frac{1}{2}kx^2, \qquad (3.27)$$

where v is the speed of the particle. We can verify that this is a conserved quantity by direct substitution of the solution (Eq. (3.23)) into the energy equation, i.e.

$$E = K + U = \frac{1}{2}m\omega^2 A^2 \cos^2\omega t + \frac{1}{2}kA^2 \sin^2\omega t = \frac{1}{2}kA^2. \qquad (3.28)$$

Example 3.2.3 *Determine the force acting on a particle that moves with a potential energy* $U(x) = \frac{1}{2}kx^2 + \alpha x^4$.

Solution 3.2.3 *We use Eq. (3.19) to get from the potential energy to an expression for the force on the particle:*

$$F(x) = -\frac{dU}{dx} = -kx - 4\alpha x^3.$$

This deviates from Hooke's Law by the addition of the term $F_a = -4\alpha x^3$. Notice that F_a is still a restoring force since it depends on an odd-power of x, i.e. it points towards the origin ($x = 0$) whatever the sign of x. As a result, we can still expect the particle to oscillate about the origin. However, Newton's Second Law now leads to the equation of motion

$$m\frac{d^2x}{dt^2} = -kx - 4\alpha x^3,$$

where m is the mass of the particle. A simple sine or cosine function with fixed frequency cannot satisfy this equation and the motion is more complicated. Such a system is referred to as an anharmonic oscillator.

3.2.3 Motion about a point of stable equilibrium

Finally, we are ready to reveal why springs are so important in physics. Consider a particle at a point of stable equilibrium $x = x_0$ in a potential $U(x)$. What motion do we expect for small departures from equilibrium? We may expand the potential as a Taylor series about x_0 as

$$U(x) \approx U(x_0) + (x - x_0)\left.\frac{dU}{dx}\right|_{x_0} + \frac{1}{2}(x - x_0)^2 \left.\frac{d^2U}{dx^2}\right|_{x_0} + \dots \tag{3.29}$$

Since the equilibrium is stable we must have $\left.\frac{dU}{dx}\right|_{x_0} = 0$ and $\left.\frac{d^2U}{dx^2}\right|_{x_0} \geq 0$. We set the constant $\left.\frac{d^2U}{dx^2}\right|_{x_0} = k$ and redefine our scale of potential energy such that $U(x_0) = 0$. Thus

$$U(x) \approx \frac{1}{2}k(x - x_0)^2 \tag{3.30}$$

and so, for small enough departures from equilibrium, the potential energy of (and hence the restoring forces acting on) the particle are identical to those of the harmonic oscillator.

The above argument may be generalised to three dimensions leading to three terms in the potential energy:

$$U(x, y, z) \approx \frac{1}{2}k_x(x - x_0)^2 + \frac{1}{2}k_y(y - y_0)^2 + \frac{1}{2}k_z(z - z_0)^2, \tag{3.31}$$

where the three spring constants, k_x k_y and k_z, correspond to the generally different restoring forces in the x, y and z directions, respectively. The equilibrium position is given by the co-ordinates (x_0, y_0, z_0).

Thus we see that although an ideal spring obeys Hooke's Law and therefore is a harmonic oscillator it is not the only harmonic oscillator. We have just seen that *any* system close to a point of stable equilibrium also constitutes a harmonic oscillator and therefore behaves like a spring. The only exception is those systems for which the second derivative of the potential just happens to vanish at the point of equilibrium, i.e. $k = 0$.

3.3 COLLISIONS

At the very heart of the Newtonian programme to understand the world is the idea that particles move around and collide with each other. Collisions will typically change the energies and momenta of the colliding particles but always such that the total energy and total momentum remain unchanged. We must be very careful though in how we use energy conservation because it is possible for some energy to "leak away", e.g. the energy carried away by a sound wave when two objects collide, and it strictly needs to be included when we come to compute the total energy after the collision. These days, collisions are exploited daily at the world's particle physics colliders in order to develop an understanding of how matter behaves at the shortest distances. Although Newton's laws do not apply in those experiments (we need relativity and the quantum theory instead) it is nevertheless true that energy and momentum remain conserved. It is also true that the methods used in solving collision problems in Newtonian mechanics are very similar to those used when it comes to tackling collisions in particle physics and that is something we will explore in more detail in Section 7.2.

3.3.1 Zero-momentum frames

We will consider the collision between two classical particles that interact by a force that goes to zero at large distances. This allows us to separate the process into three stages (see Figure 3.8): (a) the early stage, when the particles are far enough apart to be each considered isolated; (b) the interaction stage when the mutual interaction is significant and the particles are accelerating; (c) the late stage when the particles are once again isolated. The particles have masses m_1 and m_2 with initial velocities $\mathbf{v}_1$ and $\mathbf{v}_2$, respectively. In stage (c), after the collision, the velocities are $\mathbf{v}'_1$ and $\mathbf{v}'_2$ respectively. In constructing Figure 3.8 we have chosen a frame of reference in which to view the collision. Naturally enough we take this to be the frame in which the experimenter is stationary. This frame is known as the *lab frame*. While the lab frame is a familiar frame of reference to use in thinking about the collision, it turns out not to be the most useful for calculations. All isolated collisions conserve linear momentum, which is a vector quantity. So if we want to make things easy for ourselves we could choose a frame of reference in which the total momentum is zero. We will call this the *zero-momentum frame*. Let us suppose that we transform to a frame of reference travelling with velocity $\mathbf{V}$ with respect to the lab frame. How will the particles move in this new frame of reference? As discussed in Section 1.3.2, we know[5] that

---

[5] These results change in Special Relativity.

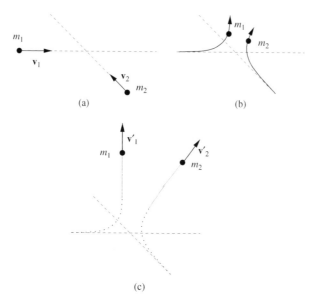

(a)   (b)

(c)

Figure 3.8   The collision between two particles in the lab frame is considered in three stages. At times long before (a) and long after (c) the collision the particles are far enough apart that we can ignore their mutual interaction and both may be treated as isolated particles. During the interaction stage (b) the particles interact and their velocity vectors are continuously changing.

$$\mathbf{v}_{1c} = \mathbf{v}_1 - \mathbf{V}$$

$$\mathbf{v}_{2c} = \mathbf{v}_2 - \mathbf{V}, \tag{3.32}$$

where $\mathbf{v}_{1c}$ and $\mathbf{v}_{2c}$ are the velocities in the new frame of reference. We assert that the total momentum in this frame is zero, i.e.

$$m_1 \mathbf{v}_{1c} + m_2 \mathbf{v}_{2c} = \mathbf{0}, \tag{3.33}$$

and solve for $\mathbf{V}$ to obtain

$$\mathbf{V} = \frac{m_1 \mathbf{v}_1 + m_2 \mathbf{v}_2}{m_1 + m_2}. \tag{3.34}$$

Notice that the form of this equation is reminiscent of the expression for the centre-of-mass vector, Eq. (2.27). In fact, for two colliding particles with position vectors $\mathbf{r}_1$ and $\mathbf{r}_2$ the time-derivative of the position of the centre-of-mass is

$$\frac{d\mathbf{R}}{dt} = \frac{1}{m_1 + m_2}\left(m_1 \frac{d\mathbf{r}_1}{dt} + m_2 \frac{d\mathbf{r}_2}{dt}\right) = \mathbf{V}. \tag{3.35}$$

Thus we see that zero-momentum frames are characterized by the fact that they are those frames in which the centre of mass is at rest. We have used the plural "frames" here because so far we did not specify the location of the origin, so there exist an infinite number of zero momentum frames all moving parallel to each other and differing only in the location of the origin. We are at liberty to specify the location of the origin and shall usually choose it to be the position of the centre of mass **R** and we refer to this particular zero-momentum frame as the centre-of-mass frame (although very often the two are used synonymously). While the distinction is not important for the discussions in this chapter, it will be so in Chapter 4 when we come to consider rotating bodies and will sometimes want to insist that the origin is coincident with the position of the centre of mass.

We can compute the momenta of the colliding particles in the centre-of-mass frame as follows:

$$\mathbf{p}_{1c} = m_1 \mathbf{v}_{1c}$$

$$= m_1 \left(\mathbf{v}_1 - \frac{m_1 \mathbf{v}_1 + m_2 \mathbf{v}_2}{m_1 + m_2} \right)$$

$$= \frac{m_1 m_2}{m_1 + m_2} (\mathbf{v}_1 - \mathbf{v}_2). \tag{3.36}$$

The result is a product of two terms, the reduced mass

$$\mu = \frac{m_1 m_2}{m_1 + m_2}, \tag{3.37}$$

and the relative velocity,

$$\mathbf{v} = \mathbf{v}_1 - \mathbf{v}_2. \tag{3.38}$$

Both μ and $\mathbf{v}$ feature heavily in two-body calculations. We could perform a similar calculation for $\mathbf{p}_{2c}$ but the result is clear from the fact that the total momentum must be zero:

$$\mathbf{p}_{1c} = \mu \mathbf{v},$$

$$\mathbf{p}_{2c} = -\mu \mathbf{v}. \tag{3.39}$$

The total kinetic energy in the lab frame can be expressed in terms of velocities in the centre-of-mass frame:

$$K = \frac{1}{2} m_1 v_1^2 + \frac{1}{2} m_2 v_2^2 = \frac{1}{2} m_1 (\mathbf{v}_{1c} + \mathbf{V}) \cdot (\mathbf{v}_{1c} + \mathbf{V}) + \frac{1}{2} m_2 (\mathbf{v}_{2c} + \mathbf{V}) \cdot (\mathbf{v}_{2c} + \mathbf{V}).$$

The simplification of this expression requires the evaluation of, e.g.

$$(\mathbf{v}_{1c} + \mathbf{V}) \cdot (\mathbf{v}_{1c} + \mathbf{V}) = v_{1c}^2 + 2\mathbf{v}_{1c} \cdot \mathbf{V} + V^2,$$

which leads to

$$K = \frac{1}{2} m_1 v_{1c}^2 + \frac{1}{2} m_2 v_{2c}^2 + (m_1 \mathbf{v}_{1c} + m_2 \mathbf{v}_{2c}) \cdot \mathbf{V} + \frac{1}{2} (m_1 + m_2) V^2. \tag{3.40}$$

Since we are working in a zero-momentum frame $(m_1 \mathbf{v}_{1c} + m_2 \mathbf{v}_{2c}) = \mathbf{0}$ and so

$$K = K_c + \frac{1}{2}(m_1 + m_2)V^2, \tag{3.41}$$

where the kinetic energy in the centre-of-mass frame is

$$K_c = \frac{1}{2}m_1 v_{1c}^2 + \frac{1}{2}m_2 v_{2c}^2. \tag{3.42}$$

So we see that the kinetic energy measured in the lab is a sum of two components: the energy of the particles as measured in the centre-of-mass frame, and a kinetic energy term $\frac{1}{2}(m_1 + m_2)V^2$ that we can ascribe to the motion of the total mass of the system at the centre-of-mass velocity. Since the collision involves a system of two particles for which there are no external forces, Eq. (2.28) implies that $\mathbf{V}$ is constant. The energy contained in the motion of the centre of mass is therefore unchanged during the collision and we can think of this contribution to the kinetic energy as being 'locked in' to the motion of the system as a whole; it cannot be used to alter the internal state of the system. Since $V^2 \geq 0$, Eq. (3.41) also shows that the zero-momentum frame is the inertial frame in which the system has the lowest possible total kinetic energy.

3.3.2 Elastic and inelastic collisions

During the interaction stage (see Figure 3.8), K_c will generally change as a result of the forces acting on the particles. If we assert that all of these forces are conservative, then we can represent them by a potential U. The mechanical energy in the centre-of-mass frame is then

$$E_c = K_c + U \tag{3.43}$$

and this is a conserved quantity. We shall define the potential energy to be zero when the particles are infinitely separated in which case U is zero both before and after the collision and hence K_c has the same value both before and after the collision (even though it can change during the collision). We distinguish quantities measured after the collision from their counterparts beforehand by the use of a prime ('). In this notation we can write that

$$K_c = K_c'. \tag{3.44}$$

Since $\frac{1}{2}(m_1 + m_2)V^2$ is a constant, this result is also true in the lab frame:

$$K = K'. \tag{3.45}$$

Furthermore, we can describe the collision by specifying the momentum vectors of the two particles in the centre-of-mass frame (both before and after the collision). They are always back-to-back since the total momentum is zero in this frame. The collision will generally alter the direction of the momentum vectors, so that $\mathbf{p}'_{1c}$ and

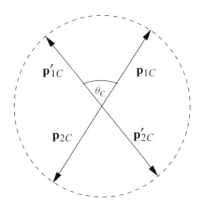

Figure 3.9   Momentum diagram for an elastic ($Q = 0$) collision.

$\mathbf{p}'_{2c}$ lie along a different axis to $\mathbf{p}_{1c}$ and $\mathbf{p}_{2c}$, but it cannot alter the lengths of the vectors (see Figure 3.9). The angle between these two axes defines the scattering angle θ_c. This type of scattering, in which the kinetic energy is conserved, is known as elastic scattering.

Frequently we want to consider collisions, not between point particles, but between objects with some internal structure. For such objects it is again possible that we observe elastic scattering as described above. Elastic scattering occurs if the internal structure of the colliding bodies remains unchanged by the collision process. This happens, to a good approximation, when we collide resilient objects such as glass or steel marbles. In general however, energy may be absorbed or released by rearrangement of the components of the bodies and such collisions are referred to as inelastic. We define the energy released by internal rearrangements as the Q-value:

$$Q = K' - K = K'_c - K_c. \qquad (3.46)$$

Typically, for macroscopic bodies the Q-value is negative, i.e. the bodies absorb energy when their internal structure changes. For example, when we drop a blob of plasticine on the floor kinetic energy goes into deformation of the blob resulting in a negative Q. Explosive collisions (i.e. those with $Q > 0$) occur when the collision causes the release of stored internal energy; the rather contrived case of the collision between two set mousetraps would be one example. Collisions with $Q > 0$ are more common on the microscopic level, e.g. electromagnetic energy stored within molecules can be released in exothermic chemical reactions.

Momentum diagrams for inelastic collisions are shown in Figure 3.10. Notice that momentum conservation still applies since there are no external forces. However the initial and final kinetic energies are now different, resulting in different magnitudes for the initial and final momentum vectors. For $Q > 0$ the momentum vectors increase in magnitude whereas for $Q < 0$ they decrease.

Example 3.3.1 *Consider a collision between two cars each of mass m on a linear air-track. The first car is initially travelling to the right with a speed v when it*

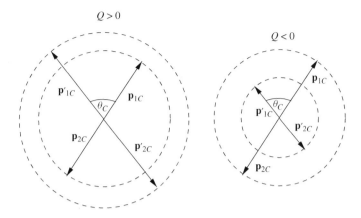

Figure 3.10   Momentum diagrams for inelastic collisions with $Q > 0$ and $Q < 0$.

collides with the second, stationary car. Calculate: (a) the final velocities of the cars in the lab frame assuming an elastic collision; (b) the Q-value of an inelastic collision if the cars stick together.

Solution 3.3.1 *(a) We shall perform the calculation in the lab frame. The total momentum is mv and the law of momentum conservation implies that the momenta after the collision must satisfy*

$$mv_1' + mv_2' = mv,$$

i.e.

$$v_1' + v_2' = v. \tag{3.47}$$

The collision is elastic so

$$\frac{1}{2}mv_1'^2 + \frac{1}{2}mv_2'^2 = \frac{1}{2}mv^2,$$

i.e.

$$v_1'^2 + v_2'^2 = v^2. \tag{3.48}$$

Squaring Eq. (3.47) and subtracting Eq. (3.48) gives

$$2v_1'v_2' = 0,$$

which means that one of the cars must have zero final velocity. Eq. (3.47) then implies that the other car must have final velocity v. Since Nature does not permit the cars to pass through each other, the car with final velocity v must be the one that was originally stationary. Thus, in terms of early- and late-stage velocities we have, e.g., $v_1 = v$ $v_2 = 0$; $v_1' = 0$ $v_2' = v$.

(b) For the case where the cars stick together we work in the centre-of-mass frame. The centre-of-mass velocity, as measured in the lab, is (using Eq. (3.34))

$$V = \frac{1}{2m}(mv + m \times 0) = \frac{v}{2}.$$

Hence

$$v_{1c} = v - \frac{v}{2} = \frac{v}{2},$$

$$v_{2c} = 0 - \frac{v}{2} = -\frac{v}{2}.$$

If the cars stick together then all of the kinetic energy in the centre-of-mass frame is lost (presumably into deforming the material that sticks the cars together) and we have

$$v'_{1c} = v'_{2c} = 0.$$

Thus

$$Q = 0 - \left(\frac{1}{2}m \left(\frac{v}{2} \right)^2 + \frac{1}{2}m \left(\frac{-v}{2} \right)^2 \right) = -\frac{1}{4}mv^2.$$

Example 3.3.2 *An experiment measures the elastic scattering of a beam of particles from a stationary target. The beam and target particles have equal mass. Show that the angle between the final velocity vectors is 90°, as long as the final velocities are both non-zero.*

Solution 3.3.2 *The initial state is similar to that described in the previous example, although we must be careful with the vector nature of the velocities. Momentum conservation now gives*

$$\mathbf{v}'_1 + \mathbf{v}'_2 = \mathbf{v},$$

where **v** *is the velocity of the beam particles and* **v**$'_1$ *and* **v**$'_2$ *are the final velocity vectors. Taking the scalar product of each side with itself ("squaring") gives*

$$v_1'^2 + 2v_1'v_2' \cos\phi + v_2'^2 = v^2,$$

where ϕ is the angle between the final velocity vectors. Invoking energy conservation for an elastic collision we have

$$v_1'^2 + v_2'^2 = v^2$$

and therefore

$$2v_1'v_2' \cos\phi = 0.$$

Since both final speeds are non-zero we must have $\cos\phi = 0$ and we conclude that the final velocity vectors form a right-angle in the lab frame.

3.4 ENERGY CONSERVATION IN COMPLEX SYSTEMS

Consider what happens when you drop a heavy stone on a sandy beach. Initially the stone has gravitational potential energy, which converts into kinetic energy as the stone falls. When the stone hits the beach, the grains of sand are given kinetic energy in the collision and they jostle against each other briefly, but in a very short time everything comes to rest with the stone embedded in the sand. Where does the energy go? We could say that friction acts between the stone and the sand, and since friction is a non-conservative force the total mechanical energy of the stone is not a constant of the motion once the stone hits the sand. But that is not to answer the question.

You might be familiar with the heating that occurs when an electric drill is used on a resilient surface, such as brick or ceramic tiles: the drill bit can become too hot to handle. This type of experience demonstrates clearly that heat can result from motion. In a series of careful experiments, Joule showed that heat could be regarded as a form of energy and that a loss of mechanical energy could be associated with a predictable temperature rise. In effect Joule demonstrated that the conservation of energy can be rescued provided we are prepared to count thermal energy when we are doing the book-keeping. So, when considering the stone that falls into the sand we can now say that the total energy is conserved but that mechanical energy is transformed into thermal energy as a result of the collision, and that we expect a rise in temperature of the sand and stone that depends on the amount of mechanical energy dissipated in the collision.

But is thermal energy really a fundamentally new type of energy? To gain a fuller understanding requires us to view the system in terms of its microscopic constituent particles. On this level, the details of the collision are very complicated. Many atoms, interacting by way of electrostatic forces, jostle each other and the original mechanical energy of the stone is dissipated into kinetic energy of the atoms within the sand and the stone itself. In this way thermal energy can be viewed as nothing more than atomic kinetic energy. Nevertheless, from a macroscopic perspective it makes much more sense to talk about heat energy, since keeping track of the kinetic energy of the individual atoms is not practicable. Figure 3.11 illustrates another way to think about the distinction between kinetic and thermal energy: in the former case there tends to be a collective, ordered, motion whilst in the latter the motion tends to be disordered.

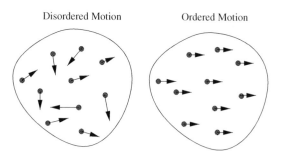

Figure 3.11   Disordered and ordered motion of the molecules in a body.

But what of the reverse process? Can disordered thermal motion spontaneously be transformed into ordered motion? Is it possible that the stone and the sand can spontaneously cool down, delivering the thermal energy back to the stone causing it to leap into the air? Such a bizarre happening is clearly contrary to our experience, but it is not impossible. It is just very unlikely to happen if the original mechanical energy is dissipated randomly.

Real systems of many constituent particles may be rigid, as is the case of a steel block, or they may be more fluid and deformable like a piece of plasticine. A little care is required when using energy conservation with deformable systems as it is often the case that the internal forces may dissipate kinetic energy into thermal energy. In addition, it is possible for a complex, deformable, system to convert internal potential energy into kinetic energy. An explosion provides a good example of the latter process at work. It is also the case with the following example of a ballerina's sauté.

Example 3.4.1 *Discuss which forces do the work when a ballerina sautés (jumps) into the air. In particular, explain why the normal reaction from the ground does no work.*

Solution 3.4.1 *It is quite sensible to first draw the free body diagrams illustrated in the dotted boxes in Figure 3.12. Figure 3.12(a) refers to the period of the jump*

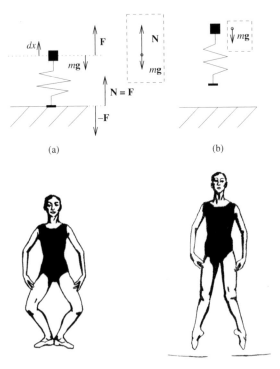

Figure 3.12 Diagram showing the mechanics of a ballerina's sauté: (a) the upwards-acceleration phase of the jump, illustrating the forces on the feet and the body; (b) the jump phase, when the acceleration is downwards. Free-body diagrams, in which the internal forces are ignored, are enclosed in dotted boxes.

when the ballerina's feet are in contact with the ground and Figure 3.12(b) refers to the period when the ballerina is in the air. In the former case, the only forces external to the ballerina's body are gravity and the normal force from the floor. Thus, motion of her centre of mass upwards is only possible if the magnitude of the normal force exceeds that of gravity. As soon as the ballerina leaves the floor the normal force is reduced to zero and the acceleration takes on a value g downwards. To view the jump in terms of energy we must identify how work is done. During the upwards acceleration phase the kinetic energy of the ballerina is increasing. Since the normal force from the floor is responsible for the upwards motion of the centre of mass it is tempting to suppose that this force does the work that is responsible for the increase in kinetic energy. This is incorrect: the force coming from the ground cannot do any work because the point of application does not move. This is not surprising because we know that the ballerina leaps into the air using her own internal energy and not energy from the Earth. At first sight, this may seem strange given that Newton's Third Law requires that all the internal forces sum to zero, i.e. $\sum_{i,j} \mathbf{F}_{ij} = \mathbf{0}$. However, this statement does not imply that $\sum_{i,j} \mathbf{F}_{ij} \cdot \mathrm{d}\mathbf{r}_i = 0$. This is the crux of the matter: not all particles are displaced by the same amount (since the ballerina constitutes a deformable system). For example, the atoms in the feet do not move at all. Figure 3.12 shows the essential features of the mechanics. To simplify matters we model the ballerina as a mass m sitting on a spring, which initially sits on the floor (i.e. the ballerina's feet are at the base of the spring). It is a crude model but one that allows us to introduce the internal forces into the problem. The internal forces that drive the upwards-acceleration phase of the body are a result of the tension $\mathbf{F}$ in the compressed spring. This corresponds to the starting point of the sauté, i.e. when the ballerina stands with her legs flexed. Newton's Third Law dictates that there is a corresponding force acting on the feet, and since the feet do not move this is also equal to the normal reaction from the ground, i.e. $\mathbf{N} = \mathbf{F}$. As the spring extends, the body moves upwards a distance $\mathrm{d}x$ so that the work done by the tension in the spring is $F\,\mathrm{d}x$ and the net work done on the body is $(F - mg)\mathrm{d}x = (N - mg)\mathrm{d}x$. Note that this is the result one would obtain upon considering the free body diagram alone but now it is clear that the work is done by the spring and not the ground. Notice also that in our model we have a simple explanation for how the ballerina's feet leave the ground: as the spring moves from compression to extension, the tension reverses direction and that leads to a net upwards force on the feet.

Conservation of energy in a complex system must therefore take account of the fact that it is possible for the internal forces to do work. If the internal forces are non-conservative then some of this work will be dissipated, e.g. it might lead to an increase in the thermal energy of the system. Of course, external forces may also do work on the system. Figure 3.13 illustrates how the different categories of force are able to contribute to the total energy of a system. If we assume that all dissipated energy becomes thermal energy, then we could write the law of conservation of energy as

$$W_{\text{ext}} = \Delta K + \Delta U_{\text{int}} + \Delta E_{\text{thermal}}, \qquad (3.49)$$

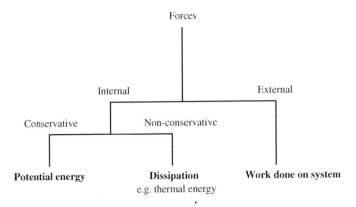

Figure 3.13   A scheme showing how different categories of force contribute to energy.

where W_{ext} is the work done on the system, ΔK and ΔU_{int} are the changes in kinetic and internal potential energy, and $\Delta E_{thermal}$ is the change in thermal energy. In the example of the ballerina's sauté, the force of gravity can be viewed as an external force that does negative work on the ballerina (i.e. it takes energy to jump) and we can therefore write

$$-mg\,dx = \Delta K - \Delta U_{int} \qquad (3.50)$$

as the ballerina increases the position of her centre of mass by a height dx. Thus the kinetic energy can only increase if the work done by internal forces (i.e. the internal potential energy lost) exceeds the magnitude of the work done by gravity. Note that since the external force is conservative, we could also think of W_{ext} as a change in external potential energy, i.e. $W_{ext} = -\Delta U_{ext} = -mg\,dx$.

We are at liberty to choose the boundary of a complex system, and this choice determines whether forces are to be considered as internal or external. Largely this is a matter of convenience. In the ballerina example we could have extended the system boundary to include the Earth, which would have resulted in a description of the sauté purely in terms of internal forces. However, one should be careful before choosing a system boundary such that there are non-conservative external forces since the thermal energy generated will probably be distributed on both sides of the system boundary, i.e. some part of $\Delta E_{thermal}$ will be external to the system. A more detailed study of thermal energy would leads us naturally to the subject of thermodynamics, but that is outside of the scope of this book.

PROBLEMS 3

3.1 A car of mass 900 kg travelling at 120 km per hour stops in 3.0 seconds. Calculate the work done on the car by the road. Calculate also the power of the braking system.

3.2 A force $\mathbf{F} = 3.0\mathbf{i} - 2.0\mathbf{j}$ (newtons) acts on a particle while it is displaced by $\Delta\mathbf{s} = -5.0\mathbf{i} - 1.0\mathbf{j}$ (cm). Calculate the work done in units of joules.

3.3 The "push-me-pull-you" is illustrated in the figure. It consists of two cars of equal mass (m) connected by a spring of negligible mass.

The cars are free to travel along a horizontal, frictionless, linear air track. The spring has an equilibrium length l, and spring constant k. At time $t = 0$ the car at x_1 has zero velocity and the car at x_2 has velocity v_0.

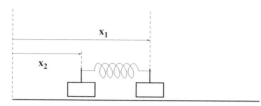

(a) Show that the centre of mass lies half-way between the two masses.
(b) Use Hooke's Law to obtain expressions for the force on each car in terms of x_1 and x_2.
(c) Use Newton's Second Law to obtain equations of motion for each of the two cars. Add these to show that a frame of reference that has its origin at the centre of mass has zero acceleration and is therefore inertial.
(d) Calculate the velocity of the centre of mass in terms of v_0.
(e) Introduce the relative coordinate $u = x_1 - x_2 - l$ and hence show that $m\ddot{u} + 2ku = 0$.
(f) Show that $u = A\sin(\omega t)$ is a particular solution to the equation you derived in (e) and hence determine an expression for ω. Show also that $A = -v_0/\omega$.
(g) Describe the motion of the two cars as seen in the lab frame.
(h) Making use of Eq. (3.28), show that the total mechanical energy of the system is fixed and equal to $mv_0^2/2$.

3.4 A uniform rope of mass per unit length λ is coiled on a table. One end is pulled straight up with constant velocity v. Consider the rate of change of momentum and show that the force exerted on the end of the rope as a function of height y is given by

$$F_a = \lambda v^2 + g\lambda y.$$

What is the total work done in lifting the end of the rope to a height y? Find an expression for the instantaneous power needed to lift the rope. Compare this with the rate of change of the total mechanical energy of the rope and comment on your result.

3.5 Two particles of mass $m_1 = 5.0\,\text{kg}$ and $m_2 = 10.0\,\text{kg}$ are moving with velocities $\mathbf{v}_1 = (2.0\mathbf{i} + 3.0\mathbf{j})\,\text{ms}^{-1}$ and $\mathbf{v}_2 = (-1.0\mathbf{i} + 4.0\mathbf{j})\,\text{ms}^{-1}$. Calculate the reduced mass for this system and determine the velocity of the centre of mass and the momentum of each of the particles as measured in the centre-of-mass frame.

3.6 The Lennard-Jones potential energy function is often used to represent the interaction between a pair of atoms:

$$U(r) = \varepsilon \left[\left(\frac{r_0}{r} \right)^{12} - 2 \left(\frac{r_0}{r} \right)^{6} \right],$$

where r is the separation of the atomic centres. Show that the equilibrium separation is r_0 and that the depth of the potential is ε. Sketch a graph of $U(r)$. If the system of two atoms has a maximum vibrational kinetic energy K_{max}, use your graph to indicate the range of allowed separations.

3.7 A proton makes a low-energy, head-on, collision with an unknown particle and rebounds straight back along its path with 4/9 of its initial kinetic energy. Assuming that the collision is elastic and that the unknown particle is originally at rest, calculate the mass of the unknown particle.

4

Angular Momentum

We have seen that the motion of a classical particle is governed by Newton's three laws, the second of which is the equation of motion $\mathbf{F} = m\mathbf{a}$. We have also seen that, by considering an extended body as a system of particles, the centre of mass also moves according to $\mathbf{F} = m\mathbf{a}$, where $\mathbf{F}$ is the sum of the external forces acting on the body and m is the sum of the masses of the constituent particles. In this way, the dynamical behaviour of an extended body is reduced to the motion of its centre of mass. However, for an extended body, centre-of-mass motion is only part of the story. An extended object may also exhibit internal motion. For example, the centre-of-mass motion of our Solar System tells us nothing about the elliptical orbits of the individual planets and moons within it and the parabolic motion of the centre of mass of a ball in flight tells us nothing about the spin of the ball. The internal motion of an extended system is usually more complicated than the motion of the centre of mass since it may well be associated with more degrees of freedom. To deal with this complexity, physicists look for conserved quantities; properties of the motion that do not change with with time, irrespective of the internal interactions that occur between the constituent particles. We have seen in the last chapter that momentum and energy are important conserved quantities. In this chapter we examine another conserved quantity, which appears both in classical and quantum physics. That conserved quantity is angular momentum.

4.1 ANGULAR MOMENTUM OF A PARTICLE

To introduce the subject of angular momentum we consider the simplest possible system, a particle with position vector $\mathbf{r}$ and momentum $\mathbf{p}$. The angular momentum $\mathbf{l}$ is defined to be

$$\mathbf{l} = \mathbf{r} \times \mathbf{p}. \tag{4.1}$$

Dynamics and Relativity   Jeffrey R. Forshaw and A. Gavin Smith
© 2009 John Wiley & Sons, Ltd

From the properties of the vector product this is a vector perpendicular to the plane containing **r** and **p**.

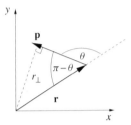

Figure 4.1   Particle travelling in the x-y plane.

Example 4.1.1 *A particle moves with momentum* **p** *and position vector* **r** *as shown in Figure 4.1. Calculate the angular momentum.*

Solution 4.1.1

$$l = |\mathbf{r} \times \mathbf{p}| = rp \sin \theta$$

but $\sin \theta = \sin(\pi - \theta)$ *so*

$$l = r_\perp p.$$

We say that $|\mathbf{l}|$ *is the moment of* **p** *about the origin and the direction of* **l** *may be determined as explained in Chapter 1. Turning the fingers of the right-hand from the direction of* **r** *to that of* **p** *causes the thumb to point out of the page, i.e. in the positive z-direction.*

Note that **l** depends on the choice of origin, as the following example demonstrates.

Example 4.1.2 *A particle is travelling with momentum p along the positive y – axis of a Cartesian co-ordinate system. Calculate the angular momentum relative to: (a) the origin; (b) the point (a,0,0).*

Solution 4.1.2 *Relative to the origin the angular momentum is*

$$\mathbf{l} = \mathbf{r} \times \mathbf{p} = \begin{vmatrix} \mathbf{i} & \mathbf{j} & \mathbf{k} \\ 0 & y & 0 \\ 0 & p & 0 \end{vmatrix} = \mathbf{0}.$$

Relative to the point (a, 0, 0) however the position vector of the particle becomes $(-a, y, 0)$ *and*

$$\mathbf{l} = \begin{vmatrix} \mathbf{i} & \mathbf{j} & \mathbf{k} \\ -a & y & 0 \\ 0 & p & 0 \end{vmatrix} = -ap\mathbf{k}.$$

Differentiating Eq. (4.1) with respect to time gives

$$\frac{d\mathbf{l}}{dt} = \frac{d\mathbf{r}}{dt} \times \mathbf{p} + \mathbf{r} \times \frac{d\mathbf{p}}{dt}$$
$$= \mathbf{r} \times \mathbf{f} \tag{4.2}$$

since $\frac{d\mathbf{r}}{dt} \times \mathbf{p} = \mathbf{0}$ (because the momentum and velocity are in the same direction). We have used Newton's Second Law to rewrite the rate-of-change of momentum as the force $\mathbf{f}$. As soon as we do so, it is to be understood that we are working in an inertial frame. The quantity $\boldsymbol{\tau} = \mathbf{r} \times \mathbf{f}$ is known as the torque. So, for a particle, we have

$$\boldsymbol{\tau} = \frac{d\mathbf{l}}{dt}. \tag{4.3}$$

Since this is a vector equation, it is independent of the choice of the origin of our co-ordinate system, a statement that we can check explicitly. We can trivially write

$$\mathbf{b} \times \mathbf{f} = \frac{d}{dt}(\mathbf{b} \times \mathbf{p}), \tag{4.4}$$

where $\mathbf{b}$ is any constant vector. Eq. (4.4) can now be added to Eq. (4.3) to obtain

$$(\mathbf{r} + \mathbf{b}) \times \mathbf{f} = \frac{d}{dt}((\mathbf{r} + \mathbf{b}) \times \mathbf{p})$$

thus

$$\boldsymbol{\tau}' = \frac{d\mathbf{l}'}{dt}, \tag{4.5}$$

where $\boldsymbol{\tau}'$ and $\mathbf{l}'$ are the torque and angular momentum calculated with respect to the point $\mathbf{r} = -\mathbf{b}$. Since $\mathbf{b}$ can be any constant vector we have shown that Eq. (4.3) is always true in an inertial frame of reference, even though the vectors $\mathbf{l}$ and $\boldsymbol{\tau}$ themselves depend on the choice of origin.

4.2 CONSERVATION OF ANGULAR MOMENTUM IN SYSTEMS OF PARTICLES

So far we have not introduced any new physics into our study of angular momentum; we have shown that, given the definitions of angular momentum and torque, Eq. (4.3) is just another way of expressing Newton's Second Law for the motion of a particle. In particular, the conservation of angular momentum seems to be nothing more than a trivial consequence of Newton's Second Law, i.e. Eq. (4.3) tells us that $d\mathbf{l}/dt = \mathbf{0}$ if $\mathbf{f} = \mathbf{0}$. The subject really starts to address new physics when we consider the rotation of systems of particles and of extended bodies.

Let us consider an extended system, which we take to be composed of N particles whose positions relative to an origin are given by the position vectors $\mathbf{r}_j$, where $j = 1, 2, 3, \ldots, N$ is an index labelling the particles. The jth particle has mass m_j and velocity $\mathbf{v}_j = \frac{d\mathbf{r}_j}{dt}$. We shall define the total angular momentum of the system $\mathbf{L}$ about the origin to be the vector sum of the particle angular momenta:

$$\mathbf{L} = \sum_{j=1}^{N} \mathbf{r}_j \times (m_j \mathbf{v}_j). \tag{4.6}$$

Let's now compute the rate of change of $\mathbf{L}$:

$$\frac{d\mathbf{L}}{dt} = \sum_{j=1}^{N} \mathbf{v}_j \times (m_j \mathbf{v}_j) + \sum_{j=1}^{N} \mathbf{r}_j \times (m_j \mathbf{a}_j), \tag{4.7}$$

where $\mathbf{a}_j = \frac{d\mathbf{v}_j}{dt}$ is the acceleration of particle j. The first term on the right-hand side is identically zero so we have

$$\frac{d\mathbf{L}}{dt} = \sum_{j=1}^{N} \mathbf{r}_j \times (m_j \mathbf{a}_j). \tag{4.8}$$

It is tempting to do as we did for the single particle and use Newton's Second Law to introduce the force on the jth particle, $\mathbf{F}_j = m_j \mathbf{a}_j$. Doing this yields

$$\frac{d\mathbf{L}}{dt} = \sum_{j=1}^{N} \boldsymbol{\tau}_j = \boldsymbol{\tau}. \tag{4.9}$$

However, such a substitution is valid only if we are working in an inertial frame of reference and we would like to be more general than that. Suppose instead that the origin of the co-ordinate system (i.e. the point about which we compute the angular momentum and torque) is accelerating relative to some inertial frame, which we generically refer to as the lab frame. Provided that this accelerating co-ordinate system is not rotating[1] we can write

$$\mathbf{a}'_j = \mathbf{A} + \mathbf{a}_j, \tag{4.10}$$

where $\mathbf{A}$ is the acceleration of the origin (of the accelerating co-ordinate system) and $\mathbf{a}'_j$ is the acceleration of the jth particle, both determined in the lab frame. Substitution for $\mathbf{a}_j$ in Eq. (4.8) gives

$$\frac{d\mathbf{L}}{dt} = \sum_{j=1}^{N} m_j \mathbf{r}_j (\mathbf{a}'_j - \mathbf{A}) \tag{4.11}$$

---

[1] Do not agonize over this caveat at this stage. We shall discuss rotating frames of reference in some detail later on.

and then

$$\frac{d\mathbf{L}}{dt} = \sum_{j=1}^{N} \mathbf{r}_j \times m_j \mathbf{a}'_j - \left(\sum_{j=1}^{N} m_j \mathbf{r}_j \right) \times \mathbf{A}. \qquad (4.12)$$

Now we can go ahead and use Newton's Second Law to simplify the first term: it is the torque calculated about the accelerating origin in the lab. The second term is generally not zero but notice that $\sum_{j=1}^{N} m_j \mathbf{r}_j$ is the position vector of the centre-of-mass. Therefore, we can eliminate it if we choose the origin to coincide with the centre of mass of the system of particles. In which case we have

$$\frac{d\mathbf{L}_c}{dt} = \sum_{j=1}^{N} \boldsymbol{\tau}_{cj} = \boldsymbol{\tau}_c \qquad (4.13)$$

and the subscript c reminds us that we are to compute the angular momentum and torque about the centre of mass.

Despite the apparently wide range of applicability of Eq. (4.13) and Eq. (4.9), these equations involve sums over the mutual interactions between constituent particles in a system to determine the net torque $\boldsymbol{\tau}$ and these sums rapidly become impossible to handle with increasing numbers of particles. To go further with systems composed of many particles we need to make the distinction between internal and external forces:

$$\mathbf{F}_j = \mathbf{F}_j^{(e)} + \sum_{k=1}^{N} \mathbf{F}_{jk}, \qquad (4.14)$$

where again $\mathbf{F}_{jk}$ is the force exerted on particle j due to particle k and $\mathbf{F}_j^{(e)}$ is the net force on particle j coming from some source outside of the system. Therefore we now have

$$\frac{d\mathbf{L}}{dt} = \sum_{j=1}^{N} \boldsymbol{\tau} = \sum_{j=1}^{N} \mathbf{r}_j \times \left(\mathbf{F}_j^{(e)} + \sum_{k=1}^{N} \mathbf{F}_{jk} \right), \qquad (4.15)$$

which, upon expanding the bracket, gives

$$\frac{d\mathbf{L}}{dt} = \sum_{j=1}^{N} \mathbf{r}_j \times \mathbf{F}_j^{(e)} + \sum_{j=1}^{N} \sum_{k=1}^{N} \left(\mathbf{r}_j \times \mathbf{F}_{jk} \right). \qquad (4.16)$$

$\boldsymbol{\tau}^{(e)} = \sum_{j=1}^{N} \mathbf{r}_j \times \mathbf{F}_j^{(e)}$ is the net external torque on the system and $\sum_{j=1}^{N} \sum_{k=1}^{N} \left(\mathbf{r}_j \times \mathbf{F}_{jk} \right)$ is the sum of all internal torques, i.e. torques exerted by particles within the system on other particles within the system. In our analogous discussion of the linear motion of extended bodies in Section 2.2.1, Newton's Third Law came to our rescue and told us that the sum of the internal *forces* is the null vector. We cannot use the same argument to show that the sum of the internal *torques* vanishes. To illustrate the issue, consider the situation depicted in Figure 4.2. Two particles have position vectors $\mathbf{r}_1$ and $\mathbf{r}_2$ and are exerting forces

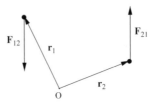

Figure 4.2   A mutual interaction between two particles that gives rise to a net torque.

F_{12} and F_{21} on each other. Although these forces are equal in magnitude and have opposite directions (consistent with Newton's Third Law) that is not sufficient to argue that the net torque is also zero. Thus, we cannot rely on Newton's laws to make a statement about the sum of internal torques. Nevertheless, it is an experimental fact that these torques sum to zero and it is at this stage that new physics enters our development. We therefore demand that

$$\sum_{j=1}^{N}\sum_{k=1}^{N}\left(\mathbf{r}_j \times \mathbf{F}_{jk}\right) = \mathbf{0}, \tag{4.17}$$

which then gives for the rate of change of angular momentum:

$$\frac{d\mathbf{L}}{dt} = \sum_{j=1}^{N}\mathbf{r}_j \times \mathbf{F}_j^{(e)} = \boldsymbol{\tau}, \tag{4.18}$$

where the total torque $\boldsymbol{\tau}$ is equal to the total torque due only to external forces acting on the body. The physics of Eq. (4.17) is not hard to understand. If it did not hold then the angular momentum of a body could change even if no external torques act upon it and that would lead to the bizarre result that isolated bodies could spontaneously start to rotate. We have thus arrived at a statement of the principle of conservation of angular momentum:

   In the absence of external torques, the total angular momentum of a system is a
   conserved quantity.

4.3   ANGULAR MOMENTUM AND ROTATION ABOUT A FIXED AXIS

We have so far succeeded in finding an equation of motion that relates the angular momentum to the net external torque. Our task in this section is to make more explicit the link between the angular momentum and the spin of the rotating body. To help simplify matters we will assert that the system of particles constitutes a rigid body. By this we mean that the relative positions of the particles that make up the body are fixed. Furthermore, we will make the restriction that the body is rotating about an axis that has a fixed direction in space. Examples of rotation of a body about a fixed axis are plentiful: a CD on a CD player rotates about a fixed axis through the centre of the CD; a yo-yo as it falls rotates about an axis whose direction is fixed, even though the yo-yo is accelerating downwards; the rear wheel of a bike shows fixed-axis rotation as long as the bike is travelling in a straight line (when the cyclist takes a bend the direction of the rotation axis is no longer fixed but changes as the direction of the bike's motion changes). We shall widen our brief to include rotations about non-fixed axes in Chapter 10.

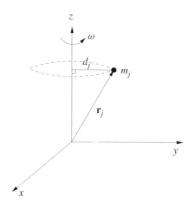

Figure 4.3   Fixed-axis rotation.

We therefore focus our attention upon the rotation of a rigid body about an axis that we define to be the z-axis. Figure 4.3 illustrates the geometry. Particle j has position vector $\mathbf{r}_j$, mass m_j and is rotating about the z-axis with angular speed ω. Since the body is rigid we can be sure that the value of ω is the same for all particles. Furthermore, each particle executes a circular orbit of radius d_j and we can use Eq. (1.24) to write

$$v_j = \omega d_j, \tag{4.19}$$

where v_j is the speed of the jth particle. This particle has angular momentum

$$\mathbf{L}_j = \mathbf{r}_j \times m_j \mathbf{v}_j = m_j \mathbf{r}_j \times \mathbf{v}_j,$$

where $\mathbf{L}_j$ is a vector perpendicular to the plane containing $\mathbf{r}_j$ and $\mathbf{v}_j$. At this stage we are only dealing with rotation about a fixed axis (the z-axis), in that case it is not usually necessary to compute the x and y components of $\mathbf{L}_j$. Thus we will consider only the projection of $\mathbf{L}_j$ on $\mathbf{k}$, the Cartesian basis vector in the z-direction, i.e.

$$L_{jz} = \mathbf{k} \cdot \mathbf{L}_j = m_j \left(\mathbf{k} \times \mathbf{r}_j \right) \cdot \mathbf{v}_j = m_j d_j v_j. \tag{4.20}$$

We have used the identity (for any vectors $\mathbf{a}$, $\mathbf{b}$ and $\mathbf{c}$):

$$(\mathbf{a} \times \mathbf{b}) \cdot \mathbf{c} = \mathbf{a} \cdot (\mathbf{b} \times \mathbf{c}) \tag{4.21}$$

as well as the result

$$\mathbf{k} \times \mathbf{r}_j = d_j \hat{\mathbf{v}}_j, \tag{4.22}$$

where $\hat{\mathbf{v}}_j$ is a unit vector in the direction of $\mathbf{v}_j$. You should take a moment to make sure that you can obtain Eq. (4.22) using the geometry shown in Figure 4.3.

The total angular momentum along the z−axis is obtained upon summing the contributions from each particle:

$$L_z = \sum_{j=1}^{N} m_j d_j v_j = \left(\sum_{j=1}^{N} m_j d_j^2 \right) \omega \tag{4.23}$$

and we have made use of Eq. (4.19). The term in brackets is a property of the rigid body and the axis of rotation. It is known as the moment of inertia about the rotation axis, and is given the symbol I.

Via Eq. (4.23) we have succeeded in achieving our goal of relating the angular momentum to the spin. Fixing the rotation axis means that we can focus our attention on the z-component of $\mathbf{L}$ but we ought not to forget that $\mathbf{L}$ is really a vector quantity. Likewise, the angular speed that appears on the right-hand side of Eq. (4.23) should be viewed as the z-component of the angular velocity. As far as this chapter is concerned we shall not really need to appreciate the vector nature of the angular velocity but it will turn out to play an important role later in the book, especially in Chapters 8 and 10. Now is a good time for us to take the trouble to define the angular velocity $\boldsymbol{\omega}$. Of course it must be defined so that its z-component is equal to the angular speed in the case of fixed-axis rotation, for that is what appears in Eq. (4.23). We choose to define $\boldsymbol{\omega}$ such that at any instant it points in the same direction as the axis of rotation of the body and with a magnitude equal to the angular speed at that instant. This definition allows for the possibility that the body wobbles around or its angular speed changes: the direction and modulus of $\boldsymbol{\omega}$ changes accordingly. Notice that since we require $\omega_z = \omega$ for rotations about the z-axis, in the sense illustrated in Figure 4.3, then we have defined that the direction of $\boldsymbol{\omega}$ should be parallel to the instantaneous axis of rotation as indicated by the right hand rule, i.e. curl the fingers of your right hand as if following the circular path of one of the particles in the body, then your thumb points in the direction of $\boldsymbol{\omega}$.

Returning to the case of rotations about the z-axis, we will save ink and write $L = L_z$, i.e. we shall write Eq. (4.23) as

$$L = I\omega. \tag{4.24}$$

Of course, the complicated bit is hidden in the symbol for the moment of inertia:

$$I = \sum_{j=1}^{N} m_j d_j^2 \tag{4.25}$$

and we turn our efforts next to showing how to compute it.

As was the case with the calculation of the centre of mass of a macroscopic object it is impossible to calculate this sum over all particles. We must instead resort to the approximation that the matter in the body is continuous and specify a function, $\rho(\mathbf{r})$, to tell us how that matter is distributed in space. We now need to write down the version of Eq. (4.25) appropriate for a continuous body. To that end, we imagining breaking the body up into an infinity of tiny volume elements. The mass of the element at $\mathbf{r} = (x, y, z)$ is

$$dm = \rho(\mathbf{r})\,dV = \rho(x, y, z)\,dx\,dy\,dz, \qquad (4.26)$$

where $dV = dx\,dy\,dz$ is the volume of the element. The sum in Eq. (4.25) now becomes a triple integral:

$$I = \int \int \int (x^2 + y^2)\rho(x, y, z)\,dx\,dy\,dz \qquad (4.27)$$

and we have used Pythagoras' Theorem to write the square of the distance from the rotation axis (the z-axis) as $d^2 = x^2 + y^2$. Performing the integral is generally not so easy and we underline the importance of looking for symmetries to make the calculation easier.

Example 4.3.1 *Calculate the moment of inertia of a uniform, thin, straight beam of length l and mass M, for rotations about an axis that passes through one end of the beam and runs perpendicular to it. Calculate also the moment of inertia when the axis runs through the centre of the beam and is perpendicular to it.*

Solution 4.3.1 *This is a one-dimensional problem with a uniform mass distribution and we can write the element of mass*

$$dm = \frac{M}{l}\,dx$$

with x as shown in Figure 4.4(a). The moment of inertia is

$$I = \int_0^l x^2 \frac{M}{l}\,dx.$$

Doing the integral gives

$$I = \frac{l^3}{3}\frac{M}{l} = \frac{1}{3}Ml^2 \qquad (4.28)$$

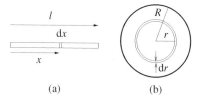

(a)   (b)

Figure 4.4 Calculating the moments of inertia of: (a) a beam and (b) a solid disc.

for rotations about an axis at one end. In the case that the axis passes through the centre, we need only change the limits of the integral:

$$I = \int_{-l/2}^{+l/2} x^2 \frac{M}{l}\, dx = \frac{1}{12}Ml^2.$$

Example 4.3.2 *Compare the moments of inertia of a uniform thin circular ring of mass M and radius R with a thin uniform disc of the same mass and radius. In both cases the rotation axis is the axis of rotational symmetry, i.e. perpendicular to the plane of the ring and disc.*

Solution 4.3.2 *For the ring all the mass lies at the same distance R from the rotation axis and so we can write down the result straightaway as $I = MR^2$. For the disc, the smart way to proceed is to realise that a disc can be built out of a series of rings. The mass of a ring of thickness dr and radius r is*

$$dm = \frac{M}{\pi R^2} 2\pi r\, dr = \frac{2M}{R^2} r\, dr.$$

Note that this is just the mass per unit area $\frac{M}{\pi R^2}$ multiplied by the area of the ring, see Figure 4.4(b). The moment of inertia is then

$$I = \int r^2\, dm = \frac{2M}{R^2} \int_0^R r^3\, dr = \frac{1}{2}MR^2.$$

The moment of inertia of the disc is smaller than that of the ring with the same mass even though the spatial extent of the two objects is the same. This is to be expected, since the d^2 term in Eq. (4.27) means that matter far from the axis of rotation has a greater contribution to the moment of inertia than matter close to the rotation axis.

Notice that in each of our calculations above the moment of inertia was of the form

$$I = Mk^2, \tag{4.29}$$

where M is the mass of the body and k is a length of the order of the spatial extent of the body known as the radius of gyration. We can therefore specify the moment of inertia by giving the mass and the radius of gyration (for a particular shape and axis of rotation). Table 4.1 gives the radii of gyration for some simple objects. For more complicated shapes the calculations become difficult, but a rough estimate of the moment of inertia may be obtained by setting k equal to the approximate size of the object (and that can be done without too much ambiguity provided the object is not too elongated).

Example 4.3.3 *An atomic nucleus with 150 nucleons and a size of around 6.4 fm may be produced in a 'high-spin' state following a nuclear fusion reaction. In such a state the nucleus rotates at $\omega \sim 10^{21}$ s^{-1}. Estimate the angular momentum of the nucleus about an axis through its centre assuming the nucleus to be a rigid body.*

TABLE 4.1   Radii of gyration for some simple uniform objects. In all cases the rotation axis is through the centre of mass. For the bar the axis is perpendicular to the bar. For the disc and the ring the axis is perpendicular to the plane in which the mass lies as shown in (c) and (d) of Figure 4.5.

Shape	k
Cylindrical shell (or thin ring) of radius R	R
Bar of length l	$\dfrac{l}{2\sqrt{3}}$
Disc of radius R	$\dfrac{R}{\sqrt{2}}$
Solid sphere of radius R	$R\sqrt{\dfrac{2}{5}}$
Spherical shell of radius R	$R\sqrt{\dfrac{2}{3}}$

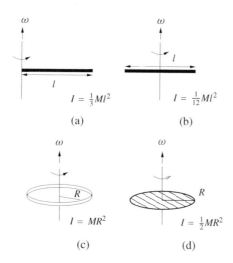

$$I = \tfrac{1}{3}Ml^2 \qquad (a)$$

$$I = \tfrac{1}{12}Ml^2 \qquad (b)$$

$$I = MR^2 \qquad (c)$$

$$I = \tfrac{1}{2}MR^2 \qquad (d)$$

Figure 4.5   Moments of inertia for some simple uniform objects for rotation about the axes shown. (a) and (b) correspond to thin rods, (c) is a thin circular ring and (d) is a flat circular disc.

Solution 4.3.3 *The mass is given in terms of the nucleon mass and is*

$$M = 150 \times 1.66 \times 10^{-27}\ \text{kg} = 2.49 \times 10^{-25}\ \text{kg}.$$

We do not know the detailed shape of the nucleus so we will approximate it by a sphere and use the radius of gyration quoted in Table 4.1, i.e.

$$I = \frac{2}{5}MR^2 = \frac{2}{5} \times 2.49 \times 10^{-25} \times 41.0 \times 10^{-30}\ \text{kg m}^2 \approx 4 \times 10^{-54}\ \text{kg m}^2.$$

The angular momentum is then

$$L = I\omega \sim 4 \times 10^{-33}\,\mathrm{kg\,m^2\,s^{-1}}.$$

As you can see, SI units are not so convenient for the description of subatomic objects. Physicists would normally express the angular momentum of a nucleus in terms of the fundamental quantity

$$\hbar = 1.054589 \times 10^{-34}\,\mathrm{kg\,m^2\,s^{-1}}$$

in which case we can write the answer as

$$L \sim 40\hbar.$$

In the above example we have used classical mechanics to obtain an order-of-magnitude estimate for a calculation that should really be carried out using quantum mechanics. While not strictly correct, such "ballpark" estimates are frequently used as a quick first approach by professional physicists to get an idea of an order of magnitude or to point the way forward to a more elaborate (and more correct) calculation.

We have shown in this section that the calculation of the angular momentum of a rigid body rotating at a known angular speed about a fixed axis boils down to being able to calculate the appropriate moment of inertia. In the last section we stated that it is an experimental fact that the sum of the internal torques is always zero. We can now bring these two results together to establish the equation of motion for the rotation of a rigid body about a fixed axis:

$$\tau = \frac{\mathrm{d}L}{\mathrm{d}t} = \frac{\mathrm{d}(I\omega)}{\mathrm{d}t} = I\frac{\mathrm{d}\omega}{\mathrm{d}t} = I\alpha, \qquad (4.30)$$

where we have introduced the angular acceleration,

$$\alpha = \frac{\mathrm{d}\omega}{\mathrm{d}t}.$$

In Eq. (4.30), τ is the component of net external torque $\boldsymbol{\tau}$ in the direction of the rotation axis. Notice that this equation is very similar in structure to Newton's Second Law with torque replacing force, angular momentum replacing linear momentum and angular acceleration replacing linear acceleration. This similarity to Newton's Second Law is handy to remember when it comes to problem solving.

Finally, we are ready to go ahead and study the motion of a particular rigid body. We shall consider the situation illustrated in Figure 4.6. The body is free to rotate about a horizontal axis in the Earth's gravitational field and the centre of mass of the object is at the point C. The perpendicular distance from the axis of rotation to C is R and P is the point on the axis of rotation that lies directly above C when the body is in equilibrium. The angle θ (see Figure 4.6) therefore specifies the extent of deviations from equilibrium. Our task is to understand the general motion of the body as it rotates about the axis and that is achieved if we can figure out how θ varies with time. To do that, we must solve Eq. (4.30) above, remembering that $\omega = \mathrm{d}\theta/\mathrm{d}t$ and $\alpha = \mathrm{d}^2\theta/\mathrm{d}t^2$. We need the torque about P that

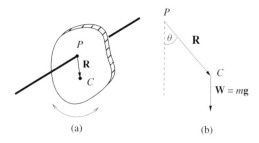

Figure 4.6   A pendulum consisting of a rigid body free to rotate about an axis.

acts on the body (more specifically we need the component of the torque along the axis of rotation) and this is provided by the weight. Now, the weight of the body acts as if it is concentrated at the centre of mass[2] and so we are able to work out the torque about P. Our particular choice of the point P guarantees that the torque is parallel to the axis of rotation, i.e.

$$\tau = -|\mathbf{R} \times \mathbf{W}| = -mg\,R\sin\theta, \tag{4.31}$$

where $W = mg$ is the weight of the body. Note the minus sign: the way we defined the angle θ means that $\boldsymbol{\omega}$ points out of the plane of the page in Figure 4.6(b) whereas $\boldsymbol{\tau}$ points into the page (from the definition of the vector product). We can now substitute τ into Eq. (4.30):

$$-mg\,R\sin\theta = I\alpha = I\frac{d^2\theta}{dt^2}. \tag{4.32}$$

As a double check that we got the sign right for the torque, you should note that the torque clearly must act so as to try and pull the body back towards equilibrium, i.e. the angular acceleration must be negative for positive θ. Note that the weight is not the only force to act upon the body. There will also be a normal reaction coming from the axis but that produces no torque about P.

We can keep things simple if we focus on the case where θ is small. Then $\sin\theta \approx \theta$ and

$$\frac{d^2\theta}{dt^2} \approx -\frac{mgR}{I}\theta. \tag{4.33}$$

Notice that this is none other than the equation for simple harmonic motion that we met in Section 3.2.2. The solution is $\theta = A\cos(2\pi ft)$ provided

$$f = \frac{1}{2\pi}\sqrt{\frac{mgR}{I}}. \tag{4.34}$$

---

[2] Can you prove this? It is easiest to consider a collection of particles. Work out the total gravitational torque about the origin and then show that this is the same torque that you would obtain if you had all the mass concentrated at the centre of mass.

This particular solution corresponds to fixing $t = 0$ when the angular displacement is at a maximum.

As an aside, we note that a measurement of the frequency of small oscillations, f, (the period is just $1/f$) provides a method to determine the moment of inertia of a body about the axis of rotation. However, that strategy runs into problems if we want to measure the moment of inertia about an axis through the centre of mass. Then there is no torque and the body doesn't behave like a pendulum. In the next section we will show how to circumvent that particular problem when we prove that the moment of inertia for rotations about one axis is sufficient to determine the moment of inertia about any parallel axis.

4.3.1   The parallel-axis theorem

Consider a rigid body, as illustrated in Figure 4.7. Two rotation axes are shown. Axis C is chosen so that it goes through the centre of mass, axis O is parallel to axis C. We imagine the body as being made up of slices perpendicular to the two axes. In any of these slices the position of C relative to O is given by the position vector $\mathbf{a}$. We first consider an element of mass dm in one of the slices. We can write

$$\mathbf{r} = \mathbf{r}_C + \mathbf{a} \tag{4.35}$$

since the moment of inertia is constructed from r^2 we take the scalar product of this equation with itself to obtain

$$r^2 = \mathbf{r} \cdot \mathbf{r} = r_C^2 + 2\mathbf{r}_C \cdot \mathbf{a} + a^2. \tag{4.36}$$

The element dm thus makes a contribution to the moment of inertia for rotations about O of

$$dI = r^2 \, dm = (r_C^2 + 2\mathbf{r}_C \cdot \mathbf{a} + a^2) \, dm. \tag{4.37}$$

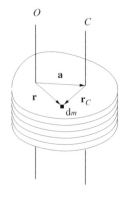

Figure 4.7   Geometry for the proof of the parallel-axis theorem.

To get the moment of inertia of the entire body for rotations about axis O we need to integrate over the whole slice and then over all the slices that make up the body, i.e.

$$I = \int \left(r_C^2 + 2\mathbf{r}_C \cdot \mathbf{a} + a^2 \right) dm,$$

$$= \int r_C^2 \, dm + 2\mathbf{a} \cdot \left(\int \mathbf{r}_C \, dm \right) + \int a^2 \, dm. \tag{4.38}$$

The term $\int \mathbf{r}_C \, dm$ vanishes due to the definition of the centre of mass. We therefore have

$$I = \int r_C^2 \, dm + a^2 \int dm,$$

i.e. $$I = I_C + Ma^2, \tag{4.39}$$

where I_C is the moment of inertia about axis C and M is the mass of the body. We can see from Eq. (4.39) that $I \geq I_C$ that is, the moment of inertia is a minimum when the rotation axis goes through the centre of mass. Note that Eq. (4.39) says nothing about axes that are *not* parallel to C.

Example 4.3.4 *Determine the moment of inertia of a thin uniform circular disc of radius R and mass M about an axis that just touches the edge and which is perpendicular to the plane of the disc (axis O in Figure 4.8).*

Solution 4.3.4 *Since we know that the moment of inertia about axis C is $\frac{1}{2}MR^2$ we can use the parallel-axis theorem to find the moment of inertia about axis O:*

$$I_O = I_C + MR^2 = \frac{3}{2}MR^2.$$

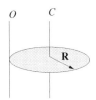

Figure 4.8   Moment of inertia about the edge of a circular disc of radius R.

4.4   SLIDING AND ROLLING

The beauty of Eq. (4.13) is that it works even when the centre of mass of a system of particles is accelerating, provided that we calculate the angular momentum and torque about the centre of mass. We have also shown that the angular momentum associated with fixed-axis rotation of a rigid body is determined by the angular speed ω and the relevant moment of inertia (Eq. (4.6)). We can combine these two results to give us

$$\tau_C = \frac{dL_C}{dt} = I_C \frac{d\omega}{dt} = I_C \alpha. \tag{4.40}$$

We therefore have a framework for describing the combined rotational and translational motion of a rigid body, which works even if the body is accelerating. The motion of the centre of mass is obtained by solving Newton's Second Law (see Section 2.2.1) and the rotation about the centre of mass may be handled using Eq. (4.40), provided the rotation axis has a fixed direction. The following example nicely illustrates how these two equations of motion are used together.

Example 4.4.1 *A bowling ball is launched across a horizontal floor with speed V_0 and no initial rotation about its centre of mass. At first it skids, then it begins to roll. What is the speed of the ball when it starts to roll?*

Solution 4.4.1 *Figure 4.9 shows the forces and velocities. Initially the ball has centre-of-mass velocity $V_C = V_0$ and angular speed $\omega = 0$ about a horizontal axis through the centre of mass (directed into the page in the figure). Kinetic friction at the point of contact between the ball and the floor will cause ω to increase while at the same time reducing V_C. The normal force N determines the friction according to*

$$F = \mu_k N = \mu_k Mg,$$

where μ_k is the coefficient of kinetic friction and M is the mass of the ball. To find the effect of this force on V_C we use Newton's Second Law:

$$-\mu_k Mg = M \frac{dV_C}{dt}.$$

Integration with respect to time gives

$$V_C = V_0 - \mu_k g t, \tag{4.41}$$

where we have used the initial condition $V_C = V_0$ at $t = 0$. Now we turn our attention to the rotation, remembering that since the ball accelerates horizonally (i.e. it slows down) we must compute the angular momentum and torque about the centre of the ball. We now use Eq. (4.40) to determine the effect of the torque on the

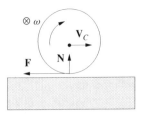

Figure 4.9   The dynamics of a bowling ball.

angular velocity. The torque is a result of the friction acting at the point of contact between the ball and the horizontal surface and has magnitude $FR = \mu_k MgR$, which causes an increase in ω. Hence

$$\tau_C = \mu_k MgR = I_C \frac{d\omega}{dt}.$$

Integrating and using the condition $\omega = 0$ at $t = 0$ we obtain

$$\omega = \frac{\mu_k MgRt}{I_C}. \tag{4.42}$$

When does this skidding phase end? At some time t_r the ball will start to roll. At this time the friction drops to almost zero since the surfaces of the ball and the floor no longer slide past each other. For rolling to occur the instantaneous velocity of the point of contact with the floor must be zero, i.e. the sum of the centre-of-mass velocity and the velocity due to rotation about the centre of mass must be zero:

$$0 = V_C - R\omega,$$

which gives

$$V_C = R\omega. \tag{4.43}$$

Eq. (4.43) is the rolling condition. We may use it to express ω in terms of V_C in Eq. (4.42) when $t = t_r$. We obtain

$$\frac{\mu_k Mg R^2 t_r}{I_C} = V_0 - \mu_k g t_r,$$

which when rearranged gives

$$t_r = \frac{V_0}{\mu_k g \left(1 + \dfrac{MR^2}{I_C}\right)}.$$

We can also use Eq. (4.41) to deduce V_r, the speed at which rolling begins:

$$V_r = \frac{V_0}{1 + \dfrac{I_C}{MR^2}}.$$

For $t > t_r$ there is much less friction and as a result a much reduced force acting on the ball. The ball's velocity will therefore be approximately constant and so V_r is approximately the final speed of the ball.

4.5   ANGULAR IMPULSE AND THE CENTRE OF PERCUSSION

Have you ever wondered why a cricket bat, a baseball bat or a tennis racquet has a sweet spot? This is the point on the bat where the ball seems to be hit most cleanly, without producing much vibration in the handle. In this section we look at sudden collisions that cause changes to both the linear and angular momenta of a rigid body, as happens when you strike a ball with a bat or when a door slams against a doorstop. In the event of such a collision there can be large forces exerted, just like in the example of the tennis player's serve in Section 2.3.5. Usually, we do not know the details of how the forces depend on time but we can nevertheless make progress if we know the impulse imparted by the collision, i.e. the change in momentum $\Delta\mathbf{p}$. Likewise, the collision will generally produce a time-dependent torque but since we don't know the time-dependence of the force we speak about the angular impulse:

$$\Delta\mathbf{L} = \mathbf{r} \times \Delta\mathbf{p} = \mathbf{r} \times \int_{t_1}^{t_2} \mathbf{F}(t)\,dt, \tag{4.44}$$

where the collision exerts a force $\mathbf{F}(t)$ at a position $\mathbf{r}$ from time t_1 to t_2.

Let us try to compute the position of the sweet spot in the collision between a bat and a ball. For simplicity we will ignore the effect of gravity by considering a bat at rest on a frictionless horizontal surface as shown in Figure 4.10. The centre of mass of the bat is at C, a distance h from the handle of the bat H. A ball strikes the bat, imparting an impulse $\Delta\mathbf{p}$ a distance b from the centre of mass as shown in the figure. We are interested in the subsequent motion of the bat and in particular the motion of the handle H immediately after the impulse has been delivered. The sweet spot ought to correspond to the special value of b such that H does not move in the split second after the impact. After the collision the bat constitutes an isolated system, so in an inertial frame the momentum of the centre of mass and angular momentum of rotation about the centre of mass are both constant. The final linear momentum is just the impulse:

$$\Delta\mathbf{p} = M\mathbf{V}_c, \tag{4.45}$$

where $\mathbf{V}_c$ is the velocity of the centre of mass and M is the total mass of the bat. Likewise, the final angular momentum (about C) is the angular impulse:

$$|\Delta\mathbf{L}| = |\mathbf{r} \times \Delta\mathbf{p}| = b\Delta p. \tag{4.46}$$

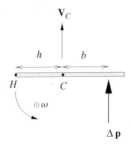

Figure 4.10   An impulse $\Delta\mathbf{p}$ causes translation and rotation of a rigid body.

We can use this expression to figure out the angular speed, ω, about a vertical axis through C, i.e.

$$b\Delta p = I_C \omega, \tag{4.47}$$

where I_C is the moment of inertia of the bat about the vertical axis. Armed with this information, we can now figure out what is happening at the handle, H. The velocity of H, immediately after the impulse, is parallel to $\Delta \mathbf{p}$ and a superposition of the velocity of the centre of mass ($\mathbf{V}_c$) with the velocity relative to the centre of mass, and the latter is the result of a pure rotation about the centre of mass. Thus we can write

$$\Delta x = \left(\frac{\Delta p}{M} - \omega h \right) \Delta t, \tag{4.48}$$

where Δx is the displacement of the handle (in the direction of $\Delta \mathbf{p}$) that occurs in a short time Δt after the impact. The sweet spot is defined such that $\Delta x = 0$, i.e.

$$\frac{\Delta p}{M} = \omega h. \tag{4.49}$$

Finally, substitute for ω using Eq. (4.47) to determine the position of the sweet spot:

$$\frac{\Delta p}{M} = \frac{bh}{I_C} \Delta p \tag{4.50}$$

and so

$$b = \frac{I_C}{hM}. \tag{4.51}$$

4.6 KINETIC ENERGY OF ROTATION

When an extended body rotates, its constituent particles will each have some kinetic energy. We can figure out the total energy of a rigid body by summing up all of these contributions. Referring to Figure 4.3, particle j has a kinetic energy

$$\frac{1}{2} m_j v_j^2 = \frac{1}{2} m_j d_j^2 \omega^2 \tag{4.52}$$

and the total kinetic energy of a system of many particles rotating about a fixed axis is therefore

$$K_{\text{rot}} = \frac{\omega^2}{2} \sum_j m_j d_j^2 = \frac{1}{2} I \omega^2. \tag{4.53}$$

If the body is only rotating about the fixed axis (i.e. the centre of mass is at rest) then this is the sole contribution to the kinetic energy. However, if we allow the body to undergo translation[3] as well as rotation then we must add together the

---

[3] Pure rotational motion arises if each and every part of the body is undergoing circular motion about some axis as determined in any inertial frame.

rotational and translational kinetic energies. In Section 3.3.1 we showed (at least for a system of two particles[4]) that the total kinetic energy is equal to

$$K = \frac{1}{2}MV_C^2 + \frac{1}{2}I_C\omega^2,$$ (4.54)

where the kinetic energy of rotation is calculated about a fixed axis through the centre of mass. Note that this expression only holds if the rotation axis passes through the centre of mass. This is so since the second term, which represents the kinetic energy of the body in the centre-of-mass frame, only reduces to $I_C\omega^2/2$ if the centre-of-mass lies on a fixed axis of rotation.

Example 4.6.1 *Determine an expression for the total kinetic energy of a solid sphere of mass M and radius R that is rolling without slipping on a flat surface at speed v.*

Solution 4.6.1 *The key to this problem is to recognise that the 'rolling without slipping' aspect of the motion implies that ω and v are connected by $v = R\omega$. This means that the rotational and translational motions are no longer independent. Thus*

$$K = \frac{1}{2}Mv^2 + \frac{1}{2}I_C\frac{v^2}{R^2}.$$

The moment of inertia of a uniform solid sphere about its centre is $\frac{2}{5}MR^2$ so we have

$$K = \frac{1}{2}Mv^2 + \frac{1}{5}Mv^2 = \frac{7}{10}Mv^2.$$

PROBLEMS 4

4.1 A turntable that rotates at a rate of $33\frac{1}{3}$ revolutions per minute has a mass of 1.00 kg and a radius of 0.13 m. Assuming the turntable to be a uniform disc, calculate the torque required if the operating speed is to be achieved in a time of 2 seconds after it is switched on. The turntable then spins freely and a lump of plasticine (mass 20 g) is dropped and sticks to the turntable 10 cm from the centre. What is the new angular frequency?

4.2 A stuntman stands on the roof of a bus, which is travelling at speed v around a circular bend of radius r. The stuntman's feet are a distance $2a$ apart, he has mass m, and his centre of mass is a height h above the roof of the bus. Obtain expressions for the normal forces acting on each of his feet. Assume that the roof of the bus remains horizontal and that his feet are equidistant from the vertical axis through his centre of mass.

4.3 A thin circular ring of radius R and mass m lies in the horizontal plane on a frictionless surface. It is free to rotate about a vertical axis fixed at some point on the circumference. A bug (mass m_b) walks from the axis around the ring at a constant speed v relative to the ring. Obtain an expression for the angular speed of the ring when the bug is directly opposite the axis.

---

[4] You may like to confirm that the result generalizes to any number of particles.

4.4 A uniform circular disc of mass m rolls without slipping such that the linear velocity of the centre of mass is v. Show that the kinetic energy is given by $K = 3mv^2/4$.

4.5 A uniform thin beam of length l and mass m is pivoted at one end and supported at the other. The beam is initially horizontal before the support is removed and the beam rotates (under gravity) in a vertical plane on the pivot.

(a) Obtain an expression for the force acting on the beam due to the pivot before the support is removed.

(b) Show that the instantaneous force acting at the pivot immediately after the support is removed is $F = mg/4$.

(c) By considering the mechanical energy of the system show that the angular speed of rotation about the pivot when the beam makes an angle θ to the horizontal is given by

$$\omega = \sqrt{\frac{3g \sin \theta}{l}}.$$

4.6 A uniform cylindrical drum of radius b and mass m rolls without slipping down a plane inclined at an angle θ to the horizontal. Find the acceleration of the centre of mass of the drum.

Use the principle of conservation of energy to determine the speed of the centre of mass after the drum has rolled a distance x down the slope. Show that this result is consistent with the expression for the linear acceleration that you just determined.

4.7 Show that the moment of inertia of a solid sphere of mass m and radius R is $\frac{2}{5}mR^2$. A spin bowler is able to impart a frictional force of $10\,N$ on the seam of a cricket ball (mass $0.15\,kg$ and radius $4.0\,cm$) for 0.1 seconds. Estimate the rotational speed of the ball.

4.8 The height of the cushion on a snooker table is chosen to be $\frac{7}{5}R$, where R is the radius of the snooker ball. This unique choice of height enables the ball to roll without slipping when it rebounds. Prove this result. You will need to use the result that the moment of inertia of the snooker ball about an axis through the centre of mass is $\frac{2}{5}MR^2$, where M is the mass of the snooker ball.

Part II

Introductory Special Relativity

5

The Need for a New Theory of Space and Time

5.1 SPACE AND TIME REVISITED

Perhaps the most astonishing idea underpinning Einstein's Special Theory of
Relativity is the rejection of the assumption that both space and time are absolute.
Since the whole of Part I of this book was built on such an assumption, it means
we will have to start all over again. Of course that is not to say that Newton's
theory is useless, for whatever Einstein's theory says, it had better be experimentally indistinguishable from Newton's theory for a very wide range of phenomena.
Before we attempt to figure out what Einstein's theory actually says, we should
first be very clear on what exactly it means to say that space and time are absolute.

Intuitively, absolute space means that we can imagine a gigantic fixed frame of
reference against which the positions of events can unambiguously be determined.
Of course the actual co-ordinates of an event will depend upon where the origin
of the reference frame is[1] but its position vector will nevertheless specify a unique
position in absolute space. Absolute time is also very intuitive. We can imagine the
Universe being filled with tiny clocks all synchronised with each other and ticking
at exactly the same rate. The time of an event can unambiguously be measured
by the time registered on a clock located close to the event (these are imaginary
clocks so we don't worry too much about the fact it isn't really practicable to put
clocks everywhere). Again, although the actual time of an event will depend upon
when we set the clocks to zero it still specifies a unique moment in time. The
consequences of absolute space and time are clear: there is no argument about how
long a body is (it is the distance between two points in absolute space), or whether

---

[1] They'll also depend upon the orientation of our axes and on our choice of co-ordinate system (e.g.
cartesian or spherical polar).

Dynamics and Relativity   Jeffrey R. Forshaw and A. Gavin Smith
© 2009 John Wiley & Sons, Ltd

or not two events occured simultaneously (which means they occured at the same absolute time).

If absolute space really existed, as Newton imagined it did, then it follows that there exists a set of very special frames of reference. Namely, all those frames which are at rest in the absolute space. Inertial frames are then those frames which are moving with some constant velocity relative to absolute space. It is interesting that long before Einstein, absolute space was under attack. Since no experiments have ever been performed that are able to identify a special inertial frame it follows that we cannot figure out which inertial frames are at rest in absolute space. Therefore, as far as physics is concerned we can dispense with the idea of absolute space in favour of the democracy of inertial frames. Physicists now take the equality of inertial frames so seriously that they have elevated it to the status of a fundamental principle: the Principle of (Special) Relativity. By postulating this relativity principle, absolute space is dismissed from physics and consigned to the realm of philosophy. Newton's theory itself obeys the relativity principle, and as such does not require the notion of absolute space. However it does assume that time is absolute.

Let's prepare the ground for later developments and gain some experience of thinking about events in space and time. Consider two inertial frames of reference, S and S' and suppose an event occurs at a time t and has Cartesian co-ordinates (x, y, z) in S. The question is, what are the corresponding co-ordinates measured in S'? To answer this we need to be more explicit and say how the S and S' move relative to each other. We'll take their relative motion to be as illustrated in Figure 5.1, i.e. S' moves at a speed v relative to S and in a direction parallel to the x-axis. Let's also suppose that their origins O and O' coincide at time $t = t' = 0$. Clearly the y and z co-ordinates of the event are the same in both frames:

$$y = y', \tag{5.1a}$$

$$z = z'. \tag{5.1b}$$

More interesting is the relationship between the co-ordinates x and x'. Common sense tells us that

$$x = x' + vt. \tag{5.2}$$

Of course this is the correct answer but only provided we assume absolute time, i.e. that $t = t'$. The proof goes like this. Firstly, we need to recognise that the

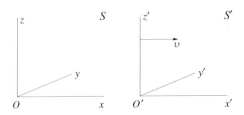

Figure 5.1   The two frames of reference S and S' moving with relative speed v in the sense shown.

relationship must be of the linear form

$$x = ax' + bt \qquad (5.3)$$

where a and b are to be determined. We'll not dwell on this, but if the relationship were not linear then it would violate the relativity principle. We also know that the point $x' = 0$ travels along the x-axis with speed v, i.e. along the line $x = vt$. From this it follows that $b = v$. The relativity principle can be used again to figure out a since one can equally well think of S' as being at rest and S as moving along the negative x-axis with speed v. This implies that

$$x' = ax - vt. \qquad (5.4)$$

Substituting for x' into Eq. (5.3) implies that $a = 1$ and we have proved the result. Notice that we did not need to invoke the idea of absolute space to derive this result: all that was needed was the relativity principle and the assumption that time is absolute. The equations (5.1) and (5.2) tell us how to relate the co-ordinates of an event in two different inertial frames and they are often referred to as the Galilean transformations.

In what follows we shall often speak of 'observers'. These are the real or fictitious people who we suppose are interested in recording the co-ordinates of events using a specified system of co-ordinates. For example, we might say that 'if an observer at rest in S measures an event to occur at the point (x, y, z) then an observer at rest in S' will measure the same event to occur at (x', y', z') where the co-ordinates in the two frames are related to each other by the Galilean transformations.'

Example 5.1.1 *A rigid rod of length 1m is at rest and lies along the x-axis in an inertial frame S. Show that if space and time are universal, the rod is also 1m long as determined by an observer at rest in an inertial frame S' which moves at a speed v relative to S in the positive x direction.*

Solution 5.1.1 *It is tempting to think that this result is so self-evident that it needs no proof but as we shall see, it is not true in Einstein's theory so it is a good idea for us to work through the proof here assuming that Eq. (5.2) holds. We shall also go very slowly and spell out explicity exactly how the length is measured in each inertial frame. For this question this level of analysis may be a little over the top but it will prepare us well for later, trickier, problems.*

We can refer to Figure 5.1 and imagine two observers, one at rest in S and the other at rest in S'. Suppose that the observer at rest in S measures the positions of each end of the rigid rod. In doing so, she records the space and time co-ordinates of two events. The first event is the measurement of one end of the rod and the second event is the measurement of the other end of the rod. To specify an event we need to specify four numbers: the three spatial co-ordinates and the time at which the event took place. The first event has co-ordinates $(x_1, 0, 0)$ and occurs at time t whilst the second event has co-ordinates $(x_2, 0, 0)$ and also occurs at time t. Obviously these two events take place at the same time since that is what we mean by making a measurement of length: we measure the positions of the ends of *the rod at an instant in time. We are told that $x_2 - x_1 = 1\,m$ and asked to find the corresponding length as measured by an observer at rest in S'.*

Our second observer makes their measurement of the length of the rod. Let's suppose they do it at a time t' (the two observers don't have to measure the length at the same time so t' does not have to equal t). Again there are two events, the measurement of one end of the rod at $(x_1', 0, 0)$ and the measurement of the other end of the rod at $(x_2', 0, 0)$. Now using the Galilean transformations it follows that $x_1' = x_1 - vt'$ and $x_2' = x_2 - vt'$ from which it follows that $x_2' - x_1' = x_2 - x_1 = 1m$, i.e. both observers agree on the length of the rod.

The relationships between measurements of events in different inertial frames under the assumption of absolute time is called 'Galilean relativity'. According to Galilean relativity, all observers will agree on things like the length of a rod or whether or not two events are simultaneous. It is now time to question the validity of this simple and intuitive relativity theory.

5.2 EXPERIMENTAL EVIDENCE

We are going to need some pretty compelling reason to give up the Galilean view of space and time. In this section we'll motivate the need for something different and we start with the 1887 experiment of Michelson and Morley.

5.2.1 The Michelson-Morley experiment

Is it possible to chase after a beam of light? In classical physics the answer seems to be a resounding 'yes'. We can even imagine running at close to the speed of light whilst shining a torch ahead of us. If we run fast enough then we might expect to see the light travelling slowly out of the front of the torch and when we reach light speed the torch is finally rendered useless. Thinking like this we are imagining that the light travels in a medium, just as every other wave we know of in Nature, and that its speed of propagation is fixed relative to the medium. The uselessness of our torch as we reach the speed of light is in this way entirely analogous to the phenomenon whereby a jet aircraft travelling at the speed of sound cannot be heard until it has passed by. This is a natural way to think, i.e. light needs a medium to support its vibrations, but it is wrong and the experiment that proves[2] it is the Michelson-Morley experiment.

If light is a wave travelling through some medium, which is historically referred to as the 'ether', then it should travel at a fixed speed c relative to the ether. This means that different observers in different inertial frames will all measure different speeds for a beam of light. Similarly an observer can tell if they are moving relative to the ether by sending out two (or more) beams of light in non-parallel directions. Only if the two beams travel at the same speed is the observer entitled to say they are at rest relative to the ether and from any difference in speeds the observer will be able to determine their speed relative to the ether. As an aside,

---

[2] Actually it does not strictly prove the absence of a medium. Rather it provides some very compelling evidence.

the idea of the ether clearly violates the relativity principle in the sense we have introduced it, because it provides a way to classify inertial frames. We need not *a priori* worry about this since the fact that Newton's laws do not allow us to classify inertial frames does not necessarily imply that the theory of light should likewise oblige.

Michelson and Morley set out to measure the anticipated difference in the speed of light in two mutually perpendicular directions. Their experiment is shown schematically in Figure 5.2. A coherent beam of light is split into two at P using a half-silvered mirror and the two subsequent beams each then travel a distance L along paths 1 and 2 before reflecting off mirrors and returning to the beam split-ter where they interfere along path 3 whence they are observed. We don't need to bother with the details of this apparatus (it is known in optics as a Michelson interferometer). All we need to concern ourselves with is the time it takes for light to travel along each path. To simplify the calculation let us assume that the whole apparatus is moving to the right with a speed v through the ether, as shown in Figure 5.2 (we could equally well assume any other orientation: our main conclu-sion won't change). From the point of view of the experimenter, it is as though the ether is moving to the left relative to the apparatus at a speed v. Let's consider path 1 first. The light beam has to travel a distance $2L$ and to work out the time this takes we need to know the speed of light travelling perpendicular to the ether. The velocity addition diagram is shown in Figure 5.3. Relative to the apparatus, the velocity of the light which travels upwards along path 1 is $\mathbf{u} = \mathbf{v} + \mathbf{c}$ where $\mathbf{v}$ is the velocity of the ether relative to the experiment and $\mathbf{c}$ is the velocity of the light relative to the ether (it might help to notice that the situation is analogous to computing the velocity of an aircraft relative to the ground given the velocity of the air and the velocity of the aircraft relative to the air)[3]. We know that $|\mathbf{c}| = c$ and $|\mathbf{v}| = v$ and hence that $|\mathbf{u}| = \sqrt{c^2 - v^2}$. The reflected light travels at the same

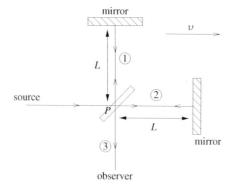

Figure 5.2   The Michelson-Morley experiment.

---

[3] One can think of the tip of the beam of light advancing through the streaming ether as if it were a particle.

Figure 5.3   The addition of velocities relevant to path 1 of the Michelson-Morley experiment.

speed and so the total time take for the round trip is

$$t_1 = \frac{2L}{c\sqrt{1 - v^2/c^2}}. \tag{5.5}$$

Path 2 is easier since the light is either parallel or antiparallel to the direction of the ether. On the outward path the light is travelling into the ether at speed $c - v$ whilst on the return path it is swept along by the ether at speed $c + v$. The time for each leg is thus different and the total time taken is

$$t_2 = \frac{L}{c - v} + \frac{L}{c + v} = \frac{2L}{c(1 - v^2/c^2)}. \tag{5.6}$$

The time taken by the light which travels along path 1 is therefore slightly shorter than for the light which travels along path 2.

The experiment of Michelson and Morley was designed to be sufficiently sensitive to this time difference that it could detect a speed through the ether comparable to the speed with which the Earth rotates around the Sun (which is about 30 km/s). In this way they hoped to be sure of seeing an effect at some time during the year since if at one instant the Earth just happened to be at rest relative to the ether, it would be unlikely to be at rest some time later and six months later one might reasonably expect it to be travelling at twice the orbital speed, i.e. 60 km/s. Of course we know that the Sun is moving at vast speeds relative to the centre of the galaxy and that the galaxy moves relative to other galaxies so one really ought to expect that the speed of the Earth through the ether is at the very least equal to its orbital speed around the Sun. However, Michelson and Morley did not measure any time difference. There was no error in their experiment, they were simply forced to conclude that the Earth does not move relative to an ether.

5.2.2   Stellar aberration

The simplest way out of this null observation might seem to be to suppose that the ether is being dragged along by the Earth, i.e. it is as if the Earth has an atmosphere of ether. However, since 1727 it was well known that this could not be so. James Bradley's observations of stellar aberration seemed to require that observers on Earth should be moving through the ether at relative speeds which differ over the course of a year by an amount equal to twice the Earth's rotational speed about the Sun. Let's go into a little more detail.

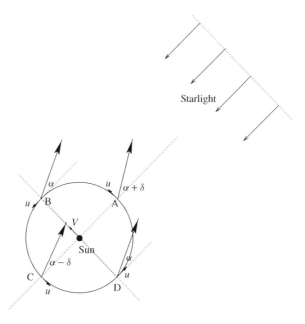

Figure 5.4   Stellar aberration.

Stellar aberration is the phenomenon whereby the positions of the stars move in ellipses over the course of each year and it is illustrated in Figure 5.4. The figure shows the motion of an observer on the Earth as they travel around the Sun. To simplify things, we assume that the observer is looking at a star which lies in the same plane as the Earth's orbit around the Sun (called the 'ecliptic'). The heavy arrows show the apparent position of the star when the Earth is at various points in its orbit. If you are having difficulties with this diagram then it might help to consider why someone running through the rain might tilt their umbrella ahead of them even though the rain is falling vertically from an overhead cloud. However this picture needs to be used carefully, for the light from a distant star is incident as plane waves on the Earth and if the wavefront normals (i.e. the vectors perpendicular to the planes of constant amplitude) are perpendicular to the ether wind then they will be unchanged by it. The situation is analogous to waves breaking on a beach: the waves may arrive parallel to the beach even if there is a strong current flowing. However it is not the wavefront normals which matter to an astronomer on Earth. The telescope they are using to view the star removes a portion of the incident wavefront. In order to focus that incident portion onto the eye of the astronomer, the telescope must be tilted to allow for the fact that the light incident into the telescope is swept by the ether wind. Using the rainfall analogy, consider a person running through vertically falling rain holding a cylindrical piece of tubing, then in order for rain entering at the top of the tube to exit at the bottom of the tube, the tube must be tilted.

Returning to the task in hand, we presume that the Sun is moving through the ether too, at a speed V, and we denote the speed of the Earth relative to the Sun u.

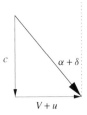

Figure 5.5   The velocity of starlight relative to the Earth at point A in its orbit around the Sun (denoted by the big arrow).

Note that if $V = 0$ it would follow that $\alpha = 0$. When the Earth is at the point A, if we suppose an ether wind is blowing at a speed $V + u$ (in the opposite direction to the direction of motion of the Earth) then adding the velocities of the ether and the velocity of the starlight relative to the ether we can obtain the velocity of the starlight relative to the Earth[4]. The addition of velocities is illustrated in Figure 5.5. From the figure, we can determine that

$$\tan(\alpha + \delta) = \frac{V + u}{c} \simeq \alpha + \delta \qquad (5.7)$$

and we assume all angles are small (which sounds reasonable since c is presumably much larger than V). Similarly, at positions B and D the apparent position of the star is given by

$$\tan \alpha = \frac{V}{c} \simeq \alpha. \qquad (5.8)$$

From these two equations it follows that $\delta \simeq u/c$. So, the star moves backwards and forwards in the night sky over the course of 1 year reaching its extreme positions when the Earth is at points A and C in its orbit. The angular size of this oscillation is just $2\delta = 2u/c$ which is independent of the Sun's velocity. If the star were not in the ecliptic then it would appear to move in an ellipse. In any case, direct observation of these stellar aberrations led Bradley to conclude that the ratio of 'the Velocity of Light to the Velocity of the Eye (which in this Case may be supposed the same as the Velocity of the Earth's annual Motion in its Orbit) as 10210 to One.'[5] Substituting for the known velocity of light gives $u \simeq 30\,\text{km/s}$ which is precisely the speed one expects given that the Earth travels around the Sun once every 365 days. In their time, Bradley's measurements had an additional significance: they provided a direct verification of Copernicus' claim that the Earth rotates around the Sun.

---

[4] More correctly it is the velocity of the light captured by the observer's telescope relative to the telescope.

[5] Letter to Edmond Halley published in Philosophical Transactions of the Royal Society of London, Vol. 35 (1727).

So where do we stand? The null (no ether) result of the Michelson-Morley experiment[6] and the apparent need of an ether to explain stellar aberration have led us into an impasse. It is an impasse which Einstein was ultimately able to overcome.

5.3 EINSTEIN'S POSTULATES

Einstein was particularly concerned with the breakdown of the principle of relativity implied by the presence of an ether. Just a few years earlier, Maxwell had written down the equations which define the classical theory of electromagnetism. The equations are beautiful and encode the idea that light is an electromagnetic wave. However, the equations taken at face value seem to predict that light travels at a speed $c = 1/\sqrt{\varepsilon_0\mu_0}$ independently of the motion of either the source which produced it or the observer who measures it[7]. This circumstance seems absurd: for a wave travelling through a medium the speed is indeed independent of the motion of the source but it certainly depends upon the motion of the observer. Of course one can sidestep this problem by supposing that Maxwell's equations are only approximately correct and that the speed c which appears in them ought to be replaced by the speed of light appropriate to the frame in which one wants to use the equations. This attempt to hold on to the ether has unpleasant consequences, for example Coulomb's Law would now be slightly different in different inertial frames. With the evidence mounting, Einstein took the dramatic step of assuming that the ether does not exist and that Maxwell's equations are correct. At a stroke he could explain the null result of Michelson-Morley, restore the principle of relativity to its central role in physics and keep the equations of Maxwell without modification[8].

In 1905, Einstein therefore made the two postulates that define his new theory of space and time, and which we can state as follows.

1st postulate: *The laws of physics are the same in all inertial frames.* This is a strong statement of the principle of special relativity which we discussed above and which was anticipated by Galileo. We talk about 'special relativity' to remind us that it is concerned only with the equivalence of inertial frames, i.e. it says nothing about accelerated frames of reference. By this Einstein insists that there should be no experiment in physics which can allow any one inertial frame to be singled out as special. It is this principle which implies that the ether does not exist.

2nd postulate: *The speed of light in vacuum is the same in all inertial frames.* This statement saves the laws of electromagnetism since if the speed of light did vary from frame to frame then Maxwell's equations would violate the 1st postulate. It also explains in a trivial manner the null result of Michelson-Morley. We stress

---

[6] Using masers, in 1959 Cedarholm & Townes constrained the speed of the ether relative to the Earth to <30 m/s.

[7] ε_0 is the permittivity, and μ_0 the permeability, of the vacuum.

[8] We shall also see that his theory is able to explain the effect of stellar aberration.

that this statement holds independently of the motion of the source or the observer. It constitutes a clean break with classical thinking and it is the source of all of the weird and wonderful physics we shall soon be encountering. As we shall see, it can ony be true if we reject the notion of absolute time.

To illustrate how weird the 2nd postulate is consider the situation illustrated in Figure 5.6. A light source sits at rest in S whilst a second frame S' moves towards the light source with a speed v. The 2nd postulate implies that:

- An observer in S measures the light to travel towards an observer in S' at a speed $c + v$. (When we say "an observer in S" we shall always mean an observer at rest in S.)
- An observer in S measures the light to travel at speed c.
- An observer in S' measures the light to travel at speed c.

The first two of these statements are reasonable but the third really does appear to be outrageous. Nevertheless it is a necessary consequence of the 2nd postulate.

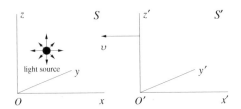

Figure 5.6   A light source sits at rest in S. A second frame S' moves towards the light source with a speed v.

PROBLEMS 5

5.1 A moving walkway moves at a speed of $0.7\,\mathrm{ms}^{-1}$ relative to the ground and is $20.0\,\mathrm{m}$ long. If a passenger steps on at one end and walks at $1.3\,\mathrm{ms}^{-1}$ relative to the walkway, how much time does she require to reach the opposite end if she walks (a) in the same direction as the walkway is moving? (b) in the opposite direction?

5.2 Convince yourself that the first of the three bulleted items listed at the end of Section 5.3 is correct.

6

Relativistic Kinematics

6.1 TIME DILATION, LENGTH CONTRACTION AND SIMULTANEITY

In the next section we shall find the new equations which will replace the Galilean transformation equations (5.1) and (5.2), but before that let us derive perhaps the two most remarkable results in Einstein's theory: the fact that time passes at different rates in different inertial frames and that it doesn't make sense to speak of the length of a metre rule without also stating the frame in which it is at rest.

Historically people have regarded distance and time as fundamental units. For example, as defined by a standard length of material and an accurate periodic device. Speed is then a derived quantity determined by the ratio of distance travelled and time taken. Nowadays, the scientific community has stopped thinking of the metre as fundamental. Instead the metre is defined to be the distance travelled in a vacuum by light in a time of exactly 1/2,9979,2458 seconds. This might look like a rather arbitrary definition but that particular sequence of numbers in the denominator means that the metre so defined corresponds to the length of the old standard metre, which was a metal bar kept locked in a vault in Paris. The advantage of defining the metre in terms of the speed of light and the unit of time means that we no longer have to worry about the fact that the metal bar is forever changing as it expands and contracts. By defining the metre this way we have chosen a value for the speed of light in a vacuum, i.e. $c = 2.99792458 \times 10^8$ m/s. There is nothing particularly special about using the speed of light here, strictly speaking one could define the metre to be the distance travelled by an average snail in 15 minutes. Then the snail speed would be fundamental. However, given the variability in snail speeds, this would not consitute a very reliable measure. Light speed is much more preferable and it has the particular advantage that it is the only speed which everyone agrees upon (by Einstein's 2nd postulate); all other speeds require the specification of an associated frame of reference.

Although this definition of distance suits most people, it isn't really the best definition for physicists who work with particles travelling close to light speed. As a result, the metre is sometimes rejected in favour of a distance measure such that 1 unit of distance is equal to the distance travelled by light in 1 second. In these units, which particle physicists prefer, $c = 1$.

6.1.1 Time Dilation and the Doppler Effect

Conversely, one could define time by specifying a speed and a distance. For example, we could make a clock by bouncing light between two mirrors spaced by a known distance, as illustrated in Figure 6.1. We can think of one 'tick' of this clock as corresponding to the time it takes the light to travel between the two mirrors and back. The time interval between any two events can then be determined by counting the number of 'ticks' of the light-clock which have elapsed between the two events. Of course there is nothing special about light here, for example we could define time by bouncing a ball between two walls.

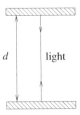

Figure 6.1   A light-clock viewed in its rest frame.

This is a good place to discuss exactly how time measurements are to be made. Consider an observer in some frame of reference S who is interested in making some time measurements. Since Einstein's theory is going to require that we drop the notion of absolute time, we need to be more careful than usual in specifying how the time of an event is determined. Ideally, the observer would like to have a set of identical clocks all at rest in S with one clock at each point in space. For convenience, the observer might choose that the clocks are all synchronised with each other. The time of an event is then determined by the time registered on a clock close to the event. Ideally the clock would be at the same place as the event otherwise we should worry about just how the information travels from the event to the clock. The observer can then determine the time of an event by travelling to the clock co-incident with the event and reading the time at which the event occured (we are imagining that the clock was stopped by the event and the time recorded). Clearly this is not a very practicable way of measuring the time of an event but that is not the point. We have succeeded in explaining in principle what we mean by the time of an event. Most importantly, the time of the event clearly has nothing to do with where the observer was when the event happened nor whether the observer actually saw the event with their eyes. We may have laboured this point to excess but that is because there is room for much confusion if these ideas are not properly appreciated.

Let us return to the light-clock of Figure 6.1. In its rest frame, the time it takes for light to do the roundtrip between the mirrors (one 'tick') is clearly

$$\Delta t_0 = \frac{2d}{c}. \tag{6.1}$$

Now let us imagine what happens if the clock is moving relative to the observer. To be specific let us put the clock in S' and an observer in S where the two frames are as usual defined by Figure 5.1. If the observer was in S' then the time for one tick of the clock would be just Δt_0. Our task is to determine the corresponding time when the observer is in S. According to this observer, the light follows the path shown in Figure 6.2. We call Δt the time it takes for the light to complete one roundtrip as measured in S. Accordingly the clock moves a distance $x_2 - x_1 = v\Delta t$ over the course of the roundtrip. Using Pythagoras' Theorem, it follows that the light travels a total distance $2(d^2 + v^2\Delta t^2/4)^{1/2}$. All of this is as it would be in Galilean relativity. Now here comes the new idea. The light is still travelling at speed c in S (in classical theory the speed would be $(c^2 + v^2)^{1/2}$ by the simple addition of velocities). As a result, the time for the roundtrip in S satisfies

$$\Delta t = \frac{2}{c}\left(d^2 + \frac{v^2\Delta t^2}{4}\right)^{1/2}. \tag{6.2}$$

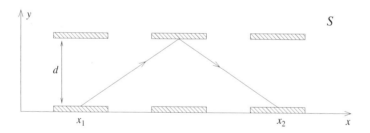

Figure 6.2   The path taken by the light in a moving light-clock.

Squaring both sides and re-arranging allows us to solve for Δt:

$$\Delta t = \frac{2d}{c} \times \frac{1}{\sqrt{1 - v^2/c^2}}. \tag{6.3}$$

The time measured in S is longer than the time measured in S' and we are forced to conclude that in Einstein's theory *moving clocks run slow*. This effect is also known as 'time dilation', and it is negligibly small if $v/c \ll 1$ but when $v \sim c$ the effect is dramatic.

The factor $1/\sqrt{1 - v^2/c^2}$ appears so often in Special Relativity that it is given its own symbol, i.e.

$$\gamma \equiv \frac{1}{\sqrt{1 - v^2/c^2}} \tag{6.4}$$

and

$$\Delta t = \gamma \Delta t_0. \tag{6.5}$$

For $v/c \leq 1$ it follows that $\gamma > 1$ and for $v/c > 1$ the theory doesn't appear to make much sense (unless we are prepared to entertain the idea of imaginary time).

To conclude this section, let us quickly check that $\Delta t = \Delta t_0$ in classical theory. Replacing c in Eq. (6.2) by $(c^2 + v^2)^{1/2}$ gives

$$\Delta t = \frac{2}{(c^2 + v^2)^{1/2}} \left(d^2 + \frac{v^2 \Delta t^2}{4} \right)^{1/2} \tag{6.6}$$

which has the solution $\Delta t = 2d/c$ as expected.

Eq. (6.5) is quite astonishing: it really does violate our intuition that time is absolute. We emphasise that this effect has nothing to do with the fact that we have considered light bouncing between two mirrors. We used light because it allows us to make use of Einstein's 2nd postulate. If we had used a bouncing ball then we would have become stuck when we had to figure out the speed of the ball in S because we are not entitled to assume that velocities add in the classical manner. When we have a little more knowledge and know how velocities add we will be able to return to the bouncing ball and we shall conclude that time is dilated exactly as for the light-clock. Clearly this must be the case for we are talking about the time interval between actual events.

The fact that time is actually different from our intuitive perception of it is no problem for physics, no matter how odd it may seem to us. There is a lesson to be learnt here. Namely, we should not expect our intuition based upon everday experiences to necessarily hold true in unfamiliar circumstances. In relativity theory, the unfamiliar circumstance is when objects are travelling close to the speed of light. The lesson also applies when tackling quantum theory. In this case common sense breaks down when we explore systems on very small length scales.

Example 6.1.1 *Muons are elementary particles rather like electrons but 207 times heavier. Unlike electrons, muons are unstable and they decay to an electron and a pair of neutrinos with a characteristic lifetime. For a muon at rest, this lifetime is 2.2 μs.*

Muons are created when cosmic rays impact upon the Earth's atmosphere at an altitude of 20 km and are observed to reach the Earth's surface travelling at close to the speed of light. (a) Use classical theory to estimate how far a typical muon would travel before it decays (assume the muon is travelling at the speed of light). (b) Now use time dilation to explain why the muons are able to travel the full 20 km without decaying.

Solution 6.1.1 *(a) Muons travelling at speed c will (on average) travel, according to classical thinking, a distance $c\Delta t_0$ before decaying where $\Delta t_0 = 2.2$ μs. Putting the numbers in gives a distance of just 660 m.*

(b) Let us suppose that the muon is travelling at a speed u towards the Earth. In the muon's rest frame its lifetime is a mere $\Delta t_0 = 2.2$ μs but from the point of view of an observer on Earth this lifetime is dilated to $\Delta t = \gamma \Delta t_0$. If γ is sufficiently large it is therefore possible that the muon could travel the 20 km and reach the Earth's surface. We can determine how large u must be using

$$\gamma \Delta t_0 > \frac{20 \ km}{u}. \qquad (6.7)$$

Since $\gamma = (1 - u^2/c^2)^{-1/2}$ we can solve this equation for $u = 0.999c$. Today, the lifetime of the muon has been measured as a function of its speed and it is found to be in excellent agreement with the prediction of time dilation.

Before leaving our discussion of time dilation we pause to consider the situation illustrated in Figure 6.3. Figure 6.3(a) shows our two frames S and S' moving relative to each other as shown. Time dilation says that, according to an observer at rest in S, clocks in S' run slow, i.e. that $\Delta t = \gamma \Delta t'$. This really does mean that all clocks run slow and so according to S an observer in S' would age more slowly. Now consider Figure 6.3(b). It represents exactly the same situation as Figure 6.3(a) since one can either think of S' moving relative to S or vice versa. Now an observer in S' will conclude that clocks in S run slow, i.e. that $\Delta t' = \gamma \Delta t$ and so from their perspective an observer at rest in S would age more slowly. At first glance these two conclusions seem to contradict each other but they do not since the observers are measuring intervals of time between different pairs of events: the observer in S is using clocks at rest in S whereas the observer in

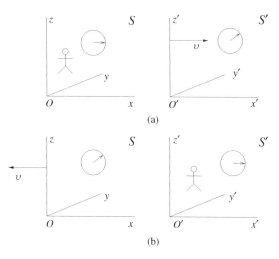

Figure 6.3   Two observers each conclude that the other is ageing more slowly than themselves. This is not a contradiction.

S' is using clocks at rest in S'. Thus it is the case that each concludes that the other is aging more slowly. Reflecting upon Einstein's 1st postulate we can see that this symmetrical situation must be correct for otherwise one could distinguish between the two inertial frames. Of course if the two observers were to meet up and compare notes then at least one of them must have undergone an acceleration. This would break the symmetry between the two and leads to the fascinating possibility that one of the observers would be genuinely older than the other upon meeting (see Section 14.1.1).

We have been very careful to explain what we mean by measurements of time and have stressed that they have nothing to do with seeing events with our eyes. Nevertheless, people do see things and it is interesting to ask how our perception of things changes in Special Relativity. Referring to Figure 5.1 we could imagine an observer situated at the origin O who is watching a clock speed away from them. We suppose that the clock is at rest at the origin O' in S'. If one tick of the clock takes a time $\Delta t'$ in S' what is the corresponding interval of time seen by the observer in S? The key word here is 'see'. Observations of events as we have hitherto been discussing them have referred explicitly to a process which does not depend upon the observer actually watching the event nor on where the observer is located when the event takes place. In contrast, the act of seeing does depend upon things like how far the observer is away from the things they are watching and the quality of the eyesight of the person doing the seeing. That distance is important when watching a moving clock becomes apparent once one appreciates that the clock is becoming ever further away and as a result light takes longer and longer to reach the observer. With this in mind, we can tackle the question in hand and attempt to work out the time interval Δt_{see} perceived by our observer at the origin O. According to all observers in S, including our observer standing at the origin, the time of one tick of the clock is given by the time dilation formula, i.e. $\Delta t = \gamma \Delta t'$. However this is not what we want. The time interval Δt_{see} is longer than Δt by an amount equal to the time it takes for light to travel the extra distance the clock has moved over the course of the tick, i.e. light from the end of the clock's tick has to travel further before it reaches the observer by an amount equal to $v\Delta t$. Therefore the perceived time interval between the start and the end of the tick is

$$\Delta t_{\text{see}} = \gamma \Delta t' + \gamma \Delta t' \frac{v}{c} = \gamma \Delta t' \left(1 + \frac{v}{c}\right) = \Delta t' \left(\frac{1 + v/c}{1 - v/c}\right)^{1/2}. \tag{6.8}$$

It is very important to be clear that this extra slowing down of the clock is an 'optical illusion', in contrast to the time dilation effect which is a real slowing down of time. To emphasise this point, if light travels at a finite speed then moving clocks will appear to run slow even in classical theory such that $\Delta t_{\text{see}} = \Delta t'(1 + v/c)$.

Eq. (6.8) leads us on nicely to the Doppler effect for light. Let us consider the situation illustrated in Figure 6.4. A light source is at rest in S' and is being watched by someone at rest in S. The time interval $\Delta t'$ could just as well be the time between the emission of successive peaks in a light wave, i.e. the frequency of the wave is $f' = 1/\Delta t'$. The person watching the light source will instead see

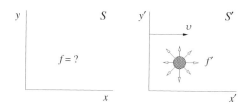

Figure 6.4   A light source of frequency f' at rest in S'.

a frequency $f = 1/\Delta t$. The two frequencies are related using Eq. (6.8):

$$f = f' \left(\frac{1 - v/c}{1 + v/c} \right)^{1/2}. \tag{6.9}$$

This is the result in the case that the light source is moving away from the observer, in which case Eq. (6.9) tells us that $f < f'$ and so the light appears shifted to shorter frequencies, i.e. it is 'red-shifted'. If the source is moving towards the observer we should reverse the sign of v in Eq. (6.9) and therefore conclude at $f > f'$, i.e. the light is now 'blue-shifted'.

Example 6.1.2 *How fast must the driver of a car be travelling towards a red traffic light ($\lambda = 675$ nm) in order for the light to appear amber ($\lambda = 575$ nm)?*

Solution 6.1.2 *In the rest frame of the car, the traffic light is moving towards them at a speed u. Our task is to determine u given the change in wavelength. We can convert wavelengths to frequencies using $c = f\lambda$ and then use Eq. (6.9) to solve for u. Because the source is moving towards the car we should use Eq. (6.9) with $v = -u$ and so*

$$\frac{c}{575 \times 10^{-9}m} = \frac{c}{675 \times 10^{-9}m} \left(\frac{1 + u/c}{1 - u/c} \right)^{1/2},$$

$$\Rightarrow \left(\frac{675}{575} \right)^2 = \frac{1 + \beta}{1 - \beta}.$$

The solution to which is $\beta = u/c = 0.159$. It is often sensible to express speeds in terms of the ratio u/c, although in this case expressing the result as a speed of just over 13 km/s makes it clear that this effect is never going to impress a court of law.

6.1.2 Length contraction

We now shift our attention to the measurement of distances in different inertial frames and to the phenomenon known as length contraction. Light bouncing between mirrors can also be used to determine distances by accurately measuring the time it takes for light to travel between the mirrors. Let us imagine a ruler of length L_0 when measured in its rest frame. Now we ask what is the length L of the ruler when it is moving? Figure 6.5 shows a ruler moving with a speed v relative to

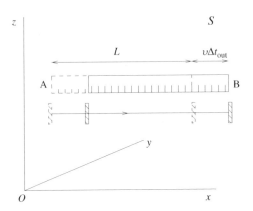

Figure 6.5   Measuring the length of a moving ruler.

S. To measure the length of the ruler we shall mount a light-clock of equal length next to it, as shown. The light-clock moves with the ruler. The light starts out from one end of the ruler and reflects from a mirror located at the opposite end of the ruler. Our strategy will be to determine the time taken for the roundtrip directly in S and equate this to the time dilation result. As a result of time dilation, the roundtrip time in S is related to the roundtrip time in the rest frame of the ruler Δt_0 by

$$\Delta t = \gamma \Delta t_0 = \gamma \frac{2L_0}{c}. \tag{6.10}$$

We shall now endeavour to determine this time interval by considering the journey of the light from the viewpoint of S. According to an observer in S, the total time is

$$\Delta t = \Delta t_{\text{out}} + \Delta t_{\text{in}}, \tag{6.11}$$

where Δt_{out} is the time taken for the light to travel on its outward journey, i.e. from A to B, and Δt_{in} is the time taken on the return journey. The figure shows explicitly the two positions of the ruler when the light starts its journey (dashed line) and when the light reaches the opposite end of the ruler (solid line). In order not to clutter the picture we have not shown the third position of the ruler, i.e. when the light finally returns back to its starting point. Since Einstein's 2nd postulate tells us the speed of light according to S, we can write

$$c\Delta t_{\text{out}} = L + v\Delta t_{\text{out}}$$

$$\Rightarrow \Delta t_{\text{out}} = \frac{L}{c - v}. \tag{6.12}$$

Each side of the first of these equations is equal to the total distance travelled by the light on its outward journey (according to S) and it takes into account the fact that the light has to travel a little further than the length of the ruler L as a result of the ruler's motion. Similarly for the return leg, the light has to travel a shorter distance than L, i.e.

$$c\Delta t_{in} = L - v\Delta t_{in}$$

$$\Rightarrow \Delta t_{in} = \frac{L}{c + v}. \tag{6.13}$$

Adding together Eqs. (6.12) and (6.13) and equating the result to Eq. (6.10) gives an equation relating L and L_0, i.e.

$$\frac{L}{c + v} + \frac{L}{c - v} = \gamma \frac{2L_0}{c}. \tag{6.14}$$

Solving for L gives

$$L = \frac{L_0}{\gamma}. \tag{6.15}$$

Again a remarkable result; for the length of the ruler is smaller when it is in motion than when it is at rest.

We could have anticipated the length contraction result knowing only the time dilation result. The argument goes as follows. Let us consider again the muons created in the upper atmosphere which we discussed in Example 6.1.1. From the viewpoint of a muon, it still lives for 2.2 μs yet has travelled all the way to the Earth's surface. However this is not such an impossible task as it would be in classical theory for the 20 km is reduced by a factor of γ. It has to be exactly the same factor of γ as before because we know that muons created at an altitude of 20 km on average just reach the Earth before decaying if they have a speed of $0.999c$ and from the viewpoint of such a muon the Earth moves towards it at that speed.

Example 6.1.3 *A spaceship flies past the Earth at a speed of 0.990c. A crew member on the ship measures its length to be 400 m. How long is the ship as measured by an observer on Earth?*

Solution 6.1.3 *This is a straightforward application of the length contraction result expressed in Eq. (6.15) with $L_0 = 400$ m. Hence*

$$\gamma = \frac{1}{\sqrt{1 - 0.990^2}} = 7.09 \tag{6.16}$$

and so $L = 400/7.09 = 56.4$ m. Perhaps the most common misuse of the length contraction formula is to confuse L and L_0.

6.1.3 Simultaneity

Classical physics, with its absolute time, has an unambiguous notion of what it means to say two events are simultaneous. However, since time is more subjective in Special Relativity, having meaning only within the context of a specified inertial frame, it may not be suprising to hear that two events that are simultaneous in one inertial frame will not in general be simultaneous in another inertial frame. Moreover, according to one observer event A may precede event B but according to a second observer event B might occur first. This last statement sounds particularly dangerous for it suggests problems with causality. Surely everyone must agree that a person must be born before they die? And indeed they must. It is a remarkable feature of Special Relativity that although the time ordering of events can be a matter for debate this is only the case for causally disconnected events, i.e. events which cannot influence each other. We shall return to this interesting discussion in Part IV. For now we content ourselves with a thought experiment which illustrates the breakdown of simultaneity.

Consider a train travelling along at a speed u relative to the platform. An observer is standing in the middle of the train. Suppose that a flashlight is attached to each end of the train and that the flashlights flash on for a brief instant. If the observer receives the light from each flashlight at the same time then she will conclude that the flashes occurred simultaneously, for the light from each flashlight had to travel the same distance (half the length of the train) at the same speed. Now consider a second observer standing on the platform watching proceedings. They must observe that our first observer does indeed receive the light from either end of the train at a particular instant in time. However, from their viewpoint the light from the front of the train has less distance to travel than the light from the rear of the train since the observer on the train is moving towards the point of emission at the front of the train and away from the point of emission at the rear of the train. None of what has been said so far is controversial; it holds in classical theory too. Here comes the difference. As a result of the 2nd postulate, the observer on the platform still sees each pulse of light travel at the same speed c. Now since both pulses arrive at the centre of the train at the same time, and the pulse from the front had less distance to travel, it follows that it must have been emitted later than the light from the rear of the train. Classical physics avoids this conclusion because although the light from the front has less distance to travel it is travelling more slowly (its speed is $c - u$) than the light from the rear (its speed is $c + u$) and the reduction in speed compensates the reduction in distance. You might like to check that this compensation is exact and that both observers agree that the pulses were emitted at the same time according to classical physics.

6.2 LORENTZ TRANSFORMATIONS

In Section 5.1 we derived the Galilean transformation equations which relate the co-ordinates of an event in one inertial frame to the co-ordinates in a second inertial frame. For their derivation we relied upon the idea of absolute time and, as the last section showed, this is a flawed concept in Special Relativity. We must therefore seek new equations to replace the Galilean transformations. These new equations are the so-called Lorentz transformations.

To derive the Lorentz transformations we shall follow the methods of Section 5.1. We shall define our two inertial frames S and S' exactly as before, and as illustrated in Figure 5.1, i.e. S' is moving along the positive x axis at a speed v relative to S. Since the motion is parallel to the x and x' axes it follows that

$$y' = y \tag{6.17}$$

$$z' = z \tag{6.18}$$

as before. Recall that we want to express the co-ordinates in S' in terms of those measured in S. Again in order for the 1st postulate to remain valid the transformations must be of the form

$$x' = ax + bt, \tag{6.19a}$$

$$t' = dx + et. \tag{6.19b}$$

Notice that we have not assumed that there exists a unique time variable, i.e. we allow for $t' \neq t$. Our goal is to solve for the coefficients a, b, d and e. As with the derivation of the Galilean transforms we require that the origin O' (i.e. the point $x' = 0$) move along the x-axis according to $x = vt$. Substituting this information into Eq. (6.19a) yields

$$-b/a = v. \tag{6.20}$$

Similarly we require that the origin O move along the line $x' = -vt'$. From Eqs. (6.19) the point $x = 0$ satisfies $x' = bt$ and $t' = et$ such that $x' = -vt'$ implies that

$$-b/e = v. \tag{6.21}$$

Eqs. (6.20) and (6.21) imply that $e = a$ and $b = -av$. Substituting these into Eqs. (6.19) gives

$$x' = ax - avt,$$

$$t' = dx + at. \tag{6.22}$$

We have two unknowns, a and d, remaining and have two postulates to implement. Let us first implement the 2nd postulate. We shall do this by considering a pulse of light emitted at the origins O and O' when they are coincident, i.e. when $t = t' = 0$. We know that this pulse must travel outwards along the x and x' axes such that it satisfies $x = ct$ and $x' = ct'$, i.e. it travels out at the same speed c in both frames. These two equations must be simultaneous solutions to Eqs. (6.22) and so we require that

$$ct' = act - avt,$$

$$t' = dct + at. \tag{6.23}$$

From which it follows directly that

$$d = -\frac{av}{c^2}. \tag{6.24}$$

It only remains to determine the value of a. Let us summarise progress so far. We have reduced Eqs. (6.19a) and (6.19b) to

$$x' = a(x - vt), \tag{6.25a}$$

$$t' = a\left(t - \frac{vx}{c^2}\right). \tag{6.25b}$$

Now it is time to make use of the 1st postulate which says that if Eqs. (6.25) are true then so necessarily are

$$x = a(x' + vt'),$$ (6.26a)

$$t = a\left(t' + \frac{vx'}{c^2}\right).$$ (6.26b)

This makes manifest the equivalence of the two frames. It can be seen by considering Figure 5.1 and swapping the primed and unprimed co-ordinate labels around whilst at the same time reversing the direction of v. We can determine the coefficient a now by substituting for x' and t' using Eqs. (6.25) into either of Eqs. (6.26), i.e.

$$x = a\left(ax - avt + avt - \frac{av^2x}{c^2}\right) = a^2x\left(1 - \frac{v^2}{c^2}\right)$$

$$\Rightarrow a = \frac{1}{\sqrt{1 - v^2/c^2}} = \gamma.$$ (6.27)

We have succeeded in deriving the Lorentz transformations:

$$x' = \gamma(x - vt),$$ (6.28a)

$$t' = \gamma(t - vx/c^2),$$ (6.28b)

$$y' = y,$$ (6.28c)

$$z' = z.$$ (6.28d)

Sometimes the inverse transformations will be more useful:

$$x = \gamma(x' + vt'),$$ (6.29a)

$$t = \gamma(t' + vx'/c^2).$$ (6.29b)

Eqs. (6.28) are perhaps the most important equations we have derived so far in this part of the book.

Example 6.2.1 *Use the Lorentz transformations to derive the formula for time dilation.*

Solution 6.2.1 *Let us consider the situation illustrated in Figure 6.6. A clock is at rest in S', let's suppose it is at position x_0'. Now consider one tick of the clock. In S', we suppose that the tick starts at time t_1' and ends at time t_2' such that $\Delta t' = t_2' - t_1'$ is the duration in the clock's rest frame. The question is: 'what is the duration of the same tick as determined by an observer in S?'*

There are two events to consider. Event 1 (start of the tick) has co-ordinates (x_0', t_1') in S' and event 2 (end of tick) which has co-ordinates (x_0', t_2') in S'. We want to know the time of each event in S. Given that we know both the location and time of the events in S' we should use Eq. (6.29b) to give us the corresponding times in S:

$$t_1 = \gamma(t_1' + vx_0'/c^2),$$

$$t_2 = \gamma(t_2' + vx_0'/c^2).$$

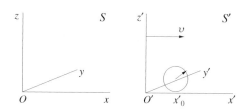

Figure 6.6   A moving clock.

Subtracting these two equations gives

$$\Delta t = t_2 - t_1 = \gamma \Delta t',$$

which is the required result. Notice that to derive this result it was crucial to be clear that the clock is at rest in S'.

Example 6.2.2 *Use the Lorentz transformations to derive the formula for length contraction.*

Solution 6.2.2 *We now consider the situation illustrated in Figure 6.7 where we have placed a ruler in S' such that it lies along the x'-axis with one end located at x'_1 and the other at x'_2. The length of the ruler in its rest frame is therefore $\Delta x' = x'_2 - x'_1$. The question now is: 'what is the length of the ruler as determined by an observer in S?'*

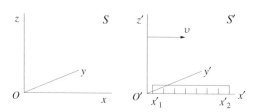

Figure 6.7   A moving ruler.

Again there are two events to consider. Event 1 (measurement of one end of the ruler) and event 2 (measurement of the other end of the ruler). The crucial point now is that both events occur at the same time in S because that is what is meant by a measurement of length. Let's call this time t_0. Given that we know the location of the two events in S' and the time of the events in S we should use Eq. (6.28a) to give us the location of the events in S:

$$x'_1 = \gamma(x_1 - vt_0),$$
$$x'_2 = \gamma(x_2 - vt_0).$$

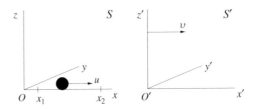

Figure 6.8   A ball bouncing back and forth between two points.

Subtracting these two equations gives

$$\Delta x' = x'_2 - x'_1 = \gamma \Delta x$$

$$\Rightarrow \Delta x = \frac{1}{\gamma} \Delta x',$$

which is the required result.

Example 6.2.3 *A ball is rolled at speed u from the point x_1 on the x-axis to the point $x_2 = x_1 + L$ at which point it is reflected back again elastically, as illustrated in Figure 6.8. In a frame moving with speed v along the positive x-axis compute:*

(a) *The spatial separation between the point where the ball starts its journey and the point where it is reflected;*
(b) *The time taken for the outward part of the ball's journey;*
(c) *The time taken for the return part of the ball's journey.*

Solution 6.2.3 *(a) Event 1 is when the ball starts on its journey and has co-ordinates (x_1, t_1) in S. Event 2 is when the ball arrives at the point of reflection. It has co-ordinates $(x_2, t_1 + L/u)$. We are asked to find $\Delta x' = x'_2 - x'_1$. Note that it is not going to be given by the length contraction formula since the two events are not simultaneous in either S or S'. We know both $\Delta x = x_2 - x_1 = L$ and $\Delta t = L/u$ and need $\Delta x'$. We therefore need to use Eq. (6.28a) which informs us that*

$$\Delta x' = \gamma(\Delta x - v\Delta t),$$

$$= \gamma L(1 - v/u) \tag{6.30}$$

and γ is of course evaluated using the relative speed of the two frames v.

(b) To get the time taken for the outward part of the journey we should use Eq. (6.28b) (we hope that by now the reader is getting the hang of selecting the correct equation to use), i.e.

$$\Delta t'_{out} = \gamma(\Delta t - v\Delta x/c^2),$$

$$= \frac{\gamma L}{u}(1 - uv/c^2). \tag{6.31}$$

(c) And for the return leg we introduce a third event corresponding to when the ball returns. In S it has co-ordinates $(x_1, t_1 + 2L/u)$. As in part (b) we use Eq. (6.28b). Notice that $x_3 - x_2 = -\Delta x$ this time whilst $t_3 - t_2 = \Delta t$ and so

$$\Delta t'_{in} = \gamma(\Delta t + v\Delta x/c^2),$$

$$= \frac{\gamma L}{u}(1 + uv/c^2). \tag{6.32}$$

Notice also that the total time for the journey is just as we would expect from time dilation, i.e $\Delta t'_{tot} = \gamma(2L/u)$ as it should be since the point of departure and point of return are one and the same place. This result confirms our earlier claim that there was nothing special about a light-clock.

6.3 VELOCITY TRANSFORMATIONS

6.3.1 Addition of Velocities

We can use the Lorentz transformations to figure out how the rules for adding velocities must change in Special Relativity. Consider an object moving with a velocity $\mathbf{v}'$ in S'. Let us determine its velocity $\mathbf{v}$ in S. The situation is illustrated in Figure 6.9. Notice that to avoid confusion the relative speed between the two frames is now u and we have simplified to the case of motion in two dimensions (in the $x - y$ plane). It is straightforward to generalize to three-dimensions. Recall that according to Galilean relativity $v_x = v'_x + u$ and $v_y = v'_y$. Neither of these holds true in Special Relativity, as we shall now see.

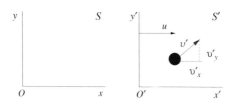

Figure 6.9 Relative velocities.

To determine the x-component of the velocity in S we make use of the Lorentz transformation formulae for x and for t:

$$v_x = \frac{dx}{dt} = \frac{\gamma(dx' + udt')}{\gamma(dt' + udx'/c^2)} = \frac{(dx'/dt') + u}{1 + u(dx'/dt')c^2}$$

$$= \frac{v'_x + u}{1 + uv'_x/c^2}. \tag{6.33}$$

Similarly we can determine the y-component of the velocity:

$$v_y = \frac{dy}{dt} = \frac{dy'}{\gamma(dt' + u\,dx'/c^2)} = \frac{dy'/dt'}{\gamma(1 + u(dx'/dt')c^2)}$$

$$= \frac{v_y'}{\gamma(1 + uv_x'/c^2)}. \qquad (6.34)$$

Notice that for $uv_x' \ll c^2$ and $u \ll c$ these results reduce to the expectation based on classical thinking. Eqs. (6.33) and (6.34) are known as the velocity transformation equations and their use is pretty straightforward. Perhaps the only place where there is room for error is when it comes to figuring out the signs. For example, if S' were moving in the negative x-direction then we should replace $u \to -u$ in the equations. We can quickly check to see that the velocity transformation equations satisfy the 2nd postulate, i.e. if $v_x' = c$ and $v_y = 0$ we have

$$v_x = \frac{c + u}{1 + uc/c^2} = c, \qquad (6.35)$$

which is as it should be.

6.3.2 Stellar Aberration Revisited

It is at this point that we can confirm that although Einstein has abolished the ether his new theory is still capable of explaining the phenomenon of stellar aberration. To understand this, let us consider the particular situation illustrated in Figure 6.10. We imagine that the Sun, Earth and star all lie in the same plane and that the Sun is at rest in S'. Suppose that light emitted from the star arrives at an angle angle α' to the vertical in S'. We shall take the relative speed between the Earth and Sun to be u and α is the angle at which the starlight arrives on Earth.

Using the velocity addition formulae with $v_x' = -c\sin\alpha'$ and $v_y' = -c\cos\alpha'$ we have that

$$v_x = \frac{u - c\sin\alpha'}{1 - \frac{u}{c}\sin\alpha'}, \qquad (6.36)$$

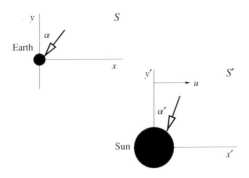

Figure 6.10   Incident starlight in the Earth and Sun rest frames.

$$v_y = \frac{-c\cos\alpha'}{\gamma(u)(1 - \frac{u}{c}\sin\alpha')}. \tag{6.37}$$

These two equations imply that

$$\tan\alpha = \frac{v_x}{v_y} = \frac{\sin\alpha' - u/c}{\cos\alpha'} \frac{1}{\sqrt{1 - u^2/c^2}}. \tag{6.38}$$

Stellar aberration is greatest when $\alpha' = 0$, in which case this result simplifies to

$$\tan\alpha = -\frac{u}{c}\frac{1}{\sqrt{1 - u^2/c^2}},$$

$$\text{i.e. } \sin\alpha = -\frac{u}{c}.$$

Now if $u \ll c$ then this is gives rise to a variation in the star's angular position of $\approx 2u/c$ over the course of one year, which is in accord with observations.

Example 6.3.1 *Consider three galaxies, A, B and C. An observer in A measures the velocities of B and C and finds they are moving in opposite directions each with a speed of $0.7c$. (a) At what rate does the distance between B and C increase according to A? (b) What is the speed of A observed in B? (c) What is the speed of C observed in B?*

Solution 6.3.1 *Again it really helps to draw a picture: we refer to Figure 6.11. (a) The relative speed between B and C according to A is just $2u = 1.4c$. We do not of course worry that this speed is in excess of c because it is not the speed of any material object. (b) According to B, A moves 'to the right' with speed u. (c) Now to determine the speed of C according to an observer in B we do need to use the addition of velocities formula since we only know the speed of C in A and the speed of A relative to B. In classical theory, the result would be $1.4c$, but this will clearly be modified to a value smaller than c in Special Relativity. The correct value is found using Eq. (6.33):*

$$\frac{u + u}{1 + u^2/c^2} = \frac{1.4c}{1.49} = 0.94c.$$

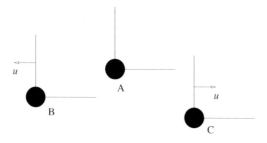

Figure 6.11   Relative motion of three galaxies viewed from an observer in A.

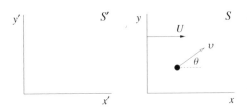

Figure 6.12   A particle moves in S at an angle θ to the x-axis at speed v.

Example 6.3.2 *A particle moves with speed v at an angle θ to the positive x-axis in the frame S. What is the direction of the particle in the frame S' given that S and S' move with relative speed U along their common x direction?*

Solution 6.3.2 *Figure 6.12 illustrates what is going on. Using the velocity addition formulae we can write down the components of the velocity of the particle in S':*

$$v'_x = \frac{v_x + U}{1 + Uv_x/c^2},$$

$$v'_y = \frac{v_y}{\gamma(U) \cdot (1 + Uv_x/c^2)}.$$

Using $\tan\theta' = v'_y/v'_x$, $v_x = v\cos\theta$ and $v_y = v\sin\theta$ gives

$$\tan\theta' = \frac{(1 - U^2/c^2)^{1/2}v\sin\theta}{U + v\cos\theta}.$$

This is an interesting result: if $U \to c$ then $\tan\theta' \to 0$ regardless of θ. This effect would only happen in classical theory as $U \to \infty$.

PROBLEMS 6

6.1 A spaceship moves relative to the Earth at a speed of $0.93c$. If a person on Earth spends 30 minutes reading the newspaper, how long have they been reading according to someone on the spaceship?

6.2 Pions are elementary particles, which decay with a half-life of 1.8×10^{-8} s as measured in a frame in which the pions are at rest. In a laboratory experiment, a beam of pions has a speed of $0.95c$. According to an observer in the lab, how long does it take for half of the pions to decay? Through what distance will they travel in that time?

6.3 An alien spacecraft is flying overhead at a great distance. You see its search-light blink on for 0.190 s. Meanwhile, on board the spacecraft, the pilot observes that the searchlight was on for 12.0 ms. What is the speed of the spacecraft relative to the Earth?

6.4 In the previous question, why was it necessary to state that the spacecraft was at a great distance overhead? Suppose that the same spacecraft is flying at $0.998c$ but this time at ground level and directly away from you. If the pilot once again turns the searchlight on for 12.0 ms, how long now does the searchlight appear to stay on according to your watch?

6.5 How long is a 1m rod according to an observer moving at speed $0.95c$ in a direction parallel to the rod?

6.6 What is the distance to the surface of the Earth as determined in the rest frame of a cosmic ray proton which is at an altitude of 12 km and which is travelling directly towards the Earth's surface at a speed of $0.97c$?

6.7 The distance to the farthest star in our galaxy is of the order of 10^5 light years. Explain why it is possible, in principle, for a human being to travel to this star within the course of a lifetime and estimate the required velocity.

6.8 As measured by an observer on Earth, a spacecraft runway has a length of 3.60 km.

   (a) What is the length of the runway as measured by the pilot of a spacecraft flying directly over the runway at a speed 4.00×10^7 ms^{-1} relative to the Earth?

   (b) An observer on Earth measures the time interval from when the spacecraft is directly over one end of the runway until it is directly over the other end. What result does she get?

   (c) According to the pilot of the spacecraft, how long does the spacecraft take to travel the length of the runway?

6.9 A pole 10 m long lies on the ground next to a barn 8 m long. An athlete picks up the pole, carries it far away, and then runs with it toward the barn at speed $0.8c$. The athlete's friend remains at rest, standing by the open door of the barn.

   (a) How long does the friend measure the pole to be as it approaches the barn?

   (b) Immediately after the pole is entirely inside the barn, the friend shuts the barn door. How long after the door is shut does it take for the front of the pole to strike the back end of the barn, as measured by the friend?

   (c) In the reference frame of the athlete what is the length of the pole and the barn?

   (d) How do you reconcile the closing of the barn door with the experience of the athlete?

6.10 Sodium light of wavelength 589 nm is emitted by a source that is moving toward the Earth with speed v. The wavelength measured by an observer on Earth is 550 nm. Find v/c.

6.11 In a frame S, event B occurs 1.3 μs after event A. Also in S the events are separated by a distance of 1.5 km along the x-axis, i.e. $x_B - x_A = 1500$ m. At what fraction of the speed of light must an observer be moving along the x-axis in order to conclude that the two events occur at the same time?

6.12 Two events occur at the same place in a certain inertial frame and are separated by a time interval of 4 s. What is the spatial separation between these two events in an inertial frame in which the events are separated by a time interval of 5 s?

6.13 Two events occur at the same time in inertial frame S and are separated by a distance of 1 km along the x-axis. What is the time difference between these two events as measured in frame S' moving with constant velocity parallel to the x-axis and in which their spatial separation is measured to be 2 km?

6.14 In a particular inertial frame, two pulses of light are emitted at a distance 4 km apart and are separated by 5 μs in time. An observer travelling at speed v along the line joining the emission points of the two pulses notes that the pulses are emitted simultaneously. Find v.

6.15 In 1994, a rather eccentric group of astronomy students wanted to celebrate the impact of the Shoemaker-Levy comet on Jupiter by holding a party of sufficiently long duration that their celebrations were simultaneous with the impact of the comet in all possible inertial frames. For how long did they need to party? How might the party end?
[The distance from the Earth to Jupiter is 8×10^{11} m and you may neglect their relative motion.]

6.16 Two particles are created in a particle physics experiment. They move apart in opposite directions, with one particle travelling at a speed of $0.70c$ and the other at a speed of $0.850c$ as measured in the laboratory. What is the speed of one particle relative to the other? Compare your answer to that which you would expect using classical ideas.

6.17 A rocket moves away from the Earth at a speed of $0.5c$ and at a later time sends out a second rocket which travels back towards the Earth at a speed of $0.8c$ relative to the parent rocket. What is the speed of the second rocket relative to the Earth?

6.18 The passage of light through a medium is characterised by a refractive index n and the velocity of light relative to the medium is c/n. Suppose that such a medium is moving with speed v parallel to the direction of propagation of the light. Derive an expression for the speed of light, V, as observed by a stationary observer. Show that your result can be used to explain Fizeau's experimental findings of 1851 which used light passing through flowing water to demonstrate that

$$V \approx \frac{c}{n} + v\left(1 - \frac{1}{n^2}\right).$$

6.19 Two rockets, A and B, start from a common point C and travel with constant speeds u in directions perpendicular to each other as observed in the rest frame of C. An observer in A measures the angle BAC to be $30°$. What is the value of u/c?

7

Relativistic Energy and Momentum

7.1 MOMENTUM AND ENERGY

In classical theory, energy and momentum are conserved quantities, and are therefore of particular significance. From a practical viewpoint, we can exploit the conservation of energy and/or momentum in order to simplify calculations. At this stage in our development of Einstein's theory we are led to contemplate just how energy and momentum are to be defined if they are to be compatible with Einstein's two postulates. *A priori* it might be that the concepts of energy and momentum are only useful in the classical regime, where speeds are small compared to light speed, in which case any attempt to extrapolate into the relativistic regime would be doomed to fail. Fortunately, this is not the case. Transcending relativity theory, it is now known that conservation laws often have their origin in symmetry. Technically speaking, the law of conservation of energy arises because physical phenomena are invariant under time translations, which means that energy is conserved because, all other things being equal, it does not matter whether one conducts an experiment today, tomorrow or at some other time in history. Similarly, momentum is conserved because physical phenomena are invariant under spatial translations. Again, in more down to earth language, momentum conservation is a consequence of the fact that, all other things being equal, it does not matter whether one conducts an experiment here or over there. That said, and since we expect the same underlying symmetries of space and time in Einstein's theory, we press on with our attempt to introduce definitions of energy and momentum that do not conflict with Einstein's postulates.

Our first guess might be to assume the familiar expressions, i.e.

$$p = mv \tag{7.1}$$

for the momentum and

$$K = \frac{1}{2}mv^2 \tag{7.2}$$

for the kinetic energy. However these will not do. To see why, we need to notice that if the total momentum or energy is conserved in one inertial frame then it cannot also be conserved in all other inertial frames, and this is a violation of the principle of relativity (Einstein's 1st postulate). To prove that the quantities defined in Eqs. (7.1) and (7.2) cannot be conserved in all inertial frames one only has to look at the velocity transformation formulae we derived earlier, i.e. Eqs. (6.33) and (6.34).

As an example consider an inertial frame S in which two particles, of mass m_1 and m_2, scatter elastically off each other. For simplicity we'll consider that the particles always move in the x-direction. Before they scatter, the particles have speeds u_1 and u_2 and afterwards v_1 and v_2. The same collision is recorded by an observer in S' which, once again, is moving with speed U along the positive x-axis. In this case the particles have speeds u_1' and u_2' before the collision and v_1' and v_2' after it. In classical theory, the total momentum before and after the collision as recorded in S is

$$m_1 u_1 + m_2 u_2 = m_1 v_1 + m_2 v_2. \tag{7.3}$$

Using the classical law of addition of velocities, i.e. $u_1 = u_1' - U$ etc., this can be rewritten as

$$m_1 u_1' + m_2 u_2' - (m_1 + m_2)U = m_1 v_1' + m_2 v_2' - (m_1 + m_2)U$$

i.e.

$$m_1 u_1' + m_2 u_2' = m_1 v_1' + m_2 v_2'. \tag{7.4}$$

Therefore we see that so long as momentum is conserved in S so it is also conserved in S'. The same can be said for kinetic energy since in S we have

$$\frac{1}{2}m_1 u_1^2 + \frac{1}{2}m_2 u_2^2 = \frac{1}{2}m_1 v_1^2 + \frac{1}{2}m_2 v_2^2, \tag{7.5}$$

which can be rewritten as

$$\frac{1}{2}m_1 u_1'^2 + \frac{1}{2}m_2 u_2'^2 - (m_1 u_1' + m_2 u_2')U + \frac{1}{2}(m_1 + m_2)U^2$$

$$= \frac{1}{2}m_1 v_1'^2 + \frac{1}{2}m_2 v_2'^2 - (m_1 v_1' + m_2 v_2')U + \frac{1}{2}(m_1 + m_2)U^2$$

and using the fact that momentum is conserved (i.e. $m_1 u_1' + m_2 u_2' = m_1 v_1' + m_2 v_2'$) this can be written as

$$\frac{1}{2}m_1 u_1'^2 + \frac{1}{2}m_2 u_2'^2 = \frac{1}{2}m_1 v_1'^2 + \frac{1}{2}m_2 v_2'^2. \tag{7.6}$$

Crucial to this argument is the fact that the law of addition of velocities is linear. However, we know that the relationship in Special Relativity is non-linear. For example,

$$u_1 = \frac{u_1' - U}{1 - u_1' U / c^2} \tag{7.7}$$

and the factor of u_1' in the denominator spoils the linearity. We are therefore forced to seek out alternatives to Eq. (7.1) and Eq. (7.2) which do satisfy the 1st postulate.

Let us aim first for a new definition of momentum. Insisting that it remain a vector quantity that lies parallel to the velocity and which is equal to the classical result in the limit $v \ll c$ dictates that the most general form available to us is

$$\boldsymbol{p} = f(v/c)m\boldsymbol{v} \tag{7.8}$$

and our task is to determine the dimensionless function[1] $f(v/c)$ which we know must satisfy $f(v/c) \simeq 1$ for $v \ll c$. The mass of the particle is labelled m and we take it to be an intrinsic property of the particle, not depending upon its state of motion, i.e. all observers will agree upon its value. To determine $f(v/c)$ we are going to focus our attention upon the very specific scattering process illustrated in Figure 7.1. Our strategy is to view this process in two different inertial frames with the momentum carefully defined so that it is conserved in both frames. Any

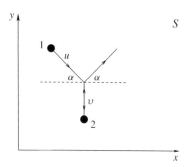

Figure 7.1   Particle 1 scatters through a vanishingly small angle α, whilst particle 2 bounces off at right angles.

---

[1] We write the argument explicitly as v/c to remind us that this is the only way to form a dimensionless function of the speed.

scattering process would in principle suffice to determine $f(v/c)$, it is only that this process provides a particularly elegant path to the answer. Having said that, the derivation we shall now present is still rather tricky and any readers wishing to avoid the details might note the key results presented in Eq. (7.15) and Eq. (7.25) and skip to the next subsection. We shall return to the topic of energy and momentum in Part IV where we shall see that there is a much more elegant way to obtain the results which we shall here work rather hard to establish.

Returning to Figure 7.1, we consider the special case where particle 1 travels in to and out of the scattering at the same angle α to the x axis, and particle 2 travels always parallel to the y axis with speed v. We shall also assume that the particles have the same mass m. Momentum conservation in the x direction then implies that if the speed of particle 1 is u before the scatter then it should remain unchanged after the scatter. This is a very symmetrical scattering and it is this symmetry which will help us arrive at a form for $f(v/c)$ without too much hard work. In order to simplify matters still further, we focus our attention upon the limit that $\alpha \to 0$. In this limit the speed $v \to 0$ too since otherwise the process would not conserve momentum in the y direction. This limit will help us a great deal since it will allow us to use the non-relativistic form for the momentum of particle 2, i.e. we shall make use of $f(0) = 1$. We have now completely specified the scattering process in inertial frame S and it clearly conserves momentum in the x and y directions independent of the actual form of $f(v/c)$.

Now let us view the same scattering from a second inertial frame, S'. In particular we choose a frame which travels along the positive x axis at a speed $u \cos \alpha$. In this frame, particle 1 travels only in the y direction, whilst particle 2 travels in the negative x direction as illustrated in Figure 7.2. The velocity addition formulae allow us to write down the x and y components of velocity of each particle in S':

$$v_{1x} = 0,$$

$$v_{1y} = -\frac{1}{(1 - (u/c)^2 \cos^2 \alpha)^{-1/2}} \frac{u \sin \alpha}{1 - (u/c)^2 \cos^2 \alpha}$$

$$= -\frac{u \sin \alpha}{\sqrt{1 - (u/c)^2 \cos^2 \alpha}}, \tag{7.9}$$

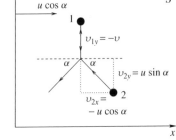

Figure 7.2   Particle 2 scatters through a vanishingly small angle α, whilst particle 1 bounces off at right angles.

and

$$v_{2x} = -u\cos\alpha,$$

$$v_{2y} = \frac{1}{(1 - (u/c)^2 \cos^2\alpha)^{-1/2}} v$$

$$= v\sqrt{1 - (u/c)^2 \cos^2\alpha}. \tag{7.10}$$

It is now that we can appreciate the advantage of picking such a symmetric scattering process. Viewed in S', the scattering looks just like that in S except that it is 'turned upside down'. This symmetry allows us to conclude that

$$v_{1y} = -v \text{ and}$$

$$v_{2y} = u\sin\alpha. \tag{7.11}$$

Either of these two equations used in conjunction with Eq. (7.9) or Eq. (7.10) implies that

$$v = \frac{u\sin\alpha}{\sqrt{1 - (u/c)^2 \cos^2\alpha}}. \tag{7.12}$$

Now we require momentum conservation in S', i.e.

$$2mf(u/c)v_{2y} = 2mf(v/c)v.$$

But in the limit of $\alpha \to 0$ we can safely take $f(v/c) \to 1$, i.e.

$$f(u/c)u\sin\alpha = v. \tag{7.13}$$

Subsituting in for v using Eq. (7.12) gives

$$f(u/c)u\sin\alpha = \frac{u\sin\alpha}{\sqrt{1 - (u/c)^2 \cos^2\alpha}},$$

which in the limit $\alpha \to 0$ reduces to

$$f(u/c) = \frac{1}{\sqrt{1 - u^2/c^2}} = \gamma(u). \tag{7.14}$$

Thus we have a new candidate for momentum in Einstein's theory. For a particle of mass m and velocity $\boldsymbol{u}$ the momentum is

$$\boldsymbol{p} = \gamma(u)m\boldsymbol{u}. \tag{7.15}$$

Although we have a candidate for momentum, we ought to be clear that we derived it by considering one very particular scattering process. If this definition is to be useful then it ought to have the property that if momentum is conserved in one inertial frame then it is also conserved in all other inertial frames and this should be true for any process. To convince ourselves that this is the case, let's consider the much more general scattering process illustrated in Figure 7.3. In inertial frame S, particles A (mass m_A) and B (mass m_B) are incident with velocities $\boldsymbol{u}_A$ and

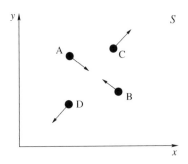

Figure 7.3   A general two-to-two scattering process: AB → CD.

u_B. These particles subsequently scatter into particles C (mass m_C) and D (mass m_D) travelling with velocities u_C and u_D. We say that the scattering process is a two-to-two process and we denote it AB → CD. It should be clear that the discussion which now follows can be generalised to include more incoming and/or outgoing particles without too much trouble.

We want to check that if momentum is conserved in S, i.e.

$$p_A + p_B = p_C + p_D \qquad (7.16)$$

then

$$p'_A + p'_B = p'_C + p'_D \qquad (7.17)$$

also holds, where the primes indicate the momenta are appropriate to the S' frame of reference, which moves along the positive x axis with a speed U. Apart from providing us with the confidence that Eq. (7.15) is not an accident of choosing the symmetric scattering process discussed above, this check has the bonus that it will also indicate the way in which we should modify the law of conservation of energy. In order to proceed, we are going to resolve the momenta into their x and y components. Let us start with the y components first (they are a little easier to deal with). Our strategy is to check directly Eq. (7.17). To do this we need to know how each of the momenta in S' can be expressed in terms of their components in S. For particle A we have[2],

$$p'_{Ay} = \gamma(v'_A)m_A v'_{Ay}$$

$$= \gamma(v'_A)m_A \cdot \frac{1}{\gamma(U)} \frac{v_{Ay}}{1 - U v_{Ax}/c^2} \qquad (7.18)$$

and to get the second line we used the velocity addition formula to relate the y component of the velocity in S' to that in S. We need to relate $\gamma(v'_A)$ to quantities defined in S. This is where the hard work resides, if you are prepared to trust our

---

[2] The same methods can be used for each of the other particles.

algebra then you might skip to the end result, Eq. (7.19), otherwise we write

$$\gamma(v_A') = \frac{1}{\sqrt{1 - v_A'^2/c^2}}$$

$$= \left(1 - v_{Ax}'^2/c^2 - v_{Ay}'^2/c^2\right)^{-1/2}$$

$$= \left(1 - \frac{(v_{Ax} - U)^2/c^2}{(1 - Uv_{Ax}/c^2)^2} - \frac{1}{\gamma(U)^2} \frac{v_{Ay}^2/c^2}{(1 - Uv_{Ax}/c^2)^2}\right)^{-1/2}$$

$$= \frac{1 - Uv_{Ax}/c^2}{\sqrt{1 + U^2 v_A^2/c^4 - v_A^2/c^2 - U^2/c^2}}$$

$$= \frac{1 - Uv_{Ax}/c^2}{\sqrt{1 - v_A^2/c^2 - U^2(1 - v_A^2/c^2)/c^2}}$$

$$= \frac{1 - Uv_{Ax}/c^2}{\sqrt{(1 - v_A^2/c^2)(1 - U^2/c^2)}}$$

$$= \gamma(v_A)\gamma(U)(1 - Uv_{Ax}/c^2). \tag{7.19}$$

Eq. (7.18) then becomes

$$p_{Ay}' = \gamma(v_A')m_A v_{Ay}'$$

$$= m_A\gamma(v_A)v_{Ay} = p_{Ay}. \tag{7.20}$$

This is an interesting result. It tells us that the y-component of momentum does not change as we move from S to S' and because of this property it is evident that if the y component of momentum is conserved in S then it must also be conserved in S'. Now we turn our attention to the x component of the momentum. Again we focus (rather arbitrarily) on particle A:

$$p_{Ax}' = \gamma(v_A')m_A v_{Ax}'$$

$$= \gamma(v_A')m_A \cdot \frac{v_{Ax} - U}{1 - Uv_{Ax}/c^2}. \tag{7.21}$$

Now we can once again use Eq. (7.19) to write

$$p_{Ax}' = m_A\gamma(U)\gamma(v_A)(v_{Ax} - U) = \gamma(U)(p_{Ax} - \gamma(v_A)m_A U). \tag{7.22}$$

Substituting this (and the corresponding expressions for the other particles) into the x component part of Eq. (7.17) gives

$$\gamma(U)(p_{Ax} + p_{Bx} - p_{Cx} - p_{Dx}) -$$

$$\gamma(U)U[\gamma(v_A)m_A + \gamma(v_B)m_B - \gamma(v_C)m_C - \gamma(v_D)m_D] = 0. \tag{7.23}$$

This is a very interesting result indeed. The first term in parentheses vanishes automatically since momentum is conserved in S. However the second term, in square brackets is not *a priori* zero. Therefore, if we are to have any chance of salvaging momentum conservation in Special Relativity we need also to insist that the term in square brackets also vanishes. This is equivalent to saying that the quantity $\gamma(v)m$ should also be conserved, i.e. we require

$$\gamma(v_A)m_A + \gamma(v_B)m_B = \gamma(v_C)m_C + \gamma(v_D)m_D. \tag{7.24}$$

Remarkably, this apparently new conservation law is nothing other than the relativistic manifestation of the law of conservation of energy. To make this more explicit, for a particle of mass m and speed u let us define the quantity

$$E = \gamma(u)mc^2. \tag{7.25}$$

Eq. (7.24) then takes the form

$$E_A + E_B = E_C + E_D.$$

This quantity has the units of energy, and it is conserved, but apart from that it is far from clear at this stage that this has anything at all to do with the kinetic energy of a classical non-relativistic particle.

7.1.1   The equivalence of mass and energy

In order the gain more insight, it makes sense for us to explore Eq. (7.25) in the limit that $u \ll c$. In this limit, we can express $\gamma(u)$ as a Taylor series about $u = 0$, i.e.

$$\gamma(u) \underset{u \ll c}{\simeq} 1 + \frac{u^2}{2c^2}. \tag{7.26}$$

Substituting this into Eq. (7.25) yields the much more revealing

$$E \underset{u \ll c}{\simeq} mc^2 + \frac{1}{2}mu^2. \tag{7.27}$$

This is simply the sum of the non-relativistic kinetic energy and a static term, i.e. mc^2 is the energy associated with a particle at rest. If we were to make the additional assumption that mass is conserved then the conservation of E is equivalent to the conservation of kinetic energy. This is what we usually do in non-relativistic mechanics, although the conservation of mass is often assumed to be self evident and is rarely explicitly stated. For example, if we go back to our process AB $\rightarrow$ CD then the conservation of E implies, in the non-relativistic limit, that

$$m_A c^2 + m_B c^2 + \frac{1}{2}m_A v_A^2 + \frac{1}{2}m_B v_B^2 = m_C c^2 + m_D c^2 + \frac{1}{2}m_C v_C^2 + \frac{1}{2}m_D v_D^2, \tag{7.28}$$

which reduces to the familiar

$$\frac{1}{2}m_A v_A^2 + \frac{1}{2}m_B v_B^2 = \frac{1}{2}m_C v_C^2 + \frac{1}{2}m_D v_D^2 \qquad (7.29)$$

if we assume $m_A + m_B = m_C + m_D$. It is very important to remember that this formula is only a good approximation if all speeds are small enough compared to the speed of light.

In contrast, Einstein's theory does not require the conservation of mass: it only requires that the sum of kinetic and static energy be conserved. We subsequently refer to the static energy as the 'rest mass energy', it is the energy possessed by a particle at rest:

$$E_{\text{rest}} = mc^2. \qquad (7.30)$$

The kinetic energy is then defined to be the difference between the total energy and the rest mass energy:

$$E_{\text{kinetic}} = K = (\gamma(u) - 1)mc^2. \qquad (7.31)$$

Example 7.1.1 *What is the rest mass energy of an electron?*

Solution 7.1.1 *The mass of an electron is 9.11×10^{-31} kg and so using Eq. (7.30) it possess an energy equal to*

$$E_{\text{rest}} = mc^2 = 8.19 \times 10^{-14} J.$$

It is almost always the case that it is more sensible and more convenient to express energies in electronvolt (eV) units rather than in joules. One electronvolt is the kinetic energy acquired by an electron that has been accelerated through a potential difference of 1 volt, i.e.

$$1\,eV = 1.60 \times 10^{-16} J.$$

The rest mass energy of an electron in these units is then given by

$$E_{\text{rest}} = \frac{8.19 \times 10^{-14}}{1.60 \times 10^{-16}} eV = 0.511\,MeV.$$

Thus an electron has a mass energy equivalent to just above one half million electron volts.

Also worth noting is that particle and nuclear physicists quite often quote particle masses in eV-based units, mainly because it is irksome to keep explicitly dividing by the speed of light squared. For example, one might say that an electron has a mass of 0.511 MeV/c^2. Similarly momenta might typically be expressed in units of MeV/c.

As we have just seen, only in the limit of small u/c is the kinetic energy given by $\frac{1}{2}mu^2$. Since only the total energy is conserved we have the intriguing possibility that mass is essentially just another form of energy and that kinetic energy might be traded for mass (and vice versa) in physical processes.

Remarkably this is just what happens in Nature. The most striking examples are to be found in particle and nuclear physics. For example, a nucleus at rest can spontaneously transform into a system of lighter particles travelling with some kinetic energy, leaving no trace of the original nucleus. In this case the total mass of the lighter particles is less than the mass of the initial nucleus by an amount that is exactly equal to the total kinetic energy of the particles (divided by c^2). Of course this phenomenon lies behind the operation of nuclear fission reactors, where an atomic nucleus breaks into two with the liberation of a significant amount of energy. In particle physics, the LEP collider at CERN (the European Centre for Particle Physics in Geneva) manufactured head-on collisions between electrons and positrons. In a single collision, all of the kinetic energy and all of the mass energy of the incoming particles was used to manufacture a single Z particle at rest. In that way, the incoming kinetic energy was converted entirely into the mass energy of a Z particle. In fact, one of the main motivations for building LEP was to produce millions of Z particles this way in order to study the detailed properties of the weak interactions and their unification with electromagnetism in the so-called 'electroweak theory'. Apart from such striking examples of the way Nature utilises the possibility to trade off mass and kinetic energy, the idea is applicable in more everyday phenomena. For example, if one would burn a mass of coal in a container sealed so that no material can enter or leave it then the mass of the container after the coal has burnt (i.e. the mass of the remaining ash plus gases) would be less than the initial mass of coal. The difference in mass being equal exactly to the total energy radiated by the container divided by c^2. This is clearly a very new idea, deviating essentially from the classical idea that there exists some immutable atomic substructure. For chemical processes, the reduction in mass is typically very small indeed[3] due to the largeness of the speed of light. That is why the mass-energy equivalence was not demonstrated in a laboratory experiment until long after Einstein's original conjecture using nuclear processes in which the energies involved are much larger and changes in mass correspondingly much more significant. Cockcroft and Walton are credited with providing the first direct evidence in their work of 1932 wherein they studied the reaction $p + Li \rightarrow \alpha + \alpha$ and showed that the reduction in mass was balanced by an increase in kinetic energy in accord with Einstein's expectations.

Before moving on, we pause to reflect upon an interesting symmetry which we have accidentally uncovered.

7.1.2 The hint of an underlying symmetry

Take a look at Eq. (7.20) and Eq. (7.22). They tell us that a particle travelling in S with energy E and momentum components (p_x, p_y) has momenta (p'_x, p'_y) in

[3] When expressed as a fraction of the total mass.

S' where[4]

$$p'_y = p_y \tag{7.32}$$

and

$$p'_x = \gamma(U)(p_x - UE/c^2). \tag{7.33}$$

Now take a closer look at these equations. They are very similar to the Lorentz transformation equations we introduced in Eq. (6.28). In fact the correspondence is exact if we were to start from the Lorentz transformations and make the replacement $x \to p_x$, $y \to p_y$ and $ct \to E/c$. The similarity is all the more striking when we write down the transformation equation for the energy E:

$$E' = \gamma(U)(E - Ucp_x). \tag{7.34}$$

The fact that energy and momentum transform between inertial frames in exactly the same way as do the time and space co-ordinates is suggestive of an underlying symmetry. Indeed such a symmetry exists, and we shall return to study it in much more detail in Part IV.

7.2 APPLICATIONS IN PARTICLE PHYSICS

In order to explore the consequences of our new formulae for energy (Eq. (7.25)) and momentum (Eq. (7.15)) we shall use some examples taken from particle physics. This choice is mainly motivated by the fact that, along with nuclear physics, this is the area of physics where the new dynamics is particularly important.

Example 7.2.1 *Find the speed of an electron that has been accelerated from rest by an electric field through a potential difference of (a) 20.0 kV (typical of a cathode ray tube in a television set); (b) 5.00 MV (typical of an X-ray machine).*

Solution 7.2.1 *The total energy of the electron after being accelerated through the potential difference V is*

$$E = mc^2 + eV$$
$$= \gamma mc^2,$$

where $e = 1.60 \times 10^{-19} C$. Hence

$$\gamma = 1 + \frac{eV}{mc^2}.$$

---

[4] We have focussed on motion in two dimensions but it should be pretty clear that $p_z = p'_z$ if we had considered a third spatial dimension.

Using $\gamma = 1/\sqrt{1 - v^2/c^2}$ gives us an expression for the speed:

$$\frac{v^2}{c^2} = 1 - \frac{1}{\gamma^2}.$$

(a) For the TV set,

$$1 + \frac{eV}{mc^2} = 1 + \frac{20.0 \times 10^3}{511 \times 10^3} = 1.039.$$

Hence the speed is given by

$$\frac{v}{c} = \sqrt{1 - \frac{1}{1.039^2}} = 0.272.$$

It is usually most convenient to express speeds as a fraction of the speed of light.

(b) For the X-ray machine,

$$1 + \frac{eV}{mc^2} = 1 + \frac{5.00 \times 10^6}{511 \times 10^3} = 10.8$$

and the speed is therefore

$$\frac{v}{c} = \sqrt{1 - \frac{1}{10.8^2}} = 0.996.$$

7.2.1 When is relativity important?

We know that when $\gamma \simeq 1$ it follows that $v \ll c$ and the formulae of non-relativistic mechanics provide a good approximation. Usually it makes sense to use the non-relativistic approach if one can be confident that it provides sufficient accuracy since it is usually easier than computing using the full apparatus of Special Relativity. For example, one really can safely neglect relativistic corrections when building a car (except for the satellite navigation system which uses the Global Positioning System (GPS)). It would certainly be an advantage if we could spot whether or not a system needs relativistic corrections before performing the necessary calculations. Clearly if we know that a particle is travelling with speed much smaller than the speed of light then we can press ahead using Newton's mechanics. But what if we are given the kinetic energy or the total energy of a particle, is there a quick way to tell if it is relativistic or not?

The answer is of course in the affirmative: if a particle has a kinetic energy which is much smaller than its rest mass energy then the particle is moving non-relativistically whereas if the kinetic energy is comparable to or greater than the rest mass energy the particle is moving relativistically. To see this we need to realise that the non-relativistic limit corresponds to $\gamma \simeq 1$ in which case the kinetic energy $(\gamma - 1)mc^2$ is much smaller than the total energy γmc^2. Another way of stating the same result is to say that the total energy is almost entirely made up of rest mass energy for a non-relativistic system.

Example 7.2.2 *For the electron in the TV set of the previous example, estimate its speed using non-relativistic mechanics.*

Solution 7.2.2 *Since the electron has a kinetic energy of 20.0 keV we might expect to be able to use non-relativistic mechanics since 20.0 keV is much smaller than the electron rest mass energy of 511 keV. Let us see how good an approximation it actually is. In the non-relativistic limit, energy conservation dictates that*

$$\frac{1}{2}mv^2 \simeq eV$$

$$\Rightarrow v \simeq \sqrt{\frac{2eV}{m}}$$

$$\simeq \sqrt{\frac{40.0}{511}}c = 0.280c.$$

This agrees with the full relativistic result to an accuracy of $\simeq 4\%$.

One final remark is in order. If you do decide to simplify a problem by working with the formulae of non-relativistic mechanics and the result is a speed which is comparable to the speed of light, or an energy which is comparable to a rest mass energy, then you were wrong to employ the non-relativistic approximation and need to start over but this time with the correct relativistic expressions. Conversely, if the speeds you obtain are small compared to the speed of light then the non-relativistic approximation was good and you can be sure of your results.

Example 7.2.3 *In particle physics experiments, physicists routinely accelerate sub-atomic particles through enormous voltages before making them collide with each other. The particle kinetic energies are often much larger than their rest mass energies which means that the situation is extremely relativistic. It also means that very many new particles can be created out of a single collision.*

For example, the Large Hadron Collider at CERN will make head-on collisions between pairs of protons. Each proton will have an energy of 7000 GeV (1 GeV = 1000 MeV).[5] The main goal of the LHC is to convert this kinetic energy into the mass of new, hitherto undiscovered, particles such as Higgs bosons or supersymmetric particles.

(a) *What is the rest energy in MeV of a proton given that each proton has a mass of 1.67×10^{-27} kg?*

(b) *What speed are the LHC protons travelling at?*

(c) *What is the momentum of an LHC proton?*

(d) *How many new protons can in principle be produced in a single collision?*

---

[5] You might like to convince yourself that this is the roughly equal to the kinetic energy of a tennis ball travelling at $\sim$5 mm/s.

Solution 7.2.3 *(a) We can compute the rest energy using $E_{rest} = mc^2$, i.e.*

$$E_{rest} = (1.67 \times 10^{-27}) \cdot (3 \times 10^8)^2 \simeq 1.5 \times 10^{-10} J$$

$$\simeq 940\, MeV.$$

(b) To get the speed we use

$$\gamma mc^2 = 7000\, GeV$$

$$\Rightarrow \gamma \simeq \frac{7 \times 10^{12}}{940 \times 10^6} \simeq 7400.$$

This is vastly greater than unity so we are certainly in the highly relativistic limit. We could have seen that right away since the proton mass is several thousand times smaller than its kinetic energy. Using

$$\frac{v}{c} = \sqrt{1 - \frac{1}{\gamma^2}} \simeq 1 - \frac{1}{2\gamma^2}.$$

In order to avoid having to evaluate the square root on a calculator (which might be a problem on some calculators) we instead have performed a Taylor expansion, which should be a good approximation since $1/\gamma^2 \ll 1$. Putting in the numbers gives

$$\frac{v}{c} \simeq 1 - 9.0 \times 10^{-9}.$$

(c) We can determine the proton momentum using $p = \gamma m v$. Since v/c is so close to unity (from part (b)) we can approximate $p \simeq E/c$ where

$$E = E_{rest} + 7000\, GeV \simeq 7001\, GeV.$$

Hence the momentum is approximately equal to $7000\, GeV/c$. If we did want to re-express this result in SI units then we would need to evaluate

$$p = \frac{(7 \times 10^{12}) \cdot (1.6 \times 10^{-19})}{3 \times 10^8}\, Ns \simeq 3.7 \times 10^{-15}\, Ns.$$

(d) The LHC protons collide head on with equal and opposite momentum, i.e. the total momentum for any given collision is zero. This means that all of the incoming energy can be used to create new particles, with all of them at rest. (If the total momentum were not zero then momentum conservation would force some of the outgoing particles to have some motion.) The total energy before the collision is $2 \times 7000\, GeV$ and each proton has a mass of $940\, MeV$, therefore the collision is capable of producing some $\simeq 14000/0.94 \simeq 15000$ new protons. This number of protons is energetically possible but in reality there are other laws of physics which prevent 15000 protons being produced (not least the conservation of electric charge). Nevertheless, it is the case that thousands of new particles can easily be produced in any given proton-proton collision at the LHC.

7.2.2 Two useful relations and massless particles

A pair of particularly useful relations can be derived using $E = \gamma mc^2$ and $p = \gamma mv$. The first of them is obtained simply by taking the ratio:

$$\frac{cp}{E} = \frac{v}{c}. \tag{7.35}$$

The usefulness of this equation lies in the fact that if we are given the energy and momentum then it is possible to compute the speed without first computing γ. The second equation takes a little more effort to derive but will turn out to be very useful indeed. Let us consider the combination $E^2 - c^2 p^2$:

$$E^2 - c^2 p^2 = (\gamma mc^2)^2 - (\gamma mv)^2 c^2$$
$$= \frac{m^2 c^4 - m^2 v^2 c^2}{1 - v^2/c^2}$$
$$E^2 - c^2 p^2 = m^2 c^4. \tag{7.36}$$

Why is this so interesting? Well the main value arises because the right-hand-side is the same in all inertial frames. We say that the combination $E^2 - c^2 p^2$ is Lorentz invariant. Quantities such as this, whose values all inertial observers agree upon, arise most naturally in the Part IV of this book where we explore more fully the symmetry alluded to at the end of the previous section. For now we note the result, its utility in helping us solve problems will be apparent when we come to tackle some of the later examples.

The formula for the total energy of a particle $E = \gamma mc^2$ tells us that a massive particle has an energy which approaches infinity as the particle's speed approaches c. Practically, this means that it is impossible to accelerate a massive particle in such a way that its speed exceeds c, for to do so would require an infinite amount of work. This is a much celebrated prediction of Einstein's theory and it is certainly in accord with experiments. For example, to accelerate the protons which will circulate at CERN's LHC to within a few metres per second of light speed requires a power input comparable to that of the whole of the city of Geneva. The power costs are in fact so prohibitive that CERN has to shut down over the winter months. Interestingly, Einstein's theory does not however exclude the possible existence of particles which travel at exactly the speed of light. According to Eq. (7.35) such particles would have an energy and momentum related by

$$E = cp. \tag{7.37}$$

You may be worrying that for these particles $\gamma \to \infty$ and therefore they have infinite energy and momentum. This problem can be avoided but only if the particles carry zero mass. In which case $E = \gamma mc^2$ and $p = \gamma mv$ are simply no longer well defined equations. The very existence of massless particles may sound like a contradiction but in Special Relativity the counter intuitive equivalence of mass and energy provides the loophole which allows for their being, providing that they travel at light speed. In fact, we now know that the wave-particle duality of quantum theory permits us to view light as being made up of particles (the study of the photoelectric effect famously providing the first direct evidence) and since these particles must necessarily travel at the speed of light then Special Relativity predicts that they should also be massless. Such massless particles are called 'photons'.

Example 7.2.4 *A neutral π meson (pion) is an elementary particle which can decay into two photons, i.e. $\pi \rightarrow \gamma\gamma$.[6] If the pion is moving with a kinetic energy equal to 1 GeV what angle is formed between the two photons if they are emitted at equal angles relative to the original direction of the pion? [The neutral pion has a mass of 135 MeV/c^2.]*

Solution 7.2.4 *It is usually a very good idea to draw a sketch in problems like this; something like that illustrated in Figure 7.4. Our goal is to compute the angle α. We have the equations for energy and momentum at our disposal, along with the laws of conservation of energy and momentum, and we must use them carefully. Let us first collect together what we know before the decay:*

$$\text{Before:} \qquad E = m_\pi c^2 + K,$$

$$p_x = (E^2 - m_\pi^2 c^4)^{1/2}/c,$$

$$p_y = 0$$

and we have made use of Eq. (7.36) to fix the momentum of the pion given its total energy. After the decay we have that

$$\text{After:} \qquad E_1 = cp_1, \qquad\qquad E_2 = cp_2,$$

$$p_{1x} = p_1 \cos\alpha, \qquad p_{2x} = p_2 \cos\alpha,$$

$$p_{1y} = p_1 \sin\alpha, \qquad p_{2y} = -p_2 \sin\alpha$$

and we have used $E = cp$ for the massless photons. We can obtain the angle α using momentum conservation in the x direction. However before that we need to figure

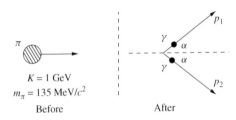

Figure 7.4   Neutral pion decay to two photons

---

[6] The symbol γ is often used to denote a photon.

out the photon momenta p_1 and p_2. This is a two step process, firstly momentum conservation in the y direction tells us that[7]

$$p_{1y} + p_{2y} = 0 \Rightarrow p_1 = p_2.$$

We choose subsequently to define $p = p_1 = p_2$ and our next task is to figure out p. We can obtain this using the conservation of energy since this implies that

$$m_\pi c^2 + K = 2cp,$$

$$\Rightarrow p = \frac{m_\pi c^2 + K}{2c}.$$

All that remains is for us to use conservation of momentum in the x direction:

$$2p \cos \alpha = (E^2 - m_\pi^2 c^4)^{1/2}/c,$$

which leads directly to (substituting for E)

$$\cos \alpha = \frac{((m_\pi c^2 + K)^2 - m_\pi^2 c^4)^{1/2}}{m_\pi c^2 + K}.$$

This is our final answer and we can substitute for the pion mass and its kinetic energy to get a numerical answer, i.e.

$$\cos \alpha = \frac{((135 + 1000)^2 - 135^2)^{1/2}}{135 + 1000} = 0.993$$

$$\Rightarrow \alpha = 6.8°.$$

As a final aside to this exercise, the decay $\pi \to \gamma\gamma$ provides a very nice and direct test of Special Relativity. Alväger et al (1962) showed that pions travelling at a speed of $0.99975c$ decayed to produce forward going photons of speed $(2.9977 \pm 0.0004) \times 10^8 \, ms^{-1}$.

Notice that the strategy for solving the last example is very similar to the one we would use in classical mechanics. The only difference is that we should use the relativistic forms for energy and momentum rather than the classical ones. It is well worth stressing that we managed to get a numerical answer without ever needing to multiply or divide by the speed of light. That happy circumstance arose because we expressed all momenta and energies in MeV based units. In fact, particle physicists often work in units where $c = 1$. It's not a bad idea to re-do the previous exercise but putting $c = 1$ everywhere, although it is a pretty trivial exercise it should help convince you that there is little point in forever writing factors of c all over the place.

---

[7] Strictly we can only conclude $p_1 = p_2$ if $\alpha \neq 0$.

7.2.3 Compton Scattering

As our final example, we shall take a close look at the theory behind an experiment performed in 1923 by Arthur Compton which provided very direct evidence that photons exist as massless particles behaving according to the ideas of Special Relativity. Compton shone short wavelength light (X-rays) onto a target and looked at the angular distribution of the scattered radiation. He was particularly interested in measuring the difference in wavelength between the incident and scattered light as a function of scattering angle. It is one of the triumphs of modern physics that Compton's results can be explained in terms of the simple process illustrated in Figure 7.5. The incoming light constitutes a source of photons and the shift in wavelength detected by Compton arose as a direct result of scattering individual photons in the source elastically off individual atomic electrons in the target.

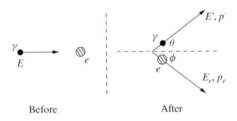

Before   After

Figure 7.5   Compton scattering.

Referring to Figure 7.5, we shall aim to compute the energy of the scattered photon in terms of the energy of the incoming photon and the photon's scattering angle θ. The photon energies and the scattering angle were measured in Compton's experiment, whereas the energy and scattering angle of the recoiling electron were not, so we must eliminate them from our theoretical analysis. Applying the conservation of momentum we can write

$$\frac{E}{c} = \frac{E'}{c}\cos\theta + p_e\cos\phi \tag{7.38}$$

$$\text{and } \frac{E'}{c}\sin\theta = p_e\sin\phi. \tag{7.39}$$

We have used $E = cp$ to write the photon momenta in terms of the incoming and outgoing photon energies. The conservation of energy tells us that

$$E + m_ec^2 = E' + E_e \tag{7.40}$$

and we have been careful not to forget that the initial electron, although it is assumed to be at rest[8], still has its rest mass energy. Given these equations our challenge is to express E' in terms of E and θ, i.e. we must eliminate all dependence upon the scattered electron momentum. It is quite easy to go around in circles in

---

[8] We do not need to worry about the fact that the electron is not actually at rest since its kinetic energy when bound in an atom ($\sim 1\,\text{eV}$) is much less than its rest mass ($\sim 10^6\,\text{eV}$).

this type of calculation, the key is to realise that we can make very good use of Eq. (7.36) to make progress, i.e. we know that

$$E_e^2 - p_e^2 c^2 = m_e^2 c^4. \tag{7.41}$$

If we can evaluate the left hand side of this expression in terms of photon variables only then we will have succeeded in our task. This we can do since Eq. (7.38) and Eq. (7.39) together imply that

$$c^2 p_e^2 = (E - E' \cos\theta)^2 + E'^2 \sin^2\theta \tag{7.42}$$

(using $\cos^2\phi + \sin^2\phi = 1$) and Eq. (7.40) implies that

$$E_e^2 = (E - E' + m_e c^2)^2. \tag{7.43}$$

Subtracting these last two equations and using Eq. (7.41) gives

$$(E - E' + m_e c^2)^2 - (E - E' \cos\theta)^2 - E'^2 \sin^2\theta = m_e^2 c^4,$$

which can be re-arranged to give

$$2EE'(1 - \cos\theta) = 2m_e c^2(E - E').$$

This equation can be easily solved for E' but we prefer to re-write it as

$$\frac{1}{E'} = \frac{1}{E} + \frac{(1 - \cos\theta)}{m_e c^2}. \tag{7.44}$$

In actual fact Compton measured the wavelength of the incoming and outgoing light and used the de Broglie relationship to relate the energy of a photon to its wavelength, i.e. $E = hc/\lambda$, where h is Planck's constant. Consequently, Eq. (7.44) becomes

$$\lambda' = \lambda + \frac{hc(1 - \cos\theta)}{m_e c^2}. \tag{7.45}$$

This is our final answer: Special Relativity and Quantum Mechanics together lead to a very definite prediction for the shift in wavelength of the scattered light as a function of the angle θ and this prediction is strikingly confirmed by Compton's original data, which we show in Figure 7.6. We plot the shift in wavelength $\Delta\lambda = \lambda' - \lambda$ as a function of $1 - \cos\theta$. The data are to be compared to the prediction of Eq. (7.45) which is also shown as the straight line.

PROBLEMS 7

7.1 A proton has mass equal to 1.673×10^{-27} kg. Use this to determine the mass of the proton in units of MeV/c^2.

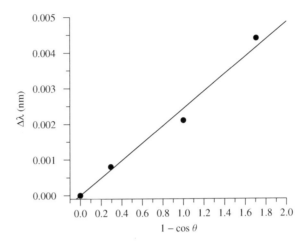

Figure 7.6   Comparison of Compton's original data with the theoretical expectations. From A.H. Compton, Phys. Rev. 21 (1923) 483.

7.2 What is the speed of a particle whose kinetic energy is equal to (a) its rest energy; (b) five times its rest energy?

7.3 The Sun produces energy at a rate of 3.8×10^{26} W. How much mass does the Sun lose each second?

7.4 Calculate the speed of an electron of kinetic energy equal to 0.1 MeV according to both classical and relativistic mechanics. [The mass of an electron is $0.511 \text{ MeV}/c^2$.]

7.5 A particle with momentum 6.0 GeV/c has a total energy of 11.2 GeV. Determine the mass of the particle and its speed.

7.6 What is the total energy of a particle of mass $80 \text{ GeV}/c^2$ which has momentum 65 GeV/c? What is its kinetic energy? Is the particle relativistic or not?

7.7 Two deuterium nuclei can fuse together to form one helium nucleus. The mass of a deuterium nucleus is $2.0136u$ and that of a helium nucleus is $4.0015u$ (u is the atomic mass unit).

(a) How much energy is released when 1 kg of deuterium undergoes fusion?
(b) The annual consumption of electrical energy in the USA is of order 10^{20} J. How much deuterium must react to produce this much energy?

7.8 A (fictitious) particle of mass 1 MeV/c^2 and kinetic energy 2 MeV collides with a stationary particle of mass 2 MeV/c^2. After the collision, the particles form a new particle. Find

(a) the speed of the first particle before the collision;
(b) the total energy of the first particle before the collision;
(c) the initial total momentum of the system;
(d) the mass of the system after the collision;
(e) the total kinetic energy after the collision.

7.9 An antiproton of kinetic energy 0.667 GeV strikes a proton which is at rest in the laboratory. They annihilate to produce two photons which emerge from the reaction travelling forward or backward on the line along which the antiproton entered.

(a) What energies do the photons have?
(b) In which direction is each photon heading?
(c) As measured in a frame attached to the incoming antiproton, what is the energy of each photon?
[The mass of an antiproton is equal to that of a proton, i.e. 938 MeV/c^2.]

7.10 Two particles, each of mass m, are moving perpendicular to each other with speeds u_1 and u_2. The particles collide and coalesce to form a single particle of mass M. Show that

$$M^2 = 2m^2(1 + \gamma(u_1)\gamma(u_2))$$

Part III
Advanced Dynamics

8

Non-inertial Frames

To this point, our attention has focused mainly on physics as viewed from inertial frames of reference. Inertial frames have the substantial advantage that Newton's laws hold within them and that Einstein's Special Relativity is formulated using them. For example, bodies not acted upon by some external force travel in straight lines (or remain at rest) and acceleration arises as a result of the action of a force. However, it is not always advantageous to work in an inertial frame. For example, a natural frame to choose when describing physics on the surface of the Earth would be a frame at rest relative to the Earth. Any such frame is not inertial because the Earth is spinning on its axis (and rotating in orbit about the Sun). In this chapter, our goal is to understand the implications of working in non-inertial frames of reference. As we shall see, Newton's laws can be rescued provided we are prepared to introduce the idea of fictitious forces. In order not to complicate matters too much we shall assume that all speeds are sufficiently small so that we can ignore the effects of relativity. We will in fact return to consider relativistic effects in accelerating frames of reference towards the end of the book.

8.1 LINEARLY ACCELERATING FRAMES

Let us start with the simplest type of acceleration, namely acceleration in a straight line. In Figure 8.1 we show two frames of reference. It looks rather similar to the pictures in the last chapter on Special Relativity except that now the frame S' is accelerating uniformly relative to S. Ignoring the relativistic effects, if a particle is located at position $\mathbf{x}(t)$ in S then its co-ordinates in S' are given by

$$\mathbf{x}'(t) = \mathbf{x}(t) - \mathbf{X}(t), \tag{8.1}$$

where $\mathbf{X}(t)$ is the position of the origin O' relative to the origin O. Differentiating twice gives us a relationship between the acceleration of the particle as it would

Dynamics and Relativity   Jeffrey R. Forshaw and A. Gavin Smith
© 2009 John Wiley & Sons, Ltd

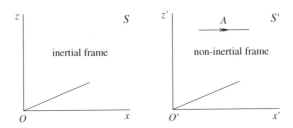

Figure 8.1   Two frames of reference S and S' which are accelerating relative to each other.

be determined in the two frames of reference, i.e.

$$a'(t) = a(t) - A(t), \qquad (8.2)$$

where $A(t)$ is the acceleration of S' relative to S (the double arrow in the figure is intended to denote acceleration).

Now we know that Newton's Second Law holds in the inertial frame and hence the acceleration of the particle in S is related to an applied force F via

$$F = ma \qquad (8.3)$$

(assuming a particle of fixed mass m). We can use Eq. (8.2) to re-write this equation as

$$F = m(a' + A). \qquad (8.4)$$

This can obviously be re-cast into the form

$$F' = ma' \qquad (8.5)$$

provided $F' = F - mA$. Thus from the point of view of an observer at rest in S' the particle moves around as though it is acted upon not only by the real force F but also by a fictitious force

$$F_{fict} = -mA. \qquad (8.6)$$

A very simple and familiar illustration of such a fictitious force occurs if one considers a ball on the floor of an accelerating car. As the car accelerates forwards so the ball rolls towards the back of the car. From the viewpoint of someone sitting in the car it is as if the ball is being pushed along. Of course there is no physical force acting upon the ball: viewed from the point of view of a person standing watching the car accelerate past, in the absence of any friction the ball would remain at rest whilst the car accelerates. The very same fictitious force is responsible for pressing the driver of the car back into their seat as the car accelerates.

Another example is provided if we consider the case of a freely falling lift (or elevator). As the lift accelerates downwards it can be used to define a non-inertial frame of reference. Any objects within the lift will experience not only their own weight but also an upwards fictitious force of magnitude mg where g is the acceleration of the lift (i.e. the acceleration due to gravity). But this is none other than the weight of the object. Hence an unfortunate passenger within the lift will feel weightless as they plummet towards the ground. We invoked the lift for dramatic effect but it should be clear that a person falling freely towards the ground will feel weightless (in the absence of any air resistance). This is a very interesting and intriguing result and in fact provides us with our first hint towards Einstein's theory of gravitation, also known as his General Theory of Relativity. It is worth our spending a moment or two to consider just why the weightlessness of free fall is a such remarkable phenomenon.

Take a look at Eq. (8.6). The mass which appears in this equation and which determines the magnitude of the fictitious force is just the mass which appears in Newton's Second Law. Now, the exact cancellation of the weight of a body in free fall only occurs if this mass is identically equal to the mass which appears in the law of gravitation and which defines the weight mg of the body. This may not at first strike us as a remarkable result but we really ought to be very impressed that the inertial mass which appears in Newton's Second Law is, as far as we can tell, identical to the mass which appears in the law of gravity. After all, these are two totally independent laws of physics. The significance of this equivalence of inertial and gravitational mass can be glimpsed if we return to the example of a lift in free fall within a uniform gravitational field and realise that physics within the lift is totally indistinguishable from physics in a lift floating in the zero gravity environment of outer space. The suggestion is that uniform gravitational fields can be eliminated if we work in freely falling frames of reference. As we shall see in Chapter 14.2, Einstein took this idea to its logical conclusion and succeeded in eliminating the force of gravity altogether in exchange for a description of the world in terms of an infinity of carefully chosen freely falling frames of reference which, as we shall later show, is equivalent to a curved spacetime.

8.2 ROTATING FRAMES

Sitting on a merry-go-round, one is in a rotating frame of reference. In order to remain at rest in that frame we feel a fictitious force called the centrifugal force which pushes outwards and balances the real centripetal force pulling us towards the centre. The centrifugal force is not the only fictitious force associated with a rotating frame. You may even have noticed the other force if you have ever attempted to play a game of 'catch' whilst riding on a merry-go-round: it is called the Coriolis force and it arises when objects are in motion in a rotating frame of reference. In this section we shall derive mathematical expressions to quantify the role of the centrifugal and Coriolis forces.

Let us consider a set of co-ordinate axes which rotate with an angular velocity ω about some axis[1] and we shall place the origin somewhere on the axis of rotation. For definiteness, you might think of such a set of axes fixed to the Earth, with the origin located at the centre of the Earth. We denote the basis unit vectors in the

---

[1] See Section 4.3 for the definition of ω.

rotating frame as $\mathbf{e}'_i$ where $i \in \{1, 2, 3\}$. In this basis, the position of some point is given by

$$\mathbf{r} = \sum_i x'_i \mathbf{e}'_i = x'_i \mathbf{e}'_i. \tag{8.7}$$

Notice that we have introduced a new and important piece of notation: we have dropped the summation sign in the final expression. This ought never to cause confusion since the summation sign really was redundant: the repeated index signals the need for a summation. We shall use this convention wherever appropriate in the remainder of the book. In an inertial frame, with basis vectors $\mathbf{e}_i$, this same vector is given by

$$\mathbf{r} = x_i \mathbf{e}_i. \tag{8.8}$$

Our task is as follows. We might imagine the point to represent the position of a particle and then we should be interested to know the velocity and acceleration of the particle. In the inertial frame the result is easy:

$$\mathbf{v} = \frac{\mathrm{d}x_i}{\mathrm{d}t} \mathbf{e}_i \tag{8.9}$$

and

$$\mathbf{a} = \frac{\mathrm{d}^2 x_i}{\mathrm{d}t^2} \mathbf{e}_i. \tag{8.10}$$

However, the result is not so simple in the case of the rotating frame because the basis vectors are time dependent. Thus we need to figure out how the $\mathbf{e}'_i$ change with time. Figure 8.2 illustrates what is going on and from it we can see that in a small interval of time, the unit vector $\mathbf{e}'_i$ changes its position from $\mathbf{e}'_i$ to $\mathbf{e}'_i + \Delta \mathbf{e}'_i$ where the modulus of $\Delta \mathbf{e}'_i$ is given by

$$|\Delta \mathbf{e}'_i| = \sin \phi \; \Delta \theta. \tag{8.11}$$

The figure also makes it clear in which direction this little vector points: it is perpendicular to both $\mathbf{e}'_i$ and $\boldsymbol{\omega}$. As a consequence of the basic properties of the vector product, it is therefore parallel to the vector $\boldsymbol{\omega} \times \mathbf{e}'_i$, i.e.

$$\Delta \mathbf{e}'_i = \Delta \theta \; \hat{\boldsymbol{\omega}} \times \mathbf{e}'_i, \tag{8.12}$$

where we use the conventional notation of putting a hat above vectors that are made into unit vectors, i.e. $\hat{\boldsymbol{\omega}} \equiv \boldsymbol{\omega}/\omega$. We can now take the limit of infinitesimal displacements, and deduce that

$$\frac{\mathrm{d}\mathbf{e}'_i}{\mathrm{d}t} = \frac{\mathrm{d}\theta}{\mathrm{d}t} \hat{\boldsymbol{\omega}} \times \mathbf{e}'_i$$

$$= \boldsymbol{\omega} \times \mathbf{e}'_i \tag{8.13}$$

since $\omega = \mathrm{d}\theta/\mathrm{d}t$. Armed with this quantity, we can go ahead and compute the velocity and acceleration of our particle at position $\mathbf{r}$. In fact we can be even

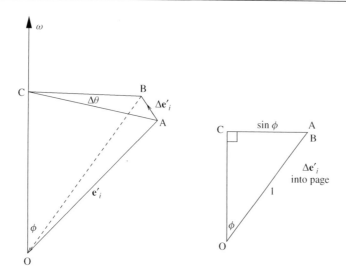

Figure 8.2   Illustrating the change in direction of a basis vector $\mathbf{e}'_i$.

more general and compute the time rate of change of any vector in terms of its components in the rotating frame. If we consider a general vector:

$$\mathbf{W} = W_i \mathbf{e}_i = W'_i \mathbf{e}'_i \tag{8.14}$$

then it follows that

$$\frac{d\mathbf{W}}{dt} = \frac{dW_i}{dt} \mathbf{e}_i$$

$$= \frac{dW'_i}{dt} \mathbf{e}'_i + W'_i \frac{d\mathbf{e}'_i}{dt}$$

$$= \frac{dW'_i}{dt} \mathbf{e}'_i + W'_i \, \boldsymbol{\omega} \times \mathbf{e}'_i. \tag{8.15}$$

The second term on the right hand side is equal to $\boldsymbol{\omega} \times \mathbf{W}$ and hence we write

$$\frac{d\mathbf{W}}{dt} = \left(\frac{d\mathbf{W}}{dt}\right)_{rot} + \boldsymbol{\omega} \times \mathbf{W}. \tag{8.16}$$

It should be clear why we chose to write the first term on the right hand side as we have done for it is the time rate of change of the vector $\mathbf{W}$ as determined in the rotating frame. The second term on the right hand side determines how the vector $\mathbf{W}$ is carried around by the rotation. We can now use this expression to determine the velocity and acceleration of our particle. For the velocity we get

$$\frac{d\mathbf{r}}{dt} = \left(\frac{d\mathbf{r}}{dt}\right)_{rot} + \boldsymbol{\omega} \times \mathbf{r}, \tag{8.17}$$

i.e.

$$\mathbf{v} = \mathbf{v}' + \boldsymbol{\omega} \times \mathbf{r}, \tag{8.18}$$

where $\mathbf{v}'$ is the velocity of the particle as determined in the rotating frame. To get the acceleration we can use Eq. (8.16) again to determine the time rate of change of $\mathbf{v}$, i.e.

$$\frac{d\mathbf{v}}{dt} = \left(\frac{d(\mathbf{v}' + \boldsymbol{\omega} \times \mathbf{r})}{dt}\right)_{\mathrm{rot}} + \boldsymbol{\omega} \times (\mathbf{v}' + \boldsymbol{\omega} \times \mathbf{r}). \tag{8.19}$$

The acceleration in the rotating frame is $\mathbf{a}' = d\mathbf{v}'/dt$ and using the product rule in the case of a vector product allows us to write

$$\mathbf{a} = \mathbf{a}' + 2\boldsymbol{\omega} \times \mathbf{v}' + \boldsymbol{\omega} \times (\boldsymbol{\omega} \times \mathbf{r}). \tag{8.20}$$

Equations (8.18) and (8.20) are our final expressions relating the velocity and acceleration in a rotating frame to the same quantities in an inertial frame. Notice that if the point is at rest in the rotating frame then $\mathbf{v}' = \mathbf{0}$ and these equations reduce to the familiar expressions which relate the velocity and acceleration to the angular velocity and position vector for a particle undergoing circular motion, i.e.

$$\mathbf{v} = \boldsymbol{\omega} \times \mathbf{r} \tag{8.21}$$

and

$$\mathbf{a} = \boldsymbol{\omega} \times (\boldsymbol{\omega} \times \mathbf{r}). \tag{8.22}$$

The second of these is none other than the equation for the centripetal acceleration of a particle undergoing circular motion which we derived in Section 1.3.4. You should certainly convince yourself that Eq. (8.22) describes a radial acceleration of magnitude equal to $\omega^2 R$, where R is the distance from the axis of rotation. However, the expressions we have just derived are more general and allow also for the case where the particle is moving in the rotating frame.

Just as we did in the case of linear acceleration, we can substitute for the acceleration $\mathbf{a}$ in Newton's Second Law in order to derive the fictitious force which acts in the non-inertial frame, i.e.

$$\mathbf{F}_{\mathrm{fict}} = -2m\boldsymbol{\omega} \times \mathbf{v}' - m\boldsymbol{\omega} \times (\boldsymbol{\omega} \times \mathbf{r}). \tag{8.23}$$

The second term on the right hand side is the centrifugal force whilst the first term is something new. It is called the Coriolis force and it acts upon objects which are moving within a rotating frame of reference.

Example 8.2.1 *Consider the rotating turntable illustrated in Figure 8.3. Show that (a) a particle rolled radially outwards from the centre will be deflected as illustrated and (b) that a particle rolled radially inwards from a point on the edge of the turntable will be deflected in the opposite direction.*

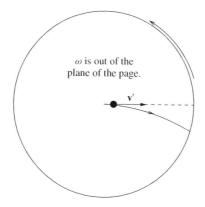

ω is out of the
plane of the page.

$\mathbf{v}'$

Figure 8.3   Trajectory of a ball rolled on a rotating disk.

Solution 8.2.1 *The first thing to realise is that the centrifugal force always acts radially outwards and hence it does not affect the general argument. Our attention therefore is focused upon the Coriolis force. In case (a), for the anti-clockwise rotation illustrated in the figure, $\boldsymbol{\omega}$ points out of the plane of the page. The velocity in the rotating frame points radially outwards and hence the vector $-\boldsymbol{\omega} \times \mathbf{v}'$ pushes the particle as illustrated. In case (b), the velocity vector points in the opposite direction and so the particle is pushed in the opposite direction. This result might at first seem counter intuitive, especially the result of part (b). There is however a simple way to understand what is happening. Viewed from an inertial frame, when the ball is released from the rim of the turntable it has both a radial and tangential component to its velocity. The tangential component is equal to ωR where R is the radius of the turntable. Now, after it has moved inwards slightly it is at a distance smaller than R from the centre but its tangential component of velocity is still (approximately) equal to ωR and this speed is faster than is needed to keep the particle travelling on a radius vector. Hence the particle moves in the direction of the rotation. Crudely stated, it is as if the particle has been thrown in the direction of motion with a speed equal to the speed of the rim of the turntable. The opposite is true for the case where the particle is rolled from the centre: it never has enough tangential speed to keep up with the rotating disk and hence it moves in the opposite direction to the rotation.*

8.2.1   Motion on the Earth

Motion in the vicinity of some region on the Earth's surface is most conveniently described by employing a system of co-ordinates fixed to the Earth. Such a system is illustrated in Figure 8.4 and this choice of basis vectors is most convenient for describing physics in the vicinity of a point at latitude λ on the Earth's surface (in principle they can of course be used to describe any physics anywhere else in the Universe). Looking at Figure 8.4, it should be clear that (for $-\pi/2 < \lambda < \pi/2$) the basis vector $\mathbf{e}_2$ points to the North, $\mathbf{e}_1$ points East (i.e. into the plane of the page) and $\mathbf{e}_3$ points upwards (i.e. radially outwards from the centre of the Earth).

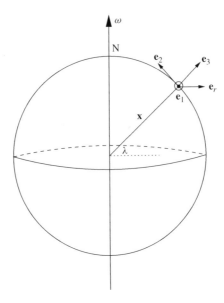

Figure 8.4   A non-inertial system of co-ordinates defined at rest relative to the Earth.

Although we have drawn the basis vectors on the surface of the Earth we should remember that the basis vectors define only directions in space. To complete our specification of the co-ordinate system we need also to specify the location of the origin and it is convenient to choose the origin to be located at the centre of the Earth. In this case the point on the surface of the Earth at which we have drawn the basis vectors is located at position

$$\mathbf{x} = R\,\mathbf{e}_3 \tag{8.24}$$

and the angular velocity of the Earth is

$$\boldsymbol{\omega} = \omega(\cos\lambda\,\mathbf{e}_2 + \sin\lambda\,\mathbf{e}_3). \tag{8.25}$$

We are now ready to explore the influence of the Earth's spin upon physics occurring in the vicinity of $\mathbf{x}$, which is a general point on the Earth's surface. We start by computing the centrifugal force acting upon a particle located at $\mathbf{x}$. Of course we already know the answer from our prior understanding of circular motion: it should be a force of magnitude $m\omega^2 R\cos\lambda$ pointing in the $\mathbf{e}_r$ direction (see Figure 8.4). To warm up, let us compute it using Eq. (8.23). First we compute the vector $\boldsymbol{\omega}\times\mathbf{x}$:

$$\boldsymbol{\omega}\times\mathbf{x} = \omega R\cos\lambda\,\mathbf{e}_1 \tag{8.26}$$

and since $\boldsymbol{\omega}\times\mathbf{e}_1 = -\omega\mathbf{e}_r$ it follows that the centrifugal force is

$$-m\boldsymbol{\omega}\times(\boldsymbol{\omega}\times\mathbf{r}) = m\omega^2 R\cos\lambda\,\mathbf{e}_r, \tag{8.27}$$

as we anticipated. Phenomenologically, the centrifugal force has the effect of slightly reducing the weight of objects on the surface of the Earth, the effect being greatest at the equator where $\cos \lambda = 1$. Note also that the centrifugal force does not acts downwards, rather it acts radially outwards from the axis of the Earth's rotation. This means that a pendulum suspended above the Earth's surface will not point exactly towards the centre of the Earth.

Example 8.2.2 *Compute the maximum deflection of a pendulum suspended close to the surface of the Earth.*

Solution 8.2.2 *The net force on a particle suspended close to the Earth's surface at a latitude λ is given by the sum of the gravitational force (i.e. the weight) and the centrifugal force:*

$$\mathbf{F}' = m\,\mathbf{g} - m\,\boldsymbol{\omega} \times (\boldsymbol{\omega} \times \mathbf{r}).$$

We already deduced the centrifugal force, but to determine the deflection of a pendulum we should express the vector $\mathbf{e}_r$ in terms of our basis vectors, i.e.

$$\mathbf{e}_r = -\sin \lambda\, \mathbf{e}_2 + \cos \lambda\, \mathbf{e}_3 \tag{8.28}$$

so that

$$\mathbf{F}' = -\omega^2 R \cos \lambda\, \sin \lambda\, \mathbf{e}_2 + (\omega^2 R \cos^2 \lambda\ - g)\,\mathbf{e}_3$$

since $\mathbf{g} = -g\,\mathbf{e}_3$.

Now for a pendulum at rest this net force is balanced by the tension in the pendulum, and so the pendulum aligns itself with this force and it is thus deflected at an angle α to the vertical. This deflection is illustration in Figure 8.5 in the case of a pendulum hanging in the northern hemisphere and is given by

$$\tan \alpha = \frac{\omega^2 R \cos \lambda\, \sin \lambda}{g - \omega^2 R \cos^2 \lambda}.$$

The maximum deflection occurs at $\lambda = 45°$ (you should convince yourself that this is indeed the case: note there is no deflection at the poles or on the equator). Since

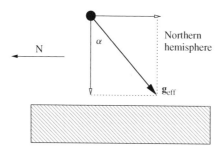

Figure 8.5   The deflection α of a pendulum suspended close to the surface of the Earth.

the deflection is very small, we can use the small angle approximation to write

$$\alpha_{\max} \approx \frac{\omega^2 R}{2g} \approx 0.1^\circ$$

and we have substituted for the angular velocity of the Earth: $\omega = 2\pi/(1\ \text{day})$ $\approx 7.3 \times 10^{-5} s^{-1}$.

Let us now turn our attention to the case of an object moving with velocity **v** on the Earth's surface. The Coriolis acceleration is given by

$$\frac{1}{m}\mathbf{F}_{\text{Cor}} = -2\,\boldsymbol{\omega} \times \mathbf{v}$$

$$= -2 \begin{vmatrix} \mathbf{e}_1 & \mathbf{e}_2 & \mathbf{e}_3 \\ 0 & \omega\cos\lambda & \omega\sin\lambda \\ v_1 & v_2 & v_3 \end{vmatrix}$$

$$= 2\omega\,(v_2\sin\lambda - v_3\cos\lambda)\,\mathbf{e}_1$$

$$-2\omega\,v_1\sin\lambda\,\mathbf{e}_2 + 2\omega\,v_1\cos\lambda\,\mathbf{e}_3. \qquad (8.29)$$

From now on we shall ignore the $\mathbf{e}_3$ component since it is negligible compared to the acceleration due to gravity, which acts also in this direction. Let us now use Eq. (8.29) to consider two different types of motion close to the surface of the Earth.

First we take a look at the case of an object dropped downwards. Initially, the object has velocity $\mathbf{v} = \mathbf{0}$. Our goal is to deduce its velocity sometime later. The precise motion is rather complicated because it is non-linear, i.e. once the particle starts to move the size of the Coriolis and centrifugal forces changes with time, but if we are happy to neglect terms in ω^2 (and higher powers of ω) then the situation is much simpler. For a start, we can therefore neglect the centrifugal acceleration. Moreover, the motion in the $\mathbf{e}_3$ direction is dominated by the force of gravity, i.e.

$$v_3 \approx -g\,t.$$

All that remains is to consider the v_1 and v_2 components of the velocity. We obtain these by integrating the accelerations, i.e.

$$v_1 \approx \int_0^t \frac{(F_{\text{Cor}})_1}{m}\,dt = 2\omega \int_0^t (v_2\sin\lambda + gt\cos\lambda)\,dt$$

$$v_2 = \int_0^t \frac{(F_{\text{Cor}})_2}{m}\,dt = -2\omega \int_0^t (v_1\sin\lambda)\,dt. \qquad (8.30)$$

These are coupled equations and as such we cannot go ahead and solve them for v_1 and v_2. However, we have not made full use of the assumption that we can neglect terms which are proportional to ω^2. Immediately we see that in this approximation $v_2 = 0$ for all t, since the first of the two equations (8.30) tells us that $v_1 \propto \omega$ and this implies, using the second equation, that $v_2 \propto \omega^2$. This is of course a major simplification and we get

$$v_1 \approx \omega g t^2 \cos \lambda. \tag{8.31}$$

The bottom line is therefore that a particle dropped from rest relative to the Earth will not fall directly towards the centre of the Earth but instead will be deflected slightly to the East. Like the motion of the ball on the turntable, this is not too hard to understand. The easterly direction is special because it points in the direction of the Earth's rotation, as can be seen in Figure 8.4. A ball dropped from a height above the ground must move this direction because, at the instant of release, its speed in the easterly direction (as viewed in an inertial frame) is too great for it to fall only along a radius vector. As a parting remark, you might like to see if you can convince yourself that the neglect of terms quadratic in ω is a good approximation if $\omega^2 R \ll g$ and if the total time of the motion is much less than 1 day.

Secondly, we shall consider the case of an object moving horizontally on the Earth's surface. In this case, we know that $v_3 = 0$ for all times and the motion is in a plane. The Coriolis acceleration is just

$$\frac{1}{m}\mathbf{F}_{\text{Cor}} = 2\omega \sin \lambda \, (v_2 \mathbf{e}_1 - v_1 \mathbf{e}_2). \tag{8.32}$$

The general situation for motion in the northern hemisphere is illustrated in Figure 8.6 and it shows the Coriolis force acting in the direction of the vector $-\boldsymbol{\omega} \times \mathbf{v}$. In the northern hemisphere $\sin \lambda > 0$ and the object is always pushed to the right. In the southern hemisphere $\sin \lambda < 0$ and the object is pushed to the left. At the equator $\sin \lambda = 0$ and there is no Coriolis force. It is a consequence of the Coriolis force that large bodies of air do not move in straight lines around the Earth. In the northern hemisphere the air swirls in a clockwise direction as viewed from above whilst in the southern hemisphere it swirls in an anti-clockwise direction. Cyclones are regions of low pressure and as such the air around the cyclone moves towards the centre. As it moves it is deflected as a result of the

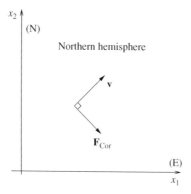

Figure 8.6   The Coriolis force acting on a body moving in the northern hemisphere.

Coriolis force[2]. The result is the menacing spiralling motion familiar from satellite pictures; the rotation always occurring in opposite directions in the northern and southern hemispheres.

Example 8.2.3 *The effect of the Coriolis force on smaller bodies moving at every-day speeds is usually negligible. In this example we are asked to compute the size of the Coriolis force on a car of mass 1500 kg travelling due North across Manchester at a speed of 100 km h^{-1}.*

Solution 8.2.3 *Let us first convert the speed into SI units, i.e.*

$$v = \frac{100 \times 10^3}{60 \times 60} \text{ms}^{-1} = 28 \text{ ms}^{-1}.$$

In addition we need to know that Manchester is at a latitude $\lambda = 53°$. Now, since the car is travelling due North, the Coriolis force is simply given by

$$\frac{1}{m} F_{\text{Cor}} = 2\omega v \sin \lambda$$

$$= 2 \times 7.3 \times 10^{-5} \times 28 \times \sin 53° \text{ ms}^{-1}$$

$$= 4.9 \text{ N.} \tag{8.33}$$

PROBLEMS 8

8.1 The centrifugal force acting on a particle of mass m at position $\mathbf{r}$ in a frame that is rotating with angular velocity $\boldsymbol{\omega}$ is

$$-m\boldsymbol{\omega} \times (\boldsymbol{\omega} \times \mathbf{r})$$

and it appears at first sight to depend upon the position of the origin. Convince yourself that this is not the case provided the origin lies somewhere on the rotation axis.

8.2 A bucket of water rotates about its symmetry axis in the Earth's gravitational field (the symmetry axis is vertical). In a frame which rotates with the bucket, determine the direction of the net force that acts on a small mass of water (which lies a distance r from the rotation axis) due to its weight and the centrifugal force which acts on it. You may assume that all elements rotate with the same angular velocity $\boldsymbol{\omega}$. Convince yourself that the tangent to the surface of the water at radius r should be orthogonal to this force and hence prove that the surface of the water forms a paraboloid of revolution.

8.3 A particle is released from rest at the top of a tall building of height 150m. If the building is at a latitude of 53°N, determine that the particle strikes the ground with a small easterly deflection and compute the size of the deflection. You may neglect air resistance.

---

[2] Anti-cyclones are regions of high pressure and the air is correspondingly pushed away from the centre.

8.4 A small bead of mass m is constrained to move on a hoop of radius R. The hoop rotates with a constant angular speed ω about a vertical axis which coincides with a diameter of the hoop. Find the angular speed Ω above which a bead located originally at the bottom of the hoop begins to slide upwards and determine the position to which it will rise.

9

Gravitation

Throughout Part I of this book, we focused our attention on the basic principles of motion as articulated by Newton. On occasion we invoked the force of gravity in order that we might consider interesting physical applications of Newton's laws. However in most cases we assumed that the motion was taking place in a sufficiently small region of space that the gravitational field could be assumed constant. However, thanks again to Newton, we now understand that gravity is a force that acts between all bodies such that the force between any two bodies is directly proportional to the product of their masses and inversely proportional to the square of the distance between them. Of course we know that gravity is not the only force at work in the natural world. Electricity and magnetism, unified by Faraday, Ampère and Maxwell into the theory of electromagnetism are also abundantly evident and are ultimately responsible for light itself. In addition, we now know that the atomic nucleus is prevented from exploding under the repulsive influence of the Coulomb force which acts between its proton constituents by a further force, the strong nuclear force, which acts only over very tiny distances. Finally, the burning of the Sun can only be understood once we recognise the existence of a fourth fundamental force: the weak force, responsible also for the process of nuclear beta decay. So it seems that physical phenomena throughout the Universe can be thought of as arising out of the interactions of matter which occur as a result of these four fundamental forces. All other forces, such as the tension in a spring or the force which drives forwards a sailing boat, are none other than complicated consequences of one or more of these fundamental forces.

Of all the forces in nature, Newton's Law of Gravity sits alongside Coulomb's Law of electrostatics (which is also described by an inverse square law and is a part of the electromagnetic theory) in being simple enough that we can make very significant progress in understanding its consequences without too much hard work. It is for that reason that we focus our attention in this chapter on gravity and over the next few pages we shall use Newton's theory of gravity in order to understand fully gravitational systems containing two bodies. That will be sufficient for us to

Dynamics and Relativity   Jeffrey R. Forshaw and A. Gavin Smith
© 2009 John Wiley & Sons, Ltd

precisely understand the elliptical orbits of planets within our solar system and the hyperbolic trajectories of comets.

Before launching into Newton's theory of gravity, it is perhaps worth recapping that, following Einstein's work on the General Theory of Relativity, we now understand that Newton's theory is only an approximation to Einstein's more accurate theory. Nevertheless, it is an excellent approximation in almost all circumstances in everyday life. For example, NASA's Apollo missions to the Moon were conducted entirely using calculations based upon Newton's theory. That said, there is one area of life where Newton's theory of gravity is inadequate. The GPS system uses a network of satellites orbiting the Earth every 12 hours (or so). Accurate position measurements require very accurate time keeping on the orbiting satellites and as a result it is necessary to account not only for Special Relativistic corrections due to the motion of the satellites but also the General Relativistic corrections which correct Newton's theory of gravity. Without these corrections, the GPS system would fail within minutes. At the end of this book we shall discuss how Einstein's General Relativity comes about and illustrate how it corrects Newton's theory, but for now we satisfy ourselves with a detailed account of Newtonian gravity.

9.1 NEWTON'S LAW OF GRAVITY

Let us start by writing down Newton's Law of Gravity. Illustrated in Figure 9.1 are two point masses, m and M, separated by a distance, r. Newton's Law tells us that a force $\mathbf{F}$ acts on the mass m such that

$$\mathbf{F} = -\frac{GMm}{r^2}\mathbf{e}_r, \tag{9.1}$$

where $\mathbf{e}_r$ is a unit vector pointing from the mass M towards the mass m, as illustrated. Moreover, a force of equal magnitude but opposite direction also acts upon the mass M. In short, the two masses attract each other with a strength described by an inverse square law. Although we have taken care to specify the law for point masses (i.e. idealized pointlike masses) we shall show in the next section that the law also applies to extended spherical bodies provided that r is the distance between the centres of the two masses.

As we shall very soon discover, the fact that the gravitational force acts along the line joining the two bodies and depends only on the distance between them

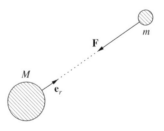

Figure 9.1 The gravitational force on a mass m due to a second mass, M.

provides us with the opportunity to solve for the motion of two gravitating bodies without too much hard work. The general motion for more than two bodies is rather more complicated and we won't address any problems of that nature, although of course no new physics is involved.

In Section 3.2, we showed that a uniform gravitational field of force is 'conservative' and hence that it can be described using a potential energy function. Specifically, we showed that the work done against gravity in moving a body around in a uniform gravitational field does not depend upon the details of the body's journey, rather it just depends on the difference in height between its starting and finishing points. Consequently, we can define a potential energy function such that for a particle moving around under the action of a conservative force the sum of the kinetic and potential energies of the particle is a constant. The fact that the law of conservation of energy can be expressed so simply is often very helpful when it comes to solving problems.

According to Newton's Law, Eq. (9.1), the gravitational field in the vicinity of a point on the Earth's surface is not exactly uniform: it decreases slightly as the distance from the centre of the Earth increases and it always points towards the centre of the Earth. Of course it is often a good approximation to assume the field is uniform but we should keep in mind that really it varies in strength and direction from point to point. Let us now show that the gravitational force described by Newton's Law is also conservative and hence that we can go ahead and define a potential energy function.

To be specific let us consider the gravitational force on the mass m due to the mass M. Let us compute the work done by this force as the particle of mass m moves from a point A to a point B in the field of the other mass M which we consider as being at some fixed point in space. Let us start by assuming that the gravitational force acting on m is conservative. It means that the work done on the particle can be written

$$\int_A^B \mathbf{F} \cdot d\mathbf{x} = -U(\mathbf{x}_B) + U(\mathbf{x}_A), \tag{9.2}$$

where $U(\mathbf{x})$ is the potential energy of the particle when it is at position $\mathbf{x}$. Notice that the sign of $U(\mathbf{x})$ is purely a matter of convention and that for any $U(\mathbf{x})$ we can also add or subtract an overall constant without changing Eq. (9.2). It is our job to introduce a potential energy function and we must be careful to interpret it correctly. The minus sign in Eq. (9.2) was chosen so that the increase in the kinetic energy of the mass m as it moves in the gravitational field (no other forces are present) from A to B is given by

$$T(\mathbf{x}_B) - T(\mathbf{x}_A) = -U(\mathbf{x}_B) + U(\mathbf{x}_B) \tag{9.3}$$

i.e.

$$T(\mathbf{x}_A) + U(\mathbf{x}_A) = T(\mathbf{x}_B) + U(\mathbf{x}_B) \tag{9.4}$$

and so the sum of the kinetic and potential energies is a constant, which we usually call the total energy. Applied locally, to infinitesimal displacements, Eq. (9.2) can be written as

$$\mathbf{F} \cdot d\mathbf{x} = -dU. \tag{9.5}$$

Using the chain rule, we can therefore write

$$\mathbf{F} \cdot d\mathbf{x} = -\left(\frac{\partial U}{\partial x_1} dx_1 + \frac{\partial U}{\partial x_2} dx_2 + \frac{\partial U}{\partial x_3} dx_3 \right)$$

$$= -\frac{\partial U}{\partial x_i} dx_i. \tag{9.6}$$

In the second line we have again made use of the summation convention introduced in Eq. (8.7). Since the left hand side is equal to $F_i \, dx_i$ and the equality is true for any infinitesimal line element it follows that we must be able to write the components of the force as

$$F_i = -\frac{\partial U}{\partial x_i} \tag{9.7}$$

which in vector notion is usually written as

$$\mathbf{F} = -\boldsymbol{\nabla} U(r), \tag{9.8}$$

where $\boldsymbol{\nabla}$ is known as the gradient operator defined by

$$\boldsymbol{\nabla} \equiv \mathbf{e}_i \frac{\partial}{\partial x_i}. \tag{9.9}$$

Thus, if the gravitational field is to be conservative then it follows that it must be possible to express it as the gradient of a scalar field, as in Eq. (9.8). Put another way, if we can find a scalar field $U(r)$ whose gradient gives the force acting upon the mass m then we will have succeeded in showing that the gravitational field is conservative.

It is not too hard to come up with the correct potential energy function. If we consider

$$U = -\frac{GMm}{r} \tag{9.10}$$

then we can go ahead and compute the corresponding components of the force using Eq. (9.7). Thus we just need to compute

$$\frac{\partial U}{\partial x_i} = \frac{dU}{dr} \frac{\partial r}{\partial x_i}. \tag{9.11}$$

If we put the mass M at the origin then $r^2 = x_1^2 + x_2^2 + x_3^2$ and so

$$\frac{\partial r}{\partial x_i} = \frac{x_i}{r}. \tag{9.12}$$

Thus

$$F_i = -\frac{GMm}{r^2}\frac{x_i}{r}. \tag{9.13}$$

Since $\mathbf{r} = x_i\mathbf{e}_i$ we can write this equation as

$$\mathbf{F} = -\frac{GMm}{r^2}\frac{\mathbf{r}}{r}, \tag{9.14}$$

which is none other than Newton's Law since $\mathbf{r}/r = \mathbf{e}_r$. Eq. (9.10) therefore provides us with a potential energy function which we can use to compute the work done on our particle of mass m as it moves in the field of the mass M. The existence of this function implies that the gravitational force is conservative.

In choosing Eq. (9.10) to define the gravitational potential energy of our mass m we have also specified that the zero of potential energy is at $r = \infty$. This choice has a nice physical interpretation. Let us consider the case where the mass m starts from rest infinitely far away from the mass M, then the conservation of energy tells us that

$$U(\infty) + 0 = U(r) + \frac{1}{2}mv^2. \tag{9.15}$$

The zero on the left hand side expresses the fact that the particle starts out at rest. Thus we see that $-U(r)$ is equal to the kinetic energy that a particle of mass m would have after falling to a distance r from the mass M starting from an initial speed of zero at infinity. Equivalently, $-U(r)$ is the work done by gravity on the particle as it falls from infinity to r.

9.2   THE GRAVITATIONAL POTENTIAL

So far we have talked about the force due to gravity acting on and the gravitational potential energy of a particle of mass m due to the field associated with a mass M. There is nothing wrong with such an approach however the idea that the mass M produces a gravitational field independent of whether there is another mass present or not is a very intuitive one and one that is conveniently expressed when we think in terms of the gravitational field strength and the gravitational potential, which are defined as follows.

The gravitational field strength, $\mathbf{g}$, is simply defined to be the force which would act on a unit mass placed in the field whilst the gravitational potential, Φ, is the gravitational potential energy of a unit mass placed in the field, i.e.

$$\mathbf{g} \equiv \frac{\mathbf{F}}{m},$$

$$\Phi \equiv \frac{U}{m} \tag{9.16}$$

and hence

$$\mathbf{g} = -\nabla\Phi. \tag{9.17}$$

Being a scalar function, the gravitational potential is usually much easier to deal with than the field strength, which is a vector quantity. Moreover, all of the information we need is contained in the potential for all we need to do is differentiate it in order the compute the field strength. Thus, the conservative nature of the gravitational force has led to a big calculational simplification. Our general task will therefore be to compute the gravitational potential for whichever problem we are faced with. Given this we can compute the other interesting quantities.

Let us next show how to compute the gravitational potential for a general distribution of mass which is described by a density $\rho(\mathbf{x})$ (this is the mass per unit volume at position $\mathbf{x}$). Figure 9.2 illustrates the general situation and our goal is to determine the potential at a general point P. To do this we must compute and sum up the potential at P arising from each and every tiny element of mass, like the one illustrated in the figure. The potential at P arising from a volume element dV located at position $\mathbf{x}'$ which has a mass equal to $\rho(\mathbf{x}')\,dV$ is given by

$$d\Phi(\mathbf{x}) = -\frac{G\rho(\mathbf{x}')\,dV}{r}. \tag{9.18}$$

Hence the potential at P arising from a general distribution of mass is the sum over all such volume elements, i.e. it is the triple integral

$$\Phi(\mathbf{x}) = -G \int_V \frac{\rho(\mathbf{x}')}{r}\,dV. \tag{9.19}$$

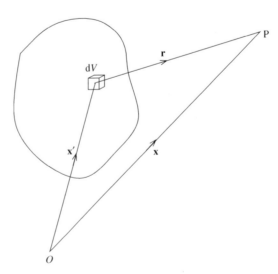

Figure 9.2 To compute the potential at P due to a general distribution of matter we sum over the infinity of volume elements dV.

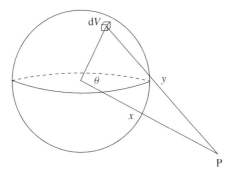

Figure 9.3   A spherical distribution of matter.

The use of this formula is perhaps best illustrated in an example. Let us use it to determine the potential arising due to a uniform sphere of radius R and of total mass M. Figure 9.3 defines the various quantities of interest. Consider breaking the sphere into lots of tiny volume elements. In spherical polar co-ordinates, the volume of a general element is

$$dV = r^2 \sin \theta \, dr \, d\theta \, d\phi \qquad (9.20)$$

and the mass of this element is just equal to

$$dM = dV \frac{M}{4\pi R^3/3}. \qquad (9.21)$$

Thus the potential at P arising from this element is equal to

$$d\Phi = -\frac{G \, dM}{y} \qquad (9.22)$$

and the potential is obtained after summing over all the elements which make up the sphere, i.e.

$$\Phi = -G \frac{3M}{4\pi R^3} \int_0^R dr \int_0^{2\pi} d\phi \int_0^{\pi} d\theta \, r^2 \frac{\sin \theta}{y}. \qquad (9.23)$$

The azimuthal ϕ integral is easy enough and just gives a factor of 2π but the other integrals are harder since the distance y varies as θ and r change. We need to express y in terms of r and θ, and we can do this using the cosine rule:

$$y^2 = r^2 + x^2 - 2xr \cos \theta. \qquad (9.24)$$

We have to choose whether to do the r or the θ integral first. Let us choose the θ integral, in which case we must evaluate

$$\int_0^{\pi} d\theta \, \frac{\sin \theta}{(x^2 + r^2 - 2xr \cos \theta)^{1/2}} \qquad (9.25)$$

for a fixed value of r. It is actually simpler if we change variables and use y instead of θ as the integration variable. In this case Eq. (9.24) gives that

$$y\,dy = xr \sin\theta\,d\theta \qquad (9.26)$$

and hence

$$\int_0^\pi d\theta\,\frac{\sin\theta}{(x^2 + r^2 - 2xr\cos\theta)^{1/2}} = \int_{x-r}^{x+r}\frac{dy}{xr} = \frac{2}{x}. \qquad (9.27)$$

We can now substitute this back into Eq. (9.23), in which case

$$\Phi = -G\frac{3M}{2R^3}\int_0^R dr\,\frac{2r^2}{x}$$

$$= -\frac{GM}{x}, \qquad (9.28)$$

which is exactly equal to the potential of a point mass M located at the origin, a result that we quoted without proof in the previous section. You might like to convince yourself that this result holds generally for any spherically symmetric distribution of matter, i.e. one for which the mass density depends only upon r and not upon θ or ϕ. The case of uniform mass density which we considered here is one example of such a distribution.

Example 9.2.1 *Show that the speed of a star orbiting in an arm of a spiral galaxy at a radius r far from the centre of the galaxy should vary as $1/\sqrt{r}$ if we assume that the mass of the galaxy is located in the spherical bulge at the centre of the galaxy.*

Solution 9.2.1 *Figure 9.4 illustrates the situation. We shall approximate the central bulge of stars by a spherically symmetric distribution of matter of total mass M and*

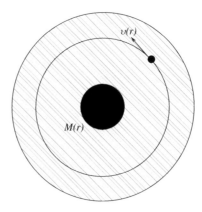

Figure 9.4   A star orbiting the centre of a galaxy. The shading denotes the presence of dark matter.

assume that the mass outside this region is negligible. In which case, a star at radius r sits in a gravitational field which, by virtue of the result we have just proven, is equivalent to that of a point mass M located at r = 0. Since the star is orbiting the centre of the galaxy, the gravitational attraction to the centre must provide the necessary centripetal acceleration, i.e.

$$\frac{mv^2}{r} = \frac{GMm}{r^2},$$

where m is the mass of the star. Thus the speed of the star is

$$v = \sqrt{\frac{GM}{r}}.$$

Remarkably, this behaviour is not what is seen in astronomical observations. Figure 9.5 shows the astronomical data for a typical galaxy and we can clearly see that the large r behaviour is approximately constant and certainly not falling as $1/\sqrt{r}$. One way to explain the data is to assume that a substantial component of the mass of the galaxy is invisible to the astronomers and that this component extends out to large distances from the centre of the galaxy (compared to the size of the central bulge). Other types of observation indicate that this mass cannot be comprised solely of large planets which are too dark to be visible and so the nature of this unseen mass is as yet unknown. Perhaps it is made up from a new type of elementary particle which is invisible to ordinary detection methods. Certainly the origin of this 'Dark Matter' is one of the big mysteries in modern physics.

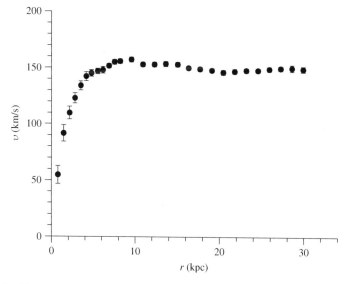

Figure 9.5   The rotation curve for stars orbiting the centre of the galaxy NGC3198 (data from K.G. Begeman, Astron. Astrophys. 223 (1989) 47).

9.3   REDUCED MASS

The remainder of this chapter will focus upon solving problems involving the motion of two bodies under their mutual gravitational interaction. There being two bodies, we need two vectors (and hence six numbers) to specify their co-ordinates at some instant in time and the motion of each particle is determined by solving the corresponding equation of motion, i.e.

$$\ddot{\mathbf{x}}_1 = -\frac{1}{m_1}\mathbf{F}(\mathbf{r}),$$

$$\ddot{\mathbf{x}}_2 = +\frac{1}{m_2}\mathbf{F}(\mathbf{r}), \tag{9.29}$$

where $\mathbf{r} = \mathbf{x}_2 - \mathbf{x}_1$ is the relative position vector and m_1 and m_2 are the masses of the two bodies[1]. We assume that the system is isolated and that the force acting upon the particles depends only upon their relative positions (this is of course true for gravitational interactions). The general configuration is illustrated in Figure 9.6. At first sight, these are two coupled second order differential equations (they are coupled since the relative position depends upon $\mathbf{x}_1$ and $\mathbf{x}_2$) and as such they might require some effort to solve. However the situation can be simplified very substantially once we appreciate that the centre of mass of the system moves with a constant velocity since no external forces are acting. As a result, we can trade off the six numbers which specify the co-ordinates of the two particles for three numbers specifying the co-ordinates of the centre of mass, $\mathbf{R}$, and three more numbers specifying the relative positions of the particles, $\mathbf{r}$. The motion of the centre of mass is easy and all of the interesting dynamics resides in the behaviour of the vector $\mathbf{r}$. Let us put this intuition into mathematical language. What we have described is a change of variables, i.e. we aim to recast Eq. (9.29) in terms

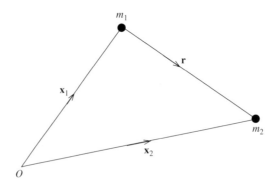

Figure 9.6   Two masses moving under the influence of their mutual gravitational interaction.

---

[1] In this equation we have introduced a shorthand notation that is very common in dynamics. We represent the derivatives with respect to time by placing 'dots' above the object being differentiated. e.g. $\dot{x} \equiv \frac{dx}{dt}$ and $\ddot{x} \equiv \frac{d^2x}{dt^2}$.

of uncoupled equations involving $\mathbf{R}$ and $\mathbf{r}$. This is similar to the approach used to study two-body collisions in Section 3.3. The non-acceleration of the centre of mass is a consequence of momentum conservation for an isolated system, i.e.

$$m_1\ddot{\mathbf{x}}_1 + m_2\ddot{\mathbf{x}}_2 = \mathbf{0} \tag{9.30}$$

and we look for a point $\mathbf{R}$ which does not accelerate, i.e. the centre of mass satisfies the equation

$$(m_1 + m_2)\ddot{\mathbf{R}} = \mathbf{0}. \tag{9.31}$$

Equations (9.30) and (9.31) have the solution

$$\mathbf{R} = \frac{m_1\mathbf{x}_1 + m_2\mathbf{x}_2}{m_1 + m_2}. \tag{9.32}$$

Subtracting the two equation in Eq. (9.29) yields

$$\ddot{\mathbf{r}} = \left(\frac{1}{m_1} + \frac{1}{m_2}\right)\mathbf{F}(\mathbf{r}). \tag{9.33}$$

Notice that equations (9.32) and (9.33) are entirely equivalent to the pair of equations (9.29) but have the virtue that they are decoupled from each other. Our attention is now focussed upon Eq. (9.33) and once we have solved it we will have solved the general motion of our two particles since

$$\mathbf{x}_1 = \mathbf{R} - \frac{m_2}{m_1 + m_2}\mathbf{r},$$

$$\mathbf{x}_2 = \mathbf{R} + \frac{m_1}{m_1 + m_2}\mathbf{r}. \tag{9.34}$$

We shall conclude this section by noting that Eq. (9.33) can be written in the form

$$\mathbf{F}(\mathbf{r}) = \mu\ddot{\mathbf{r}}, \tag{9.35}$$

where

$$\mu = \frac{m_1 m_2}{m_1 + m_2} \tag{9.36}$$

is none other than the 'reduced mass' of the system that we encountered in Section 3.3.1. This way of writing the equation of motion makes explicit the fact that the motion of this two body system is mathematically equivalent to the motion of a single particle of mass μ under the action of the force $\mathbf{F}$. That is, the problem in hand reduces to a problem whose mathematical analysis is exactly the same as the analysis of the motion of a single particle. Of course we should always remember that there are really two particles and that we are solving for their relative position.

9.4   MOTION IN A CENTRAL FORCE

Equation (9.35) certainly constitutes progress in solving for the general motion of two isolated bodies. We shall now make two further assumptions which will allow us to solve the problem completely. Firstly, we assume that the force is 'central', which means that it acts along the line joining the particles, i.e. $\mathbf{F} \propto \mathbf{e}_r$. Secondly, we assume that the force is conservative which, for a central force, means that it depends only on the distance between the particles and not on their orientation, i.e. $\mathbf{F} \propto f(r)\mathbf{e}_r$ where $f(r)$ is related to the corresponding potential via

$$\frac{dU}{dr} = f(r).$$

These are not very restrictive assumptions and the gravitational interaction between two particles satisfies them both.

Our task is to solve

$$\mu\ddot{\mathbf{r}} = -\frac{dU}{dr}\mathbf{e}_r \tag{9.37}$$

which is still a system of three coupled second order differential equations. Now, since the force is conservative we know that the sum of the kinetic and potential energies must be conserved. In addition, the central nature of the force leads also to the conservation of angular momentum, as we shall now show.

The total angular momentum of the system about some origin is

$$\mathbf{L} = m_1\mathbf{x}_1 \times \dot{\mathbf{x}}_1 + m_2\mathbf{x}_2 \times \dot{\mathbf{x}}_2. \tag{9.38}$$

Substituting using Eq. (9.34) and choosing to work in the centre-of-mass frame (i.e. $\dot{\mathbf{R}} = \mathbf{0}$) implies that

$$\mathbf{L} = \mu\mathbf{r} \times \dot{\mathbf{r}}. \tag{9.39}$$

This is easily seen to be a constant vector since

$$\frac{d\mathbf{L}}{dt} = \mu\dot{\mathbf{r}} \times \dot{\mathbf{r}} + \mu\mathbf{r} \times \ddot{\mathbf{r}} = \mathbf{0} \tag{9.40}$$

and we used the fact that $\ddot{\mathbf{r}}$ is parallel to $\mathbf{r}$ for a central force. Outside of the centre-of-mass frame, angular momentum is of course still conserved (this is just a result of the central nature of the force) but it is only in that frame that the angular momentum takes on the form written in Eq. (9.39). The existence of these two conserved quantities will help us greatly.

Indeed, the conservation of angular momentum implies immediately that the motion must be planar. Generally speaking the position vector $\mathbf{r}$ is constrained always to lie in the plane which is perpendicular to the angular momentum vector. This follows immediately from Eq. (9.39) since the nature of the vector product implies that $\mathbf{L}$ is always perpendicular to the position vector $\mathbf{r}$. Now since the angular momentum vector is constant, it follows that the the position vector lies always in the same plane and hence the motion is planar.

For motion in a plane, we can use polar co-ordinates to write (see Section 1.3.4[2])

$$\dot{\mathbf{r}} = \dot{r}\mathbf{e}_r + r\dot{\theta}\mathbf{e}_\theta \tag{9.41}$$

thus, using Eq. (9.39),

$$\mathbf{L} = \mu r^2\dot{\theta}\ \mathbf{e}_r \times \mathbf{e}_\theta. \tag{9.42}$$

Example 9.4.1 *Show that the radius vector* **r** *sweeps out area at a constant rate.*

Solution 9.4.1 *Figure 9.7 illustrates how the radius vector sweeps out area in a plane. For an infinitesimal displacement* $\mathrm{d}r = r\,\mathrm{d}\theta$, *the area swept out is*

$$\mathrm{d}A = \frac{1}{2}r^2\,\mathrm{d}\theta$$

and hence

$$\frac{\mathrm{d}A}{\mathrm{d}t} = \frac{1}{2}r^2\dot{\theta}.$$

Since $|\mathbf{L}| = L = \mu r^2\dot{\theta}$ *it follows that*

$$\frac{\mathrm{d}A}{\mathrm{d}t} = \frac{L}{2\mu} \tag{9.43}$$

which is a constant of the motion. This result is often known as Kepler's Second Law.

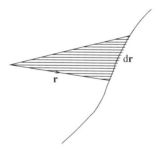

Figure 9.7   Kepler's Second Law informs us that the radius vector **r** sweeps out area at a constant rate.

---

[2] Note that we have changed to the notation more commonly used for basis vectors in advanced dynamics, i.e. $\mathbf{e}_r \equiv \hat{\mathbf{r}}$ and $\mathbf{e}_\theta \equiv \hat{\boldsymbol{\theta}}$.

Armed with the conservation of energy we can write that

$$E = \frac{1}{2}\mu\dot{\mathbf{r}}^2 + U(r)$$

$$= \frac{1}{2}\mu\dot{r}^2 + \frac{1}{2}\mu r^2\dot{\theta}^2 + U(r) \tag{9.44}$$

and we have used Eq. (9.41). Furthermore, we can eliminate the dependence upon θ by introducing the angular momentum L, i.e.

$$E = \frac{1}{2}\mu\dot{r}^2 + \frac{L^2}{2\mu r^2} + U(r). \tag{9.45}$$

This is a very powerful equation for it depends only upon the variables r and t, all other quantities being constants. Re-arranging we have

$$\frac{dr}{dt} = \sqrt{\frac{2}{\mu}\left(E - U(r) - \frac{L^2}{2\mu r^2}\right)} \tag{9.46}$$

and thus once we are given a particular potential $U(r)$ we can go ahead and integrate to obtain $r(t)$. Notice also that the motion looks exactly like the motion in one-dimension of a particle of mass μ in a potential

$$U_{\text{eff}} = U(r) + \frac{L}{2\mu r^2}. \tag{9.47}$$

That is as far as we shall take the general development. Let us now consider the particular case of a gravitational field.

9.5 ORBITS

Equation (9.46) already allows us to make some very general statements about the types of solution we expect. Figure 9.8 shows a plot of the effective potential U_{eff}. At large r the Newtonian $1/r$ term dominates whereas at small r the $1/r^2$ term dominates, we shall call this term the centrifugal barrier term. Now $E = K + U_{\text{eff}}$ where $K = \frac{1}{2}\mu\dot{r}^2 > 0$ and it follows that $E > U_{\text{eff}}$. Thus for motion occurring with total energy E, only those values of r for which $U_{\text{eff}} < E$ are accessible.

Figure 9.8 shows the three possible scenarios. In scenario (a) $E > 0$ and only the region $r < r_0$ is inaccessible, i.e. there is insufficient energy for the system to access this region. This corresponds to a motion where there is a distance of closest approach to the point $r = 0$ but no maximum distance. This type of motion is illustrated in Figure 9.9(a). It helps to have a particular system in mind, so we might consider the case where one of the bodies is much lighter than the other, e.g. a comet moving under the influence of the Sun's gravity. In this case the centre of mass of the two body system is virtually co-incident with the centre of the Sun and the reduced mass is equal to the comet's mass to a very good approximation. Scenario (a) then corresponds to an unbound orbit where the comet is deflected by

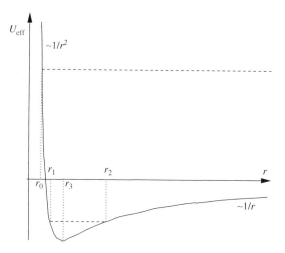

Figure 9.8   The effective potential U_{eff}.

the Sun. Scenario (b) occurs when the energy $E < 0$. In this case, the trajectory is bound to lie in the region $r_1 < r < r_2$. In the comet example, this corresponds to a bound orbit with the comet orbiting the Sun in an ellipse, as illustrated in Figure 9.9(b). The orbits of the planets around the Sun also correspond to this type of $E < 0$ bound orbit. Finally, scenario (c) occurs when the total energy is equal to the value of the effective potential at its minimum. In this case the orbit is constrained to a single value of $r = r_3$ which corresponds to circular motion, as illustrated in Figure 9.9(c). We see now why the $1/r^2$ contribution to the effective potential was called the centrifugal barrier term. In the case of circular motion it is the term which exactly balances the gravitational pull towards the centre. These

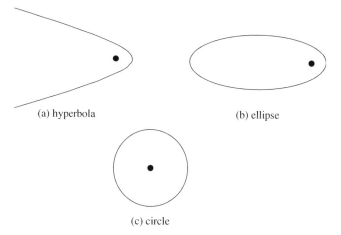

Figure 9.9   The three types of orbit allowed for general two-body motion in a conservative and central field of force.

remarks provide us with a good qualitative insight into the solutions. However it is now time to compute the precise form of the solutions mathematically.

Let us begin by re-writing Eq. (9.46) in the case that the force acting is gravitational:

$$\dot{r} = \sqrt{2\left(\eta + \frac{G\tilde{M}}{r}\right) - \frac{\lambda^2}{r^2}} \qquad (9.48)$$

and we have introduced the more convenient energy per unit mass $\eta \equiv E/\mu$ and angular momentum per unit mass $\lambda \equiv L/\mu$. We have also introduced the mass $\tilde{M} \equiv Mm/\mu$. In the case that $M \gg m$, which is often the case, we can assume $\tilde{M} = M$ without introducing any significant error. Integrating this equation gives us $r(t)$, however we would rather have r as a function of the polar angle θ since such a functional dependence will directly describe the spatial trajectory. The dependence upon t can easily be traded for a dependence upon θ since we know that $\dot{\theta} = \lambda/r^2$ and hence

$$\dot{r} = \dot{\theta}\frac{dr}{d\theta}$$

$$= \frac{\lambda}{r^2}\frac{dr}{d\theta}. \qquad (9.49)$$

Thus we can write

$$\frac{dr}{d\theta} = \frac{r^2}{\lambda}\sqrt{2\left(\eta + \frac{G\tilde{M}}{r}\right) - \frac{\lambda^2}{r^2}} \qquad (9.50)$$

and hence

$$\theta = \int \frac{dr}{r^2}\left[\frac{2}{\lambda^2}\left(\eta + \frac{G\tilde{M}}{r}\right) - \frac{1}{r^2}\right]^{-1/2}. \qquad (9.51)$$

Although this integral looks rather foreboding it is in fact one that we can perform. Let us first change variables to $u \equiv 1/r$, in which case

$$\theta = -\int du \left[\frac{2}{\lambda^2}\left(\eta + G\tilde{M}u\right) - u^2\right]^{-1/2}. \qquad (9.52)$$

By completing the square, this integral can be manipulated into the form $\int dw/\sqrt{c^2 - w^2}$ and this is a standard integral. Thus we write

$$\frac{2}{\lambda^2}\left(\eta + G\tilde{M}u\right) - u^2 = -\left(u - \frac{G\tilde{M}}{\lambda^2}\right)^2 + \left[\frac{G^2\tilde{M}^2}{\lambda^4} + \frac{2\eta}{\lambda^2}\right]. \qquad (9.53)$$

Putting

$$w \equiv u - \frac{G\tilde{M}}{\lambda^2}$$

and

$$c^2 \equiv \frac{G^2\tilde{M}^2}{\lambda^4} + \frac{2\eta}{\lambda^2}$$

we have

$$\theta = -\int \frac{dw}{\sqrt{c^2 - w^2}}$$

$$= \cos^{-1}\left(\frac{w}{c}\right) + \theta_0, \tag{9.54}$$

where θ_0 is a constant of integration which we are free to choose equal to zero (because it corresponds only to a shift in what we call the zero on the polar angle scale). Thus we have the solution that $w = c\,\cos\theta$ and it is time to change back to more familiar variables, i.e.

$$\frac{1}{r} - \frac{G\tilde{M}}{\lambda^2} = \frac{G\tilde{M}}{\lambda^2}\sqrt{\left(1 + \frac{2\eta\lambda^2}{G^2\tilde{M}^2}\right)}\cos\theta. \tag{9.55}$$

This is our final answer, for it tells us how r varies with θ. However, it is somewhat cluttered with symbols and for that reason let us introduce two more quantities

$$\alpha \equiv \frac{\lambda^2}{G\tilde{M}} \tag{9.56}$$

and

$$\varepsilon \equiv \sqrt{\left(1 + \frac{2\eta\lambda^2}{G^2\tilde{M}^2}\right)}. \tag{9.57}$$

Notice that these are both constants of the motion. Thus the polar equation describing the spatial trajectory $r(\theta)$ is simply

$$\frac{1}{r} - \frac{1}{\alpha} = \frac{1}{\alpha}\varepsilon\cos\theta \tag{9.58}$$

which can be re-arranged to read

$$r = \frac{\alpha}{1 + \varepsilon\cos\theta}. \tag{9.59}$$

This equation is familiar to mathematicians, for it is none other than the polar equation for what are called the 'conic sections'. Although we shall not prove it, these are so named because they are the curves that are generated upon slicing through a right circular cone in the various possible ways. The parameter ε determines which type of curve we are dealing with. For $\varepsilon > 1$ the curve is a hyperbola, for $\varepsilon < 1$ it is an ellipse, for $\varepsilon = 0$ it is a circle and for $\varepsilon = 1$ a parabola. These are the curves which were plotted in Figure 9.9.

We can connect the results we have just derived to the qualitative statements we made above. In particular, notice that $\varepsilon > 1$ corresponds to $E > 0$ and so the unbound orbit we described earlier can now be seen to correspond to motion along a hyperbola. Similarly, $\varepsilon < 1$ corresponds to $E < 0$ and we have an ellipse whilst $\varepsilon = 1$ corresponds to $E = 0$ and a circle. Notice that we have traded off the energy and angular momentum of the orbit for the parameters ε and α respectively.

To conclude this section we shall spend a little time exploring the content of Eq. (9.59) in the case of elliptical orbits, i.e. $\varepsilon < 1$. Figure 9.10 shows an elliptical orbit with the semi-major axis a and semi-minor axis b marked. One of the masses is located at $r = 0$ whilst the other follows the elliptical orbit, sweeping out area at a constant rate in accord with Kepler's Second Law. Note how the mass at the origin is not at the centre of the ellipse. In fact, the point $r = 0$ is known as the focus of the ellipse and it is displaced by a distance Δ from the centre of the ellipse. The fact that all of the planets in the solar system orbit around the Sun in ellipses with the Sun at one focus was first established by Johannes Kepler (1571-1630) and it is known as Kepler's First Law.

Let us compute the semi-major axis a. It is defined such that

$$2a = r(0) + r(\pi) = \frac{\alpha}{1+\varepsilon} + \frac{\alpha}{1-\varepsilon}$$

$$\therefore a = \frac{\alpha}{1 - \varepsilon^2}. \tag{9.60}$$

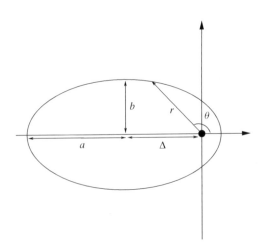

Figure 9.10   The various parameters that define an elliptical orbit.

The focus is a distance Δ from the centre, where

$$\Delta = a - r(0)$$
$$= a - \frac{\alpha}{1 + \varepsilon}$$
$$= a - (1 - \varepsilon)a$$
$$= \varepsilon a. \tag{9.61}$$

Thus the energy variable ε also tells us how squashed the ellipse is. For example, $\Delta = 0$ corresponds to a circular orbit whilst $\Delta = a$ corresponds to an ellipse with one focus at a point on the orbit and an infinite semi-major axis, i.e. it is an ellipse which closes at infinity. This special type of ellipse is more commonly known as a parabola. For obvious reasons, the parameter $\varepsilon = \Delta/a$ is called the 'eccentricity' of the ellipse. From a geometrical point of view it makes more sense to describe an elliptical orbit in terms of the eccentricity and the length of the semi-major axis and we might like to think of these as the two independent variables which specify the orbit (rather than the energy and angular momentum).

The semi-minor axis can also be computed upon realising that

$$\cos\theta = -\frac{\Delta}{r} \tag{9.62}$$

at the point on the orbit which lies a distance b from the centre of the ellipse. We can substitute this value of $\cos\theta$ into Eq. (9.59) in order to determine the value of r at this point and then use Pythagoras' Theorem to determine b, i.e.

$$r = \frac{\alpha}{1 - \varepsilon\Delta/r}. \tag{9.63}$$

Substituting for Δ and α in terms of ε and a and re-arranging gives $r = a$ and Pythagoras' Theorem then gives

$$b^2 = a^2 - \Delta^2,$$
$$\therefore b = a(1 - \varepsilon^2)^{1/2}. \tag{9.64}$$

The total energy of the orbit is $E = \eta m$. Let us express it in terms of the parameters that define the geometry of the ellipse, a and ε. Using Eq. (9.57) we can write

$$\eta = -(1 - \varepsilon^2)\frac{G\tilde{M}}{2\alpha} \tag{9.65}$$

but $\alpha = a(1 - \varepsilon^2)$ and $\tilde{M}\mu = Mm$ hence

$$E = -\frac{GMm}{2a}. \tag{9.66}$$

This is a somewhat surprising result for it tells us that the total energy in an elliptical orbit depends only upon the length of the semi-major axis and not upon the eccentricity of the orbit. For example, we can compute the total energy of an elliptical orbit by computing instead the energy in a circular orbit whose radius is equal to a.

Example 9.5.1 *Prove that the velocity of a particle in an orbit described by Eq. (9.59) is purely tangential at the pericentre and the apocentre of the orbit.*

Solution 9.5.1 *The pericentre of the orbit is the point of closest approach and the apocentre is the point of farthest approach. From Figure 9.10 we can see that these occur at $\theta = 0$ and $\theta = \pi$ respectively. The motion is tangential when $\mathrm{d}r/\mathrm{d}t = 0$. Differentiating Eq. (9.59) gives*

$$\frac{\mathrm{d}r}{\mathrm{d}t} = \frac{\alpha\varepsilon}{(1 + \varepsilon\cos\theta)^2}\sin\theta\,\frac{\mathrm{d}\theta}{\mathrm{d}t}$$

and this vanishes whenever $\sin\theta = 0$, i.e. when $\theta = 0$ or π.

Finally we shall derive Kepler's Third Law by computing the period of the orbit, T. We already know from Kepler's Second Law, Eq. (9.43), that area is swept out at a constant rate equal to $L/(2\mu)$. Integrating over one cycle (and remembering that the area of an ellipse is equal to πab) we have

$$\pi ab = \frac{L}{2\mu}T. \tag{9.67}$$

Re-arranging and substituting for L/μ and b gives

$$T = \frac{2\pi a^2}{\lambda}(1 - \varepsilon^2)^{1/2},$$

$$= \frac{2\pi a^2}{\lambda}\left(\frac{-2\eta\lambda^2}{G^2\tilde{M}^2}\right)^{1/2}. \tag{9.68}$$

We now substitute for the total energy $\eta = -G\tilde{M}/(2a)$ to get

$$T = \frac{2\pi a^{3/2}}{\sqrt{G\tilde{M}}}. \tag{9.69}$$

This result, namely that the period $T \propto a^{3/2}$ was also discovered by Kepler and it constitutes his Third Law. Kepler thought that the constant of proportionality was the same for all planets but as we have derived it, it is not the same since the mass $\tilde{M}$ is not exactly equal to the mass of the Sun. Nevertheless, $\tilde{M} = M$ is a good approximation in practice, leading to deviations from a single constant of proportionality of no more than 0.05% for all the planets.

Note that if time is measured in years and distances in astronomical units (AU) then Eq. (9.69) becomes[3]

$$T^2 = a^3. \tag{9.70}$$

---

[3] Neglecting the reduced mass correction.

This is clear since the Earth is defined to orbit the Sun such that $T = 1$ year when $a = 1$ AU and hence the constant of proportionality $4\pi^2/(G\tilde{M}) = 1$ year^2/AU^3.

Example 9.5.2 *The comet Hale-Bopp is (just about) in orbit around the Sun. The orbit is very eccentric, with $\varepsilon = 0.99511$ and the distance of closest approach to the Sun (the perihelion distance) is 0.9141 AU (it was last at perihelion in 1997). Determine the period of the comet's orbit and its farthest distance from the Sun (the aphelion distance).*

Solution 9.5.2 *To compute the period of the orbit is a straightforward application of Kepler's Third Law once we have the distance of the semi-major axis, a. The perihelion distance is given by*

$$0.9141 \, AU = a(1 - \varepsilon)$$

which can be solved to give $a = 187$ AU. Since the distance is provided in AU we need not work too hard, i.e. we can use Eq. (9.70) to get the period:

$$T = a^{3/2} = 2560 \text{ years.}$$

The distance of farthest approach is given by

$$a(1 + \varepsilon) = 373 \, AU.$$

Example 9.5.3 *The day before the 1969 moonlanding, the Apollo 11 spacecraft was put into orbit around the Moon. The spacecraft had a mass of 9970 kg and the period of the orbit was 119 minutes. In addition, the pericentre and apocentre of the orbit were 1838 km and 1861 km. Use these data to determine the mass of the Moon. Also determine the maximum and minimum speeds of Apollo 11 when it was in lunar orbit.*

Solution 9.5.3 *Using Eq. (9.69), we can determine the mass of the Moon if we have the period of the orbit and the length of the semi-major axis a. Since we anticipate that the mass of Apollo 11 is much smaller than the mass of the Moon we do not need to worry about the reduced mass. Our task is therefore to deduce a. We know that*

$$a(1 - \varepsilon) = 1838 \, km$$

$$a(1 + \varepsilon) = 1861 \, km$$

and adding these two equation together gives

$$a = 1849.5 \, km.$$

Re-arranging Eq. (9.69) yields the mass of the Moon:

$$M = \frac{4\pi^2 a^3}{GT^2} = 4\pi^2 \frac{(1849.5 \times 10^3)^3}{6.67 \times 10^{-11} \times (119 \times 60)^2} = 7.35 \times 10^{22} \, kg.$$

To determine the maximum and minimum speeds we need to compute the speeds at the pericentre and apocentre of the orbit. We can do this using the conservation of energy since we know the total energy of the orbit and, from Example 9.5.1, that the velocity is tangential to the orbit at the extrema of the orbit. Thus

$$-\frac{GM}{2a} = \frac{1}{2}u^2 - \frac{GM}{r},$$

where r is 1861 km at the apocentre or 1838 km at the pericentre. Re-arranging allows us to determine that

$$u^2 = \frac{GM}{a}\left(\frac{2a}{r} - 1\right).$$

Substituting for a = 1849.5 km, the maximum and minimum speeds are 1.64 km/s and 1.62 km/s.

PROBLEMS 9

9.1 Prove that the gravitational potential inside a hollow spherical shell is a constant.

9.2 Suppose a hole were drilled straight through the Earth along a diameter. Show that a body dropped into the hole would execute simple harmonic motion.

9.3 The gravitational self-energy of an object is the total work done against gravity in order to assemble its constituent parts from infinity. Show that the self-energy of a uniform sphere of mass M is

$$E = -\frac{3}{5}\frac{GM^2}{R}.$$

If the sphere rotates about a diameter such that the sum of its rotational and gravitational energies is zero what is its angular velocity?

9.4 A comet orbits the Sun such that the perihelion distance is 7.48×10^{10} m and its speed at perihelion is 5.96×10^4 ms^{-1}. (i) Determine the magnitude of the angular momentum of the comet divided by its mass; (ii) Determine the kinetic energy and gravitational potential energy of the comet (both divided by the comet's mass) at the point of closest approach. Is the orbit bound or unbound? (iii) Using the conservation of energy, in conjunction with fact that the angular momentum is always proportional to the tangential component of the comet's velocity, determine the radial and tangential components of the velocity of the comet, and hence its speed, when it is a distance 1.50×10^{11} m from the Sun.

9.5 A star of mass M is located at the centre of a spherical dust cloud of uniform density ρ. A planet of mass m orbits around the star in a circular orbit of radius r within the cloud. Show that the period of the planet's orbit is given by

$$T = 2\pi \left(\frac{r^3}{G\left(M + \frac{4}{3}\pi\rho r^3\right)}\right)^{1/2}.$$

If the star emits a short burst of radiation which drives away the dust cloud but leaves the planet's position and velocity unchanged show that the planet remains bound to the star if the initial radius is smaller than $(3M/(4\pi\rho))^{1/3}$. Determine the semi-major axis, eccentricity and period of the new orbit.

9.6 In Example 9.5.3, we investigated the motion of the Apollo 11 spacecraft as it orbited the Moon. The astronauts boarded the Eagle lunar module in order to descend to the Moon's surface. The Eagle detached from the command module when the spacecraft was at a height of 110 km above the Moon's surface and immediately fired its rockets to reduce its speed so that it transferred into an elliptical orbit with a perigee of 15 km above the Moon's surface. By how much did the lunar module need to reduce its speed in order to shift to this orbit? [The radius of the Moon is 1740 km.]

9.7 The Earth's orbit about the Sun has an eccentricity of 0.017 and mid-winter in the northern hemisphere occurs when the Earth is close to perihelion. Use this information in conjunction with Kepler's Second Law to estimate by how many days summer is longer than winter in the northern hemisphere. Compare your answer to that which you would get by counting the days between the relevant equinoxes.

10

Rigid Body Motion

In Chapter 4, when we considered the motion of rigid bodies we always made the assumption that the axis of rotation was fixed. This simplification allowed us to deal only with the components of the angular momentum and the angular velocity along the rotation axis. Now we will treat the more general situation in which the axis of rotation may not have a fixed direction in space; this will generally bring into play all three components of $\mathbf{L}$ and $\boldsymbol{\omega}$. To motivate the discussion let us first look at a simple example: a light rigid rod with a mass m at either end, rotating with the midpoint fixed and with the rod making a fixed angle θ with the x_3 axis of a Cartesian coordinate system (see Figure 10.1). The two masses each describe circular motion of the same frequency about the x_3 axis, hence the masses have an equal angular velocity $\boldsymbol{\omega}$, which is parallel to the x_3 axis. Recall that $\boldsymbol{\omega}$ is defined always to be parallel to the axis of rotation, as discussed in Section 4.3. Now let us compute the total angular momentum. We ignore the contribution of the rod, assuming its mass to be negligible, and sum the angular momenta of the two masses to obtain

$$\mathbf{L} = m\,\mathbf{r} \times \mathbf{v} + m(-\mathbf{r}) \times (-\mathbf{v}) = 2m\,\mathbf{r} \times \mathbf{v}. \tag{10.1}$$

$\mathbf{L}$ is perpendicular to both $\mathbf{r}$ and $\mathbf{v}$ and is composed of a component

$$L_3 = 2mr^2\omega \sin^2\theta, \tag{10.2}$$

which is parallel to $\boldsymbol{\omega}$ and a component

$$L_\perp = 2mr^2\omega \sin\theta \cos\theta, \tag{10.3}$$

which rotates about the x_3 axis with angular speed ω. We have used the relationship $v = \omega r \sin\theta$ to write Eq. (10.2) and Eq. (10.3) in terms of ω. Note that the very

Dynamics and Relativity   Jeffrey R. Forshaw and A. Gavin Smith
© 2009 John Wiley & Sons, Ltd

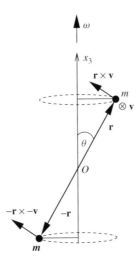

Figure 10.1   A light rod with a mass at either end rotating about the centre of the rod.

fact that $L_\perp$ is non-zero implies that the angular momentum vector about O is *not* parallel to the angular velocity ω. In our previous discussion of the angular momentum of rigid bodies in Chapter 4 we only needed the component L_3 in the direction of ω and ignored terms like $L_\perp$. Since torque is equal to the time rate of change of angular momentum, we can spot straight away that the rotating $L_\perp$ component implies that an external torque has to be acting on the masses in order for them to perform the uniform circular motion about the x_3 axis. Notice that we have just discussed the physics in two stages. We first talked about the connection between ω and $\mathbf{L}$ and then we linked the rate of change of $\mathbf{L}$ with a torque. This chapter mirrors that procedure and so first we shall concentrate on how $\mathbf{L}$ and ω are connected for a general rigid body. Then we will turn our attention to relating the rate of change of the angular momentum to the applied torque, i.e. we will solve for the motion of the body.

10.1   THE ANGULAR MOMENTUM OF A RIGID BODY

A body is considered rigid if we can think of it as being composed of particles whose relative positions do not change with time. Even if we are considering a body with an apparently smooth matter distribution, the 'particles' may be thought of as being infinitesimal volume elements each containing a tiny bit of matter, the position of which we can track precisely (see Figure 10.2). If these bits of matter do not move relative to each other the body is rigid. For such a body we can always find a frame of reference in which the particles are always stationary. This is a frame which therefore moves with the body and it is known as a body-fixed frame. It immediately follows that the position vectors of all the particles are constants in a body-fixed frame. When the body is rotating in an inertial frame of reference, the body-fixed frame is clearly non-inertial. To construct equations of motion we

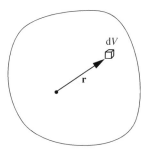

Figure 10.2   A continuous body with a small element of volume dV.

will find it convenient to make a transformation to a non-rotating inertial frame, which we shall call the lab frame. Our previous discussion, in Section 8.2, of transformations between rotating and non-rotating frames led to the result

$$\frac{d\mathbf{r}}{dt} = \left(\frac{d\mathbf{r}}{dt}\right)_{rot} + \boldsymbol{\omega} \times \mathbf{r}, \tag{10.4}$$

where $\mathbf{r}$ is the position vector of a particle relative to an origin chosen to lie somewhere on the axis of rotation.

Let us now obtain the angular momentum of the body by adding together the contributions from its constituent particles, in a similar way to that used for the previous example of two masses on a rod. Applying Eq. (10.4) when the rotating frame is a body-fixed frame gives

$$\mathbf{v} = \frac{d\mathbf{r}}{dt} = \boldsymbol{\omega} \times \mathbf{r} \tag{10.5}$$

since, by definition, $\left(\dfrac{d\mathbf{r}}{dt}\right)_{rot} = \mathbf{0}$ for any particle in the body-fixed frame. We now construct the total angular momentum by summing over particles:

$$\mathbf{L} = m_\alpha \mathbf{r}_\alpha \times \mathbf{v}_\alpha = m_\alpha \mathbf{r}_\alpha \times (\boldsymbol{\omega} \times \mathbf{r}_\alpha), \tag{10.6}$$

where we have used the summation convention on repeated indices (see Section 8.2) and α labels the particles. Rewriting the triple vector product using the identity

$$\mathbf{a} \times (\mathbf{b} \times \mathbf{c}) = (\mathbf{a} \cdot \mathbf{c})\mathbf{b} - (\mathbf{a} \cdot \mathbf{b})\mathbf{c}$$

gives

$$\mathbf{L} = m_\alpha [r_\alpha^2 \boldsymbol{\omega} - (\mathbf{r}_\alpha \cdot \boldsymbol{\omega})\, \mathbf{r}_\alpha]. \tag{10.7}$$

Alternatively, we can write the ith component of $\mathbf{L}$ in a Cartesian co-ordinate system as

$$L_i = m_\alpha \left(r_\alpha^2 \omega_i - r_{\alpha j}\, \omega_j\, r_{\alpha i}\right), \tag{10.8}$$

where there is also an implicit summation on the index j that comes from the scalar product (see Eq. (8.7)). If you are uncomfortable with the summation convention, we can make things more transparent by putting the summations back in:

$$L_i = \sum_\alpha m_\alpha \left(r_\alpha^2 \omega_i - r_{\alpha i} \sum_{j=1}^{3} r_{\alpha j} \omega_j \right). \qquad (10.9)$$

Notice that **L** depends linearly on the rotational frequency, i.e. if we multiply ω by a scalar factor then **L** increases by the same factor. However, **L** is not generally parallel to ω. We can make the connection between the two vectors clearer by rewriting Eq. (10.9) as

$$L_i = \sum_j \sum_\alpha m_\alpha \left(r_\alpha^2 \delta_{ij} - r_{\alpha i} r_{\alpha j} \right) \omega_j, \qquad (10.10)$$

where we have introduced the Kronecker delta symbol, δ_{ij}, which is defined to be

$$\delta_{ij} = 0 \ (i \neq j),$$
$$= 1 \ (i = j). \qquad (10.11)$$

One way to make sense of Eq. (10.10) is to view it as the multiplication of the column vector ω, (whose elements are ω_j) by a matrix with elements

$$I_{ij} = \sum_\alpha m_\alpha \left(r_\alpha^2 \delta_{ij} - r_{\alpha i} r_{\alpha j} \right), \qquad (10.12)$$

i.e. we can write Eq. (10.10) as

$$L_i = I_{ij} \, \omega_j. \qquad (10.13)$$

To see that this equation really represents the multiplication of a vector by a matrix you should remember that there is an implicit summation on the index j and that this summation runs over the columns of the matrix (whose components are I_{ij}), e.g. $L_2 = I_{21}\omega_1 + I_{22}\omega_2 + I_{23}\omega_3$ is the component of **L** in the $\mathbf{e}_2$ direction. Notice also that I_{ij} has the dimensions of a moment of inertia, i.e. mass times the square of a length. It links the angular velocity and the angular momentum, but crucially, unlike the scalar moment of inertia used in Chapter 4, I_{ij} possesses directional information. It is the co-ordinate representation of a geometric object known as the moment of inertia tensor.

10.2   THE MOMENT OF INERTIA TENSOR

In the previous section we obtained the result

$$L_i = I_{ij} \, \omega_j,$$

which expresses the fact that the components of the vectors **L** and ω are linked by a 3×3 matrix with elements I_{ij}. We can associate the matrix elements I_{ij} with the moment of inertia tensor $\mathbf{I}$ and write

$$\mathbf{L} = \mathbf{I}\boldsymbol{\omega}, \tag{10.14}$$

which is now an expression completely independent of our choice of co-ordinate system. The tensor $\mathbf{I}$ may be thought of as a geometrical object that is able to map the vector $\boldsymbol{\omega}$ on to the vector $\mathbf{L}$. The operation that $\mathbf{I}$ performs generally changes both the length and the direction of the vector $\boldsymbol{\omega}$ on which it is operating. That $\mathbf{I}$ can change the direction of $\boldsymbol{\omega}$ means that $\mathbf{I}$ itself has directional properties, e.g. it tells us how the $\mathbf{e}_1$ component of $\mathbf{L}$ depends upon the $\mathbf{e}_2$ component of $\boldsymbol{\omega}$, through I_{12}. The moment of inertia tensor has two spatial indices and it is accordingly known as a rank 2 tensor. Incidentally, scalars and vectors may also be thought of as tensors: a vector has one spatial index and is a rank 1 tensor; a scalar has no directional dependence and is a rank 0 tensor. Tensors of higher rank also exist (and are used in General Relativity) but in this chapter we will need nothing higher than rank 2. Note that we do not use any special typesetting to distinguish $\mathbf{I}$ as a rank 2 tensor, but that should not cause any confusion. It is important to realize that both the angular momentum and the moment of inertia tensor are defined *relative to an origin* and so we should always speak of "the moment of inertia about a point". This is in contrast to the simpler treatment in Chapter 4, where we were only ever interested in the moment of inertia for rotations about some axis, and the component of the angular momentum about the same axis. As we shall very soon see, the moment of inertia about an axis is something that lives within the moment of inertia tensor – the latter being the more general object.

As soon as we choose a co-ordinate system with associated basis vectors we can express the tensor $\mathbf{I}$ as a matrix. Thus the moment of inertia tensor $\mathbf{I}$ is represented in the basis $(\mathbf{e}_1, \mathbf{e}_2, \mathbf{e}_3)$ as the matrix

$$\mathbf{I} = \begin{pmatrix} I_{11} & I_{12} & I_{13} \\ I_{21} & I_{22} & I_{23} \\ I_{31} & I_{32} & I_{33} \end{pmatrix}. \tag{10.15}$$

In the same basis you might like to check that we can express Eq. (10.13) as multiplication of the column vector $\boldsymbol{\omega}$ by the matrix $\mathbf{I}$ to give the column vector $\mathbf{L}$:

$$\begin{pmatrix} L_1 \\ L_2 \\ L_3 \end{pmatrix} = \begin{pmatrix} I_{11} & I_{12} & I_{13} \\ I_{21} & I_{22} & I_{23} \\ I_{31} & I_{32} & I_{33} \end{pmatrix} \begin{pmatrix} \omega_1 \\ \omega_2 \\ \omega_3 \end{pmatrix}. \tag{10.16}$$

Let us for a moment consider the diagonal elements of the moment of inertia tensor. The first of these is

$$I_{11} = \sum_\alpha m_\alpha \left(r_\alpha^2 \delta_{11} - r_{\alpha 1}^2 \right). \tag{10.17}$$

By Pythagoras' Theorem, $r_\alpha^2 = r_{\alpha 1}^2 + r_{\alpha 2}^2 + r_{\alpha 3}^2$, also $\delta_{11} = 1$, and we can write

$$I_{11} = \sum_\alpha m_\alpha \left(r_{\alpha 2}^2 + r_{\alpha 3}^2 \right). \tag{10.18}$$

Notice that $r_{\alpha 2}^2 + r_{\alpha 3}^2$ is the square of the distance of particle α from the x_1 axis. Thus I_{11} is the moment of inertia for rotation about the x_1 axis. Similarly, the other diagonal elements represent the moments of inertia for rotation about the x_2 and x_3 axes, respectively:

$$I_{22} = \sum_{\alpha} m_{\alpha} \left(r_{\alpha 1}^2 + r_{\alpha 3}^2 \right),$$

$$I_{33} = \sum_{\alpha} m_{\alpha} \left(r_{\alpha 1}^2 + r_{\alpha 2}^2 \right). \tag{10.19}$$

These are none other than the moments of inertia for rotation about the three co-ordinate axes, i.e. the objects that we already met in Chapter 4. Notice also that **I** forms a symmetric real matrix, i.e. $I_{ij} = I_{ji}$ and $I_{ij}^* = I_{ij}$, so there are only three independent off-diagonal elements. These are known as the products of inertia and they have the form:

$$I_{12} = I_{21} = -\sum_{\alpha} m_{\alpha}\, r_{\alpha 1}\, r_{\alpha 2},$$

$$I_{23} = I_{32} = -\sum_{\alpha} m_{\alpha}\, r_{\alpha 2}\, r_{\alpha 3}, \tag{10.20}$$

$$I_{13} = I_{31} = -\sum_{\alpha} m_{\alpha}\, r_{\alpha 1}\, r_{\alpha 3}.$$

Thus far we have expressed the elements of the moment of inertia tensor as discrete sums over all the particles in the body, but as usual the body may be better described by a continuous density function $\rho(\mathbf{r})$, where an element of mass dm at position $\mathbf{r}$ is contained within a volume dV such that $dm = \rho(\mathbf{r})dV$ (see Figure 10.2). In which case the sums in Eq. (10.12) are replaced by integrals over the continuous mass distribution and Eq. (10.12) should be written

$$I_{ij} = \int_V dm\, [r^2 \delta_{ij} - r_i r_j] = \int_V dV\, \rho(\mathbf{r})[r^2 \delta_{ij} - r_i r_j]. \tag{10.21}$$

In this way, our picture of a body as being made up of particles is replaced by a picture in which the body is made up of infinitesimal elements of volume dV and mass dm.

We are certainly free to use whatever co-ordinate system we choose for the evaluation of the matrix elements I_{ij}. However, using a co-ordinate system in the lab frame of reference will immediately introduce the problem that the matrix elements I_{ij} will, in general, change as the body rotates. Alternatively we can choose a co-ordinate system in the body-fixed frame and this has the virtue that I_{ij} are constants in time. This provides an important simplification and we will therefore tend to calculate the moment of inertia matrix in the body-fixed frame. The price that we will pay for using the body-fixed frame is that we will have be careful with the dynamical equations of motion, since this frame of reference is rotating and is therefore non-inertial.

10.2.1 Calculating the moment of inertia tensor

The components of the moment of inertia tensor are calculated using Eq. (10.12) or Eq. (10.21) and they depend on the co-ordinate system used. In this section we look at some cases where the rigid body possesses a degree of symmetry that aids in the calculation. In particular, where there is an axis of rotational symmetry there is no change in the moment of inertia in the lab as long the body rotates only about that axis. In which case it does not matter whether we calculate the moment of inertia in the body-fixed frame or the lab frame, the result will be the same.

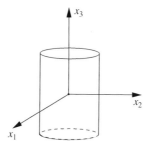

Figure 10.3   A cylinder with its axis of symmetry along the x_3 axis.

For example, consider a solid cylinder with its symmetry axis along the x_3 axis as shown in Figure 10.3. If the cylinder rotates about the symmetry axis then the angular velocity vector is $(0, 0, \omega)$ and the angular momentum is

$$\begin{pmatrix} L_1 \\ L_2 \\ L_3 \end{pmatrix} = \begin{pmatrix} I_{11} & I_{12} & I_{13} \\ I_{21} & I_{22} & I_{23} \\ I_{31} & I_{32} & I_{33} \end{pmatrix} \begin{pmatrix} 0 \\ 0 \\ \omega \end{pmatrix} = \omega \begin{pmatrix} I_{13} \\ I_{23} \\ I_{33} \end{pmatrix}. \tag{10.22}$$

Also, since there is rotational symmetry about the x_3 axis, the products of inertia $I_{13} = I_{31}$ and $I_{23} = I_{32}$ are identically zero. To see why this is so let us consider the product of inertia I_{23}. Notice that for each term proportional to x_2 in the sum I_{23} in Eq. (10.20), the symmetry of the mass distribution ensures that there is a term of equal magnitude, but opposite sign, corresponding to position $-x_2$ (see Figure 10.4). These matching terms always cancel giving rise to $I_{23} = 0$. Thus Eq. (10.22) becomes

$$\begin{pmatrix} L_1 \\ L_2 \\ L_3 \end{pmatrix} = \omega \begin{pmatrix} I_{13} \\ I_{23} \\ I_{33} \end{pmatrix} = \omega \begin{pmatrix} 0 \\ 0 \\ I_{33} \end{pmatrix} \tag{10.23}$$

or $L_3 = I_{33} \omega$ and $L_1 = L_2 = 0$. So in this case the moment of inertia about the x_3 axis is the only element of the moment of inertia matrix that matters. We could therefore dispense with the fancy notation and write the equation $L = I\omega$ just as we did in Chapter 4.

Now let us now consider another special case. Namely, that of a planar object, i.e. something thin and flat that we can approximate as being two dimensional. As such it can be described by an area mass density distribution $\sigma(\mathbf{r})$. For such a

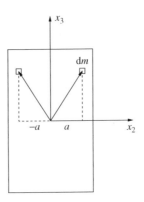

Figure 10.4   Cross-section of a cylinder showing the symmetry with respect to a change of sign of the x_2 co-ordinate.

planar object an element of mass $dm = \sigma(\mathbf{r})\,dA$, where dA is an element of area. The components of the moment of inertia are given by

$$I_{ij} = \int_A dA\,\sigma(\mathbf{r})[r^2\delta_{ij} - r_i r_j]. \tag{10.24}$$

Since the object is flat and thin we can choose its position and orientation such that it lies in the plane where $x_3 = 0$, as indicated in Figure 10.5. This simplifies the calculation of the moment of inertia tensor because two of the products of inertia, I_{13} and I_{23}, are automatically zero. Calculation of the diagonal elements is also simplified by choosing $x_3 = 0$:

$$I_{11} = \int_A dA\,\sigma(\mathbf{r})r_2^2,$$

$$I_{22} = \int_A dA\,\sigma(\mathbf{r})r_1^2,$$

$$I_{33} = \int_A dA\,\sigma(\mathbf{r})(r_1^2 + r_2^2) = I_{11} + I_{22}. \tag{10.25}$$

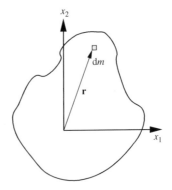

Figure 10.5   Moment of inertia of a planar object. The x_3 axis is out of the page.

The moment of inertia for a planar object can therefore always be written in the form

$$\mathbf{I} = \begin{pmatrix} I_{11} & I_{12} & 0 \\ I_{21} & I_{22} & 0 \\ 0 & 0 & I_{11} + I_{22} \end{pmatrix}. \tag{10.26}$$

Provided we choose the axis of rotation so that it is parallel to x_3, i.e. $\boldsymbol{\omega} = (0, 0, \omega)$, we can use Eq. (10.26) to show that once again we are in a situation where the motion is governed by a single moment of inertia, $I_{33} = I_{11} + I_{22}$, since $\mathbf{L} = \mathbf{I}\boldsymbol{\omega} = I_{33}\,\omega\,\mathbf{e}_3$. Incidentally, we have also shown that the moment of inertia about an axis perpendicular to the plane of a planar object can be expressed as the sum of the moments of inertia about two perpendicular axes lying in the plane. This result is known as the Perpendicular Axis Theorem.

Let us now take a look at another example in which symmetry helps in the calculation of the moment of inertia tensor, and which gives us some results that we will use later in the chapter. We will consider the rotation of a solid cube. We will be interested in rotations about an axis through the centre of the cube and about an axis along an edge of the cube. Remember that we must always choose the origin of our co-ordinate system to lie on the rotation axis since our derivation of Eq. (10.12) starts with Eq. (10.4), which is valid only for rotations about the origin. However, once we have chosen an origin somewhere on the rotation axis, we are then free to choose the directions of our co-ordinate axes to make the calculation of $\mathbf{I}$ as simple as possible.

Example 10.2.1 *Calculate the moment of inertia tensor for a uniform cube of mass M and side b that is suitable for rotations about any axis through: (a) its centre; (b) a corner.*

Solution 10.2.1 *(a) To make the calculation easier, it makes sense to use the symmetry of the cube and to choose a body-fixed Cartesian co-ordinate system with axes parallel to the edges of the cube, and the origin at the centre of the cube (Figure 10.6). Recall that the moment of inertia tensor is defined relative to an origin and that, to be useful, the origin ought to lie on the intended axis of rotation. With these choices, all of the products of inertia vanish and we have*

$$I_{11} = I_{22} = I_{33}.$$

Now

$$I_{11} = \int_V dV \frac{M}{b^3}(x_2^2 + x_3^2) = 2\int_V dV \frac{M}{b^3}x_3^2$$

since the density $\rho = M/b^3$ is uniform. Putting $dV = dx_1\,dx_2\,dx_3$ and integrating over x_3 we obtain

$$I_{11} = 2\frac{M}{b^3}\int_{-\frac{b}{2}}^{\frac{b}{2}}\int_{-\frac{b}{2}}^{\frac{b}{2}}\left[\frac{x_3^3}{3}\right]_{-\frac{b}{2}}^{+\frac{b}{2}} dx_1\,dx_2$$

$$= \frac{1}{6}Mb^2.$$

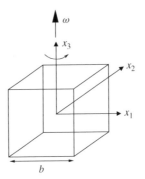

Figure 10.6   A cube rotating about an axis through its centre.

Hence

$$\mathbf{I} = \frac{1}{6}Mb^2 \begin{pmatrix} 1 & 0 & 0 \\ 0 & 1 & 0 \\ 0 & 0 & 1 \end{pmatrix}. \tag{10.27}$$

(b) When the moment of inertia tensor is calculated about a corner the products of inertia do not vanish. However the symmetry of the problem with respect to the interchange of the coordinate axes still helps us, giving $I_{11} = I_{22} = I_{33}$ and $I_{12} = I_{23} = I_{31}$. Then

$$I_{11} = \int_0^b \int_0^b \int_0^b \frac{dx_1 dx_2 dx_3}{b^3} M(x_2^2 + x_3^2) = \frac{2}{3}Mb^2 \text{ and}$$

$$I_{12} = -\int_0^b \int_0^b \int_0^b \frac{dx_1 dx_2 dx_3}{b^3} M x_1 x_2 = -\frac{1}{4}Mb^2.$$

So that

$$\mathbf{I} = Mb^2 \begin{pmatrix} \frac{2}{3} & -\frac{1}{4} & -\frac{1}{4} \\ -\frac{1}{4} & \frac{2}{3} & -\frac{1}{4} \\ -\frac{1}{4} & -\frac{1}{4} & \frac{2}{3} \end{pmatrix} = \frac{1}{12}Mb^2 \begin{pmatrix} 8 & -3 & -3 \\ -3 & 8 & -3 \\ -3 & -3 & 8 \end{pmatrix}.$$

Example 10.2.2 *Calculate the angular momentum relative to the origin when the cube of the previous example rotates about an edge parallel to the x_3 axis with an angular speed ω.*

Solution 10.2.2 *Since the cube rotates about an edge we can use the moment of inertia tensor from part (b) of the previous example to obtain*

$$\mathbf{L} = Mb^2 \begin{pmatrix} \frac{2}{3} & -\frac{1}{4} & -\frac{1}{4} \\ -\frac{1}{4} & \frac{2}{3} & -\frac{1}{4} \\ -\frac{1}{4} & -\frac{1}{4} & \frac{2}{3} \end{pmatrix} \begin{pmatrix} 0 \\ 0 \\ \omega \end{pmatrix} = Mb^2\omega \begin{pmatrix} -\frac{1}{4} \\ -\frac{1}{4} \\ \frac{2}{3} \end{pmatrix}.$$

*Note that the angular momentum and the rotational velocity are not parallel in this case (see Figure 10.7). Also note that the direction of **L** is fixed in the body-fixed frame in which we computed **I** but that in the lab frame it rotates with the cube, i.e. about the x_3 axis with angular frequency ω.*

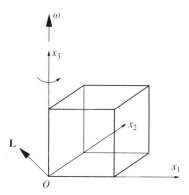

Figure 10.7   A cube rotating about an axis along one edge.

10.3   PRINCIPAL AXES

We have looked at some situations where an obvious symmetry helps us to simplify the moment of inertia tensor. In the general case it is still possible to simplify things through a good choice of the coordinate axes. Although we do not prove it here, it is a theorem of linear algebra that as long as the inverse of **I**, written $\mathbf{I}^{-1}$, exists (i.e. $\mathbf{II}^{-1}$ is the identity) then there must also exist a set of orthogonal basis vectors in which **I** takes on the diagonal form

$$\mathbf{I} = \begin{pmatrix} I_1 & 0 & 0 \\ 0 & I_2 & 0 \\ 0 & 0 & I_3 \end{pmatrix}. \tag{10.28}$$

The axes defined by this choice of basis are known as principal axes. Since there is no ambiguity, we use only one index for the principal axis elements of **I**, i.e. $I_1 \equiv I_{11}$ etc. The diagonal elements I_1, I_2 and I_3 are known as the principal moments of inertia.

For the special case of rotation about a principal axis, e.g. the x_1 axis, we have

$$L = \begin{pmatrix} I_1 & 0 & 0 \\ 0 & I_2 & 0 \\ 0 & 0 & I_3 \end{pmatrix} \begin{pmatrix} \omega \\ 0 \\ 0 \end{pmatrix} = \begin{pmatrix} I_1\omega \\ 0 \\ 0 \end{pmatrix} \tag{10.29}$$

or $\mathbf{L} = I_1\boldsymbol{\omega}$, which means that **L** and $\boldsymbol{\omega}$ are parallel. In general, when $\boldsymbol{\omega}$ has components along more than one principal axis, **L** will not be parallel to $\boldsymbol{\omega}$.

The task of determining the principal axes and the principal moments of inertia is an exercise in linear algebra. If $\hat{\boldsymbol{\alpha}}$ represents a principal axis, and I_α is the principal moment of inertia about that axis, then

$$\mathbf{L} = \mathbf{I}\boldsymbol{\omega} = \omega\,\mathbf{I}\hat{\boldsymbol{\alpha}} = I_\alpha\omega\,\hat{\boldsymbol{\alpha}}, \tag{10.30}$$

i.e.

$$\mathbf{I}\hat{\boldsymbol{\alpha}} = I_\alpha\hat{\boldsymbol{\alpha}}. \tag{10.31}$$

So, the vector $\hat{\boldsymbol{\alpha}}$ is special in that operation on it by $\mathbf{I}$ multiplies $\hat{\boldsymbol{\alpha}}$ by a constant but doesn't alter its direction. Eq. (10.31) is known as an eigenvalue equation and $\hat{\boldsymbol{\alpha}}$ is said to be an eigenvector of $\mathbf{I}$. I_α is the principal moment of inertia about the axis $\hat{\boldsymbol{\alpha}}$, and it is the corresponding eigenvalue. In matrix form Eq. (10.31) is written

$$\begin{pmatrix} I_{11} & I_{12} & I_{13} \\ I_{21} & I_{22} & I_{23} \\ I_{31} & I_{32} & I_{33} \end{pmatrix} \begin{pmatrix} \alpha_1 \\ \alpha_2 \\ \alpha_3 \end{pmatrix} = I_\alpha \begin{pmatrix} \alpha_1 \\ \alpha_2 \\ \alpha_3 \end{pmatrix}, \tag{10.32}$$

where $(\alpha_1, \alpha_2, \alpha_3)$ are the components of the $\boldsymbol{\alpha}$. Rearranging Eq. (10.32) gives us

$$\begin{pmatrix} I_{11} - I_\alpha & I_{12} & I_{13} \\ I_{21} & I_{22} - I_\alpha & I_{23} \\ I_{31} & I_{32} & I_{33} - I_\alpha \end{pmatrix} \begin{pmatrix} \alpha_1 \\ \alpha_2 \\ \alpha_3 \end{pmatrix} = 0. \tag{10.33}$$

A non-trivial[1] solution to this equation exists only if the matrix multiplying the column vector has no inverse, that is if

$$\begin{vmatrix} I_{11} - I_\alpha & I_{12} & I_{13} \\ I_{21} & I_{22} - I_\alpha & I_{23} \\ I_{31} & I_{32} & I_{33} - I_\alpha \end{vmatrix} = 0. \tag{10.34}$$

Equating the determinant to zero[2] generally gives rise to a cubic equation in I_α, known as the characteristic equation, which can be solved to obtain three possible values of I_α. The solutions to the characteristic equation are the principal moments of inertia. Then, each solution for I_α may be substituted in turn into Eq. (10.33) to obtain simultaneous linear equations that can be solved to give the eigenvector components $\alpha_1, \alpha_2, \alpha_3$. Let us now look at an example of this procedure.

Example 10.3.1 *Determine the principal axes of a solid cube of side b and mass M for rotations about a corner.*

---

[1] We can get what is called a trivial solution by setting $\hat{\boldsymbol{\alpha}} = \mathbf{0}$ but this does not determine a direction for a principal axis.

[2] We are assuming a certain familiarity with linear algebra and refer to any number of mathematics textbooks for further details.

Solution 10.3.1 *We have already calculated the moment of inertia tensor for a cube about a corner. It is*

$$\frac{1}{12}Mb^2 \begin{pmatrix} 8 & -3 & -3 \\ -3 & 8 & -3 \\ -3 & -3 & 8 \end{pmatrix}.$$

So that the characteristic equation for the eigenvalues may be written

$$\begin{vmatrix} 8-I' & -3 & -3 \\ -3 & 8-I' & -3 \\ -3 & -3 & 8-I' \end{vmatrix} = 0. \tag{10.35}$$

We have simplified notation with the substitution $I' = \frac{I}{\frac{1}{12}Mb^2}$, *where I denotes an eigenvalue of* **I**. *Expansion of the determinant in Eq. (10.35) gives*

$$(8-I')[(8-I')^2 - 9] + 3[-3(8-I') - 9] - 3[9 + 3(8-I')] = 0.$$

This cubic equation can be put into the form

$$(2-I')(11-I')(11-I') = 0$$

from which we have $I' = 2$ *or* 11 *(twice). The principal moments of inertia are* $I = \frac{1}{6}Mb^2$ *and* $\frac{11}{12}Mb^2$. *Our next task is to figure out the corresponding eigenvectors. We start with the eigenvalue equation for* $I' = 2$:

$$\begin{pmatrix} 8-2 & -3 & -3 \\ -3 & 8-2 & -3 \\ -3 & -3 & 8-2 \end{pmatrix} \begin{pmatrix} \alpha_1 \\ \alpha_2 \\ \alpha_3 \end{pmatrix} = 0$$

from which we obtain two independent equations:

$$6\alpha_1 - 3\alpha_2 - 3\alpha_3 = 0,$$

$$-3\alpha_1 + 6\alpha_2 - 3\alpha_3 = 0. \tag{10.36}$$

Note that we need three equations to completely determine the eigenvector and we have only two. That, however, is not surprising since any eigenvector can be multiplied by an arbitrary constant and it will remain a solution to the eigenvalue equation. Thus we are free to fix the overall normalization of the eigenvectors. It is not difficult to show that the solution to Eq. (10.36) must satisfy $\alpha_1 = \alpha_2 = \alpha_3$. *Now if we insist that* $\hat{\alpha}$ *is a unit vector then* $\alpha_1^2 + \alpha_2^2 + \alpha_3^2 = 1$ *and we can determine the direction of one of the principal axes, i.e.*

$$\hat{\alpha} = \frac{1}{\sqrt{3}} \begin{pmatrix} 1 \\ 1 \\ 1 \end{pmatrix}; I_\alpha = \frac{1}{6}Mb^2.$$

We must now address the solutions corresponding to $I' = 11$. *The fact that this solution occurs twice results in only one independent linear equation upon*

*substituting the eigenvalue back into the eigenvalue equation. So, if we denote an
eigenvector by $\hat{\boldsymbol{\beta}}$ then*

$$\begin{pmatrix} -3 & -3 & -3 \\ -3 & -3 & -3 \\ -3 & -3 & -3 \end{pmatrix} \begin{pmatrix} \beta_1 \\ \beta_2 \\ \beta_3 \end{pmatrix} = 0,$$

which yields the equation

$$\beta_1 + \beta_2 + \beta_3 = 0. \tag{10.37}$$

*This is the equation of a plane that lies perpendicular to the direction of the first prin-
cipal axis. To see this notice that we can write Eq. (10.37) as $\boldsymbol{\beta} \cdot \hat{\boldsymbol{\alpha}} = 0$. Eq. (10.37)
admits an infinity of solutions, all corresponding to vectors that lie in the plane
perpendicular to $\hat{\boldsymbol{\alpha}}$. However, only two vectors are needed in order to form a basis
in the plane (i.e. any other vector can be written as a linear combination of the
original two). We are free to choose any such pair of vectors and, rather arbitrarily,
we pick the first to be*

$$\hat{\boldsymbol{\beta}} = \frac{1}{\sqrt{2}} \begin{pmatrix} 1 \\ -1 \\ 0 \end{pmatrix} ; I_\beta = \frac{11}{12} M b^2.$$

*This vector clearly has components that satisfy Eq. (10.37) and we have set the
length to unity through the choice of the factor of $\frac{1}{\sqrt{2}}$. The third principal axis is
now fixed (up to an overall sign) by the requirement that it is perpendicular to the
other two (in order that the principal axes should form an orthonormal basis). You
should be able to show that*

$$\hat{\boldsymbol{\gamma}} = \frac{1}{\sqrt{6}} \begin{pmatrix} 1 \\ 1 \\ -2 \end{pmatrix} ; I_\gamma = \frac{11}{12} M b^2.$$

To finish off we can write the moment of inertia tensor in the basis $(\hat{\boldsymbol{\alpha}}, \hat{\boldsymbol{\beta}}, \hat{\boldsymbol{\gamma}})$ as

$$\mathbf{I} = \frac{1}{12} M b^2 \begin{pmatrix} 2 & 0 & 0 \\ 0 & 11 & 0 \\ 0 & 0 & 11 \end{pmatrix}.$$

The directions of the principal axis vectors are shown in Figure 10.8.

Notice that when we considered the moment of inertia of the cube about its
centre, we showed that the moment of inertia tensor, Eq. (10.27), was diagonal in
a co-ordinate system aligned with the edges of the cube. Thus the basis vectors $\mathbf{e}_1$,
$\mathbf{e}_2$ and $\mathbf{e}_3$ already defined a set of principal axes and we left the matter there. On
the other hand, when we considered rotation about a corner, one principal axis lay
along the diagonal and the other two lay anywhere in the plane perpendicular to
the diagonal. But one ought really to recognize that for the cube rotating about its

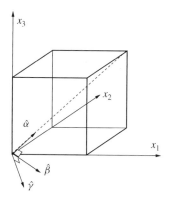

Figure 10.8   The principal axes of a solid cube for rotations about a corner.

centre, the three eigenvalues are degenerate. In such a case, any linear sum of the principal axis basis vectors is also an eigenvector of **I**, i.e. given the eigenvectors $\hat{\boldsymbol{\alpha}}, \hat{\boldsymbol{\beta}}, \hat{\boldsymbol{\gamma}}$ we can always construct a new vector, $a\hat{\boldsymbol{\alpha}} + b\hat{\boldsymbol{\beta}} + c\hat{\boldsymbol{\gamma}}$, that also satisfies the eigenvalue equation:

$$\mathbf{I}(a\hat{\boldsymbol{\alpha}} + b\hat{\boldsymbol{\beta}} + c\hat{\boldsymbol{\gamma}}) = I_0(a\hat{\boldsymbol{\alpha}} + b\hat{\boldsymbol{\beta}} + c\hat{\boldsymbol{\gamma}}), \tag{10.38}$$

where $I_0 = Mb^2/6$ is the degenerate eigenvalue. Thus we were really free to choose *any* set of mutually perpendicular axes as principal axes for rotations of the uniform cube about its centre.

Fortunately, it is not always necessary to solve a cubic equation in order to figure out the principal axes of rotation of a rigid body. There are two circumstances under which the process simplifies quite considerably. Namely, when the object is flat (i.e. planar) and when the object possesses an axis of symmetry. If a body is both flat and symmetric then no calculation is needed and one can write down the principal axes directly, as we shall see in the following example.

Example 10.3.2 *Determine a set of principal axes for a square plate that rotates about a corner.*

Solution 10.3.2 *Well we already worked out the moment of inertia tensor of a general planar object, see Eq. (10.26) and the first thing to notice is that it is already partially diagonal after picking the* $\mathbf{e}_3$ *axis to lie perpendicular to the plane of the body and through the point of rotation. That means that* $\mathbf{e}_3$ *is a principal axis because it satisfies the eigenvalue equation*

$$\mathbf{I}\mathbf{e}_3 = (I_{11} + I_{22})\mathbf{e}_3.$$

We therefore need only find the other two principal axes. Generally, that would mean we would need to find the eigenvectors of the 2×2 *submatrix*

$$\begin{pmatrix} I_{11} & I_{12} \\ I_{21} & I_{22} \end{pmatrix}$$

and that leads to a quadratic rather than cubic characteristic equation that is easier to solve. However, for the square we do not need to do even that because the diagonal of the square is an axis of reflection symmetry. We have seen that, if we choose a basis such that one of the basis vectors lies along the symmetry axis then the corresponding products of inertia vanish (e.g. see Example 10.2.2) and that fact alone is sufficient to guarantee that the axis is also a principal axis. In the case of the square plate it therefore follows that the diagonal is a principal axis, call it e_1. The third axis now comes for free, because it must be orthogonal to the other two axes. In a right-handed co-ordinate system the three principal axes of a square plate are thus as shown in Figure 10.9.

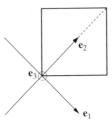

Figure 10.9   The principal axes of a square plate for rotations about a corner. The e_3 vector points out of the page.

10.4   FIXED-AXIS ROTATION IN THE LAB FRAME

All that was rather technical but now we are ready to start analysing the general motion of rotating bodies. By the end of this chapter, we shall have succeeded in understanding what happens to an object thrown through the air (it wobbles and spins), how a gyroscope works and why some rotations of a tennis racquet are safer than others! In this section we "warm up" by re-examining the simpler instance of fixed-axis rotation.

First we shall consider the case where the fixed axis just happens to also be a principal axis, e_3 say. In this case we can write

$$\mathbf{L} = I_3\,\omega\,\mathbf{e}_3, \qquad (10.39)$$

and crucially **L** is also parallel to the x_3 axis and the problem maps onto the more familiar one-dimensional one we met in Chapter 4, i.e.

$$L = I_3\,\omega, \qquad (10.40)$$

where

$$I_3 = \int_V dV\,\rho(\mathbf{r})\,(x_1^2 + x_2^2). \qquad (10.41)$$

Note that in order to maintain rotation only in the e_3 direction any torque τ that acts must be parallel to the x_3 axis, then we can write

$$\boldsymbol{\tau} = \tau\,\mathbf{e}_3 \qquad (10.42)$$

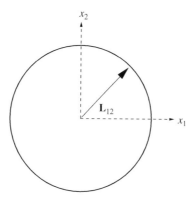

Figure 10.10   Rotation of a solid cube about an edge results in a rotating angular momentum vector in the lab frame. The component of **L** in the $x_1 - x_2$ plane, L_{12}, describes a circle about the x_3 axis.

and the vector equation Eq. (4.9) reduces to

$$\frac{\mathrm{d}L}{\mathrm{d}t} = I_3 \frac{\mathrm{d}\omega}{\mathrm{d}t} = \tau. \tag{10.43}$$

There is a subtlety here that is well worthy of a mention. Eq. (4.9) is derived only in the case that the centre-of-mass does not accelerate but we may be interested in cases other than that. However, in this special case $\mathbf{e}_3$ is a fixed direction in space and so Eq. (10.43) is still valid. If $\mathbf{e}_3$ were not fixed in an inertial frame then we would need to remember to include the term $I_3\omega \, \mathrm{d}\mathbf{e}_3/\mathrm{d}t$ when we compute $\mathrm{d}L/\mathrm{d}t$. Notice also that for fixed ω, it is possible to have rotations about a principal axis without the need for a torque. This is a special feature of rotations about a principal axis.

The other type of fixed-axis rotation occurs when a body rotates about an axis other than a principal axis. In this case, a torque is needed in order to sustain the rotation and this might ordinarily be provided by a fixed axle. We already investigated an example of rotation about an axis other than a principal axis when we considered a cube rotating about one side, in Example 10.2.2. In the following example we will return to that scenario and determine now the torque required to sustain the rotation.

Example 10.4.1 *A solid uniform cube rotates at constant angular speed ω about a fixed axle attached to one edge of the cube. Calculate the magnitude of the torque provided by the fixed axle.*

Solution 10.4.1 *To calculate the torque we need to work out the rate of change of* **L**. *In Example 10.2.2 we showed that*

$$\mathbf{L} = Mb^2\omega \begin{pmatrix} -\frac{1}{4} \\ -\frac{1}{4} \\ \frac{2}{3} \end{pmatrix}. \tag{10.44}$$

This equation gives **L** *in the body-fixed frame. In this frame* **L** *does not change with time, but this does not mean that there will be no torque since the frame is non-inertial. In the lab frame, which is inertial, the body rotates about the x_3 axis and* **L** *will rotate with the body, giving rise to a torque (see Figure 10.10). While the L_3 component clearly doesn't change with time, the components of* **L** *in the $x_1 - x_2$ plane form a time-dependent vector* $\mathbf{L}_{12}$ *with magnitude L_{12}:*

$$L_{12}^2 = L_1^2 + L_2^2 = (Mb^2\omega)^2 \left(\left(\frac{1}{4}\right)^2 + \left(\frac{1}{4}\right)^2 \right) = \frac{1}{8}(Mb^2\omega)^2.$$

Since $\mathbf{L}_{12}$ *sweeps out a circle in the $x_1 - x_2$ plane with constant angular speed ω we can write*

$$\frac{d\mathbf{L}_{12}}{dt} = -L_{12}\,\omega\,\hat{\mathbf{r}}_{12},$$

where $\hat{\mathbf{r}}_{12}$ *is a radial unit vector in the $x_1 - x_2$ plane. Now that we have gone to the effort of shifting to an inertial frame, we are in a position to calculate the torque:*

$$\boldsymbol{\tau} = \frac{d\mathbf{L}}{dt} = \frac{d\mathbf{L}_{12}}{dt} = -\frac{1}{\sqrt{8}} Mb^2\omega^2 \,\hat{\mathbf{r}}_{12}. \tag{10.45}$$

Although we managed to solve the last example, we had to figure out that $\mathbf{L}_{12}$ *precesses in a circle in the lab frame. It would be useful if we had a more general, algebraic, way of solving this problem and the formalism that we shall develop in the next section will allow us to do just that. Moroeever, it will permit us to finally move away from the special case of rotation about a fixed axis.*

10.5 EULER'S EQUATIONS

To depart from the special case of fixed-axis rotation and deal with the general rotational motion of a rigid body our starting point is Eq. (4.9), used in conjunction with Eq. (10.14):

$$\frac{d\mathbf{L}}{dt} = \frac{d}{dt}(\mathbf{I}\,\boldsymbol{\omega}) = \boldsymbol{\tau}. \tag{10.46}$$

This equation is valid in the lab frame (i.e. our generic inertial frame) or in a non-rotating, accelerating frame provided we work relative to the centre of mass. But it is not generally valid in the body fixed frame, which is a rotating frame. Thus to solve for the motion of the body we might consider starting with a set of axes that are fixed in the lab frame. However, as we have previously stressed, to do that necessitates the use of a time-dependent moment of inertia matrix. As a result, it is usually more convenient to work within a body-fixed frame of reference and modify the equations of motion accordingly, i.e. we can no longer use Eq. (10.46) directly but must transform it into the body-fixed frame.

Using the general rule for transforming the time-derivative of a vector between the lab and a rotating frame (Eq. (8.16)), we have

$$\frac{d\mathbf{L}}{dt} = \left(\frac{d\mathbf{L}}{dt}\right)_{body} + \boldsymbol{\omega} \times \mathbf{L} = \boldsymbol{\tau} \tag{10.47}$$

and also

$$\left(\frac{d\mathbf{L}}{dt}\right)_{body} = \left(\frac{d(\mathbf{I}\,\boldsymbol{\omega})}{dt}\right)_{body} = \mathbf{I}\left(\frac{d\boldsymbol{\omega}}{dt}\right)_{body}, \tag{10.48}$$

since in this frame any matrix representation of $\mathbf{I}$ is time-independent. The time-derivative of the angular velocity is actually the same vector in both frames, because

$$\frac{d\boldsymbol{\omega}}{dt} = \left(\frac{d\boldsymbol{\omega}}{dt}\right)_{body} + \boldsymbol{\omega} \times \boldsymbol{\omega} = \left(\frac{d\boldsymbol{\omega}}{dt}\right)_{body} = \dot{\boldsymbol{\omega}}. \tag{10.49}$$

We now have all the elements we need to write the equations of motion in the body-fixed frame. Using Eq. (10.46)–(10.49) we arrive at

$$\boldsymbol{\tau} = \mathbf{I}\,\dot{\boldsymbol{\omega}} + \boldsymbol{\omega} \times (\mathbf{I}\,\boldsymbol{\omega}) = \mathbf{I}\,\dot{\boldsymbol{\omega}} + \boldsymbol{\omega} \times \mathbf{L}. \tag{10.50}$$

Even though we have dropped our explicit denotation of the frame of reference, it is crucial to remember that this equation is generally valid only in a body-fixed frame since $\mathbf{I}$ must be independent of time.

Note that for the special case of rotation at constant angular velocity, i.e. $\dot{\boldsymbol{\omega}} = 0$, Eq. (10.50) gives

$$\boldsymbol{\tau} = \boldsymbol{\omega} \times \mathbf{L}. \tag{10.51}$$

If the body is also rotating about a principal axis then we have seen that $\mathbf{L}$ and $\boldsymbol{\omega}$ will be parallel, in which case $\boldsymbol{\tau} = 0$ and so no torque is required (which confirms the result from the last section). With Eq. (10.51) we now have a more direct method to address problems like the one posed in Example 10.4.1. You might like to check that you can obtain Eq. (10.45) by computing $\boldsymbol{\omega} \times \mathbf{L}$.

At this point we are still free to choose a set of coordinate axes in the body-fixed frame. Taking the basis vectors to be along the principal axes we have

$$\boldsymbol{\omega} \times \mathbf{L} = \begin{vmatrix} \mathbf{e}_1 & \mathbf{e}_2 & \mathbf{e}_3 \\ \omega_1 & \omega_2 & \omega_3 \\ I_1\omega_1 & I_2\omega_2 & I_3\omega_3 \end{vmatrix}. \tag{10.52}$$

Expanding the determinant and taking components of Eq. (10.50) gives us the three equations:

$$I_1\dot{\omega}_1 + (I_3 - I_2)\omega_2\omega_3 = \tau_1,$$

$$I_2\dot{\omega}_2 + (I_1 - I_3)\omega_3\omega_1 = \tau_2,$$

$$I_3\dot{\omega}_3 + (I_2 - I_1)\omega_1\omega_2 = \tau_3. \tag{10.53}$$

These are known as Euler's equations[3]. Notice that they are coupled, non-linear, differential equations of motion, which give the time dependence of ω under the influence of a net external torque τ. By *coupled*, we mean that each equation contains variables that appear in the other two equations and by *non-linear* we mean that the equations involve products of the components of ω. You should compare this structure with that of Newton's Second Law, which yields three linear, uncoupled differential equations in a Cartesian basis. This additional complexity means that we cannot hope to provide a general solution to Euler's equations, rather we shall use them to examine some interesting special cases.

10.6   THE FREE ROTATION OF A SYMMETRIC TOP

10.6.1   The body-fixed frame

Consider a rigid body tossed into the air: the body experiences a net external force due to gravity, but if the gravitational field is uniform there will be no net torque about the centre of mass and the gravitational field will not induce any rotation of the body. In this section we investigate this kind of torque-free rotation. We simplify matters by focussing on an object that is a solid of revolution about one axis (often called a symmetric top). To be entitled to use Euler's equations (Eq. (10.53)), we must be certain that Eq. (10.46) is valid and since we would like to consider the possibility that the centre of mass is accelerating (e.g. as it is for an object tossed through the air) that means we must always compute the angular momentum and torque about the centre of mass.

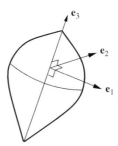

Figure 10.11   A free symmetric top with body-fixed axes.

We need first to identify the principal axes and here the symmetry of the top helps us since the axis of symmetry is automatically a principal axis; we will label this as e_3. The other two principal axes are labelled e_1 and e_2 (see Figure 10.11) and symmetry with respect to rotations about e_3 means that any pair of orthogonal unit vectors lying in the plane perpendicular to e_3 will suffice. The symmetry also dictates that the corresponding principal moments of inertia are the same, i.e.

$$I_1 = I_2 = I. \tag{10.54}$$

---

[3] After Leonhard Paul Euler (1707–1783).

These are in general different from the principal moment of inertia I_3. Thus the third Euler equation gives

$$I_3 \dot{\omega}_3 = 0 \tag{10.55}$$

from which we immediately conclude that

$$\omega_3 = \text{constant} \equiv \omega_t. \tag{10.56}$$

We call ω_t the top (or spin) frequency; it is the angular speed at which the top spins about its symmetry axis. We now have the task of solving for ω_1 and ω_2 from the first two of the Euler equations, which are now linear since ω_t is constant:

$$I \dot{\omega}_1 + (I_3 - I)\omega_2 \omega_t = 0, \tag{10.57}$$

$$I \dot{\omega}_2 + (I - I_3)\omega_1 \omega_t = 0. \tag{10.58}$$

Introducing the frequency

$$\Omega = \frac{I_3 - I}{I} \omega_t \tag{10.59}$$

these equations can be rewritten as

$$\dot{\omega}_1 + \Omega \omega_2 = 0, \tag{10.60}$$

$$\dot{\omega}_2 - \Omega \omega_1 = 0. \tag{10.61}$$

Differentiating Eq. (10.60) and substituting for $\dot{\omega}_2$ using Eq. (10.61) leaves us with a second-order ordinary differential equation:

$$\ddot{\omega}_1 + \Omega^2 \omega_1 = 0. \tag{10.62}$$

In a similar fashion we can obtain the corresponding equation for ω_2:

$$\ddot{\omega}_2 + \Omega^2 \omega_2 = 0. \tag{10.63}$$

Eq. (10.62) and Eq. (10.63) should be immediately recognisable as equations for simple harmonic motion of frequency Ω in each of the variables ω_1 and ω_2. Hence, the general solution for ω_1 is

$$\omega_1 = A \cos(\Omega t + \phi). \tag{10.64}$$

ω_2 is governed by a similar equation, except that the amplitude and phase of this second equation are *not* independent of the constants A and ϕ. Rather, Eq. (10.60) implies that

$$\omega_2 = -\frac{\dot{\omega}_1}{\Omega} = A \sin(\Omega t + \phi). \tag{10.65}$$

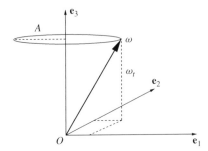

Figure 10.12   The angular velocity of the free symmetric top as observed in the body-fixed frame. The vector $\boldsymbol{\omega}$ precesses about the $\mathbf{e}_3$ direction.

We are thus led to a solution in which the $\boldsymbol{\omega}$ vector sweeps out a circle in the plane defined by the body-fixed vectors $\mathbf{e}_1$ and $\mathbf{e}_2$ as shown in Figure 10.12. The total angular velocity $\boldsymbol{\omega}$ in the body-fixed frame is thus

$$\boldsymbol{\omega} = A\cos(\Omega t + \phi)\,\mathbf{e}_1 + A\sin(\Omega t + \phi)\,\mathbf{e}_2 + \omega_t\,\mathbf{e}_3. \qquad (10.66)$$

Note that the magnitude of $\boldsymbol{\omega}$ is a constant since

$$\omega^2 = \omega_1^2 + \omega_2^2 + \omega_3^2 = A^2 + \omega_t^2. \qquad (10.67)$$

In the body-fixed frame $\boldsymbol{\omega}$ precesses about the symmetry axis with frequency Ω. It is important to keep in mind that although we are working in the frame of reference of the body, $\boldsymbol{\omega}$ describes the rotation of the body as seen in the lab frame. However, because the lab and body-fixed basis vectors do not coincide, the components of $\boldsymbol{\omega}$ are different in the two frames.

10.6.2   The lab frame

The motion of the free symmetric top is described by Eq. (10.64) in the body-fixed frame. In this frame $\boldsymbol{\omega}$, which gives the instantaneous angular velocity of the body in the lab, precesses about the body-fixed symmetry axis. Admittedly this is a bit of a mind bender! Can we understand what the motion looks like in the lab? To do so we will identify some conserved quantities that will turn out to simplify the job of making the transformation between the body-fixed and the lab frames. In order to discuss this transformation mathematically we must first specify the co-ordinate system to be used in the lab frame. We choose Cartesian co-ordinates defined by the basis vectors $\mathbf{i}, \mathbf{j}, \mathbf{k}$. Since there is no external torque, $\mathbf{L}$ is a constant vector when viewed from the lab frame and so it is convenient to fix our co-ordinate system to be aligned with the direction of $\mathbf{L}$. We therefore define $\mathbf{k}$ such that

$$\mathbf{L} = L\mathbf{k}, \qquad (10.68)$$

where L is constant. However, in the body-fixed frame the components of $\mathbf{L}$ are not constant, since the body, and hence the vectors $\mathbf{e}_1, \mathbf{e}_2, \mathbf{e}_3$, rotate with respect to the lab axes. Using Eq. (10.66) and Eq. (10.14) we obtain

$$\mathbf{L} = \mathbf{I}\boldsymbol{\omega},$$

$$= IA\cos(\Omega t + \phi)\,\mathbf{e}_1 + IA\sin(\Omega t + \phi)\,\mathbf{e}_2 + I_3\omega_t\,\mathbf{e}_3. \qquad (10.69)$$

Now $\mathbf{L} \cdot \boldsymbol{\omega}$ is also a constant since, using Eq. (10.66) and Eq. (10.69), we have that

$$\mathbf{L} \cdot \boldsymbol{\omega} = IA^2 + I_3\omega_t^2 \qquad (10.70)$$

and the right-hand side is manifestly constant. Since Eq. (10.70) involves the scalar product of two vectors it is independent of the co-ordinate system, and so must also be true in the lab, even though we have calculated it in the body-fixed frame.

As an aside, we can show that $\mathbf{L} \cdot \boldsymbol{\omega}$ is also constant for free rigid bodies even when they do not have an axis of symmetry. To prove this, we work in the centre-of-mass frame and write the rotational kinetic energy of the particles making up the rigid body as

$$T = \sum_\alpha \frac{1}{2}m_\alpha v_\alpha^2,$$

$$= \sum_\alpha \frac{1}{2}m_\alpha(\boldsymbol{\omega} \times \mathbf{r}_\alpha) \cdot \mathbf{v}_\alpha. \qquad (10.71)$$

Rearranging the triple scalar product we obtain

$$T = \sum_\alpha \frac{1}{2}m_\alpha\,\boldsymbol{\omega} \cdot (\mathbf{r}_\alpha \times \mathbf{v}_\alpha),$$

$$= \frac{1}{2}\boldsymbol{\omega} \cdot \sum_\alpha m_\alpha(\mathbf{r}_\alpha \times \mathbf{v}_\alpha),$$

$$= \frac{1}{2}\,\boldsymbol{\omega} \cdot \mathbf{L}. \qquad (10.72)$$

If there is no net torque, there is no work done to rotate the body about its centre of mass and the rotational kinetic energy must therefore be conserved. Hence, $\mathbf{L} \cdot \boldsymbol{\omega}$ is constant.

Returning to the free symmetric top, we have already shown that the magnitudes of $\boldsymbol{\omega}$ and $\mathbf{L}$ are both constant. Constant T implies that there must be a fixed angle between $\boldsymbol{\omega}$ and $\mathbf{L}$. Since $\mathbf{L}$ is a constant vector in the lab frame the most $\boldsymbol{\omega}$ can do in the lab is rotate about $\mathbf{L}$ maintaining a constant angle to it.

The result that there is a constant angle between $\mathbf{L}$ and $\boldsymbol{\omega}$ is not quite enough to tell us exactly what the body is doing. What we really need to figure out is what happens to the vector $\mathbf{e}_3$ (the symmetry axis of the body) in the lab frame. Fortunately, we can show that $\mathbf{e}_3$ also lies in the plane defined by $\mathbf{L}$ and $\boldsymbol{\omega}$ by constructing the vector

$$\mathbf{L} - I\boldsymbol{\omega} = I\omega_1\mathbf{e}_1 + I\omega_2\mathbf{e}_2 + I_3\omega_t\mathbf{e}_3 - I\boldsymbol{\omega},$$

$$= (I_3 - I)\omega_t\mathbf{e}_3. \tag{10.73}$$

Since ω_t is a constant, Eq. (10.73) describes a constant relationship between the three vectors $\mathbf{L}$, $\boldsymbol{\omega}$ and $\mathbf{e}_3$ such that the three vectors lie in a plane. This certainly does not mean that the plane formed by the three vectors is itself fixed in space; although we do know that the direction of $\mathbf{L}$ is constant it is still possible for $\boldsymbol{\omega}$ and $\mathbf{e}_3$ to rotate at the same rate about $\mathbf{L}$. This is what happens, as we will now show by examining the time dependence of $\mathbf{e}_3$. To do this we transform the time-derivative of $\mathbf{e}_3$ from the body-fixed frame to the lab. We have

$$\frac{d\mathbf{e}_3}{dt} = \left(\frac{d\mathbf{e}_3}{dt}\right)_{\text{body}} + \boldsymbol{\omega} \times \mathbf{e}_3 = \boldsymbol{\omega} \times \mathbf{e}_3, \tag{10.74}$$

since $\mathbf{e}_3$ is a constant vector in the body-fixed frame. Quite generally, we can express $\mathbf{e}_3$ in terms of the basis vectors $\mathbf{i}, \mathbf{j}, \mathbf{k}$ in the lab as

$$\mathbf{e}_3 = \cos\Theta\,\mathbf{k} + \sin\Theta(\cos\Phi\,\mathbf{i} + \sin\Phi\,\mathbf{j}), \tag{10.75}$$

where Θ represents the fixed angle between $\mathbf{e}_3$ and $\mathbf{L}$, and Φ is the angle between the projection of $\mathbf{e}_3$ into the $\mathbf{i}$-$\mathbf{j}$ plane, and the $\mathbf{i}$ axis (see Figure 10.13). Now Eq. (10.73) can be rearranged to give $\boldsymbol{\omega}$:

$$\boldsymbol{\omega} = \frac{L}{I}\mathbf{k} - \frac{I_3 - I}{I}\omega_t\,\mathbf{e}_3 \tag{10.76}$$

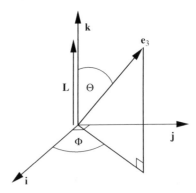

Figure 10.13   The $\mathbf{e}_3$ vector relative to the lab coordinate axes.

and we can use this in Eq. (10.74) to obtain

$$\frac{d\mathbf{e}_3}{dt} = \left(\frac{L}{I}\mathbf{k} - \frac{I_3 - I}{I}\omega_t\,\mathbf{e}_3\right) \times \mathbf{e}_3 = \frac{L}{I}\mathbf{k} \times \mathbf{e}_3. \tag{10.77}$$

Combining this with Eq. (10.75) gives

$$\frac{d\mathbf{e}_3}{dt} = \frac{L}{I}\sin\Theta\,(\cos\Phi\,\mathbf{j} - \sin\Phi\,\mathbf{i}). \qquad (10.78)$$

This can be compared with the time derivative of Eq. (10.75):

$$\frac{d\mathbf{e}_3}{dt} = \frac{d\Phi}{dt}\sin\Theta\,(\cos\Phi\,\mathbf{j} - \sin\Phi\,\mathbf{i}) \qquad (10.79)$$

to give

$$\frac{d\Phi}{dt} = \frac{L}{I}, \qquad (10.80)$$

provided that $\Theta \neq 0$. Thus we have shown that $\boldsymbol{\omega}$ and $\mathbf{e}_3$ both precess about $\mathbf{L}$ in the lab frame at a constant frequency

$$\omega_p \equiv \frac{d\Phi}{dt} = \frac{L}{I}. \qquad (10.81)$$

If $\Theta = 0$, then we are back to the situation of fixed-axis rotation about a principal axis (the symmetry axis), and $\mathbf{e}_3$, $\mathbf{L}$, and $\boldsymbol{\omega}$ are all parallel. For $\Theta \neq 0$, Eq. (10.81) takes on a more revealing form when we write it in terms of the top frequency ω_t. Using Eq. (10.69) we have

$$\mathbf{L} \cdot \mathbf{e}_3 = \omega_t I_3 = L\cos\Theta, \qquad (10.82)$$

which together with Eq. (10.81) implies that

$$\omega_p = \frac{I_3}{I\cos\Theta}\,\omega_t. \qquad (10.83)$$

Thus, in the lab frame, a free symmetric top spins about the symmetry axis (with angular speed ω_t) while the symmetry axis precesses (with angular speed ω_p) about the fixed $\mathbf{L}$ vector. This mode of motion is often referred to as 'wobbling' because of the rotating orientation of the symmetry axis. The relationship between the co-planar vectors $\mathbf{L}$, $\boldsymbol{\omega}$ and $\mathbf{e}_3$ is presented in Figure 10.14 for the case that the top is prolate, i.e $I_3 < I$. The precession of $\boldsymbol{\omega}$ around $\mathbf{e}_3$ in the body-fixed frame describes what is labelled as the body cone. In the lab frame $\boldsymbol{\omega}$ precesses about $\mathbf{L}$ to produce the space cone. The space and body cones intersect along a line defined by the vector $\boldsymbol{\omega}$ and as the motion progresses the body cone rolls around the space cone. For an oblate top ($I_3 > I$), the diagram is similar, except that $\boldsymbol{\omega}$ lies on the other side of $\mathbf{L}$ and the space cone sits inside of the body cone[4].

We now have all the bits and pieces that we need to fully describe the translational and rotational motion of a free symmetric top. Remember that in this context 'free' means free of a net external torque, but there may well be external forces that produce no net torque. Let us examine an example of free rotation that caught

---

[4] You should be able to prove this.

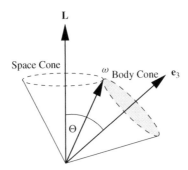

Figure 10.14   Space and body cones for a symmetric top.

the attention of Richard Feynman[5] while he was sitting the cafeteria of Cornell University[6]. Feynman told his biographer, Jagdish Mehra that he was watching a student playing with a plate, tossing it into the air and catching it. If you have ever tried this you will know that it is important to give the plate some angular momentum in order to keep its orientation stable and thereby make it easier to catch. The plate would have been spinning about its symmetry axis, and as we have shown, the symmetry axis would have been simultaneously precessing about the angular momentum vector, making the plate wobble in flight. Feynman noticed that an emblem printed on the plate rotated at about half the frequency of the wobble, i.e. $\omega_p \approx 2\omega_t$. We will obtain this result in the following example.

Example 10.6.1 *Show that the precession frequency of Feynman's plate is twice the top frequency, provided the plate doesn't wobble too much.*

Solution 10.6.1 *By saying that the plate does not wobble too much it is meant that there is only a small angle between the $\mathbf{e}_3$ axis and the $\mathbf{L}$ vector. In fact, you might like to use Figure 10.14 to help picture the wobbling motion by noting that the shaded area directly specifies the spatial orientation of the plate. For small angles, we can set $\cos\Theta \approx 1$ in Eq. (10.83) and obtain*

$$\omega_p \approx \frac{I_3}{I}\omega_t.$$

To make the calculation of the ratio of moments of inertia easier we treat the plate as a perfectly flat disc. Since this is a planar object we can use the perpendicular axis theorem to write

$$I_3 \equiv I_{33} = I_{11} + I_{22} = 2I$$

hence, $\omega_p \approx 2\omega_t$.

---

[5] Richard Phillips Feynman (1918–1988).

[6] Jagdish Mehra, *The Beat of a Different Drum: The Life and Science of Richard Feynman*, Clarendon Press.

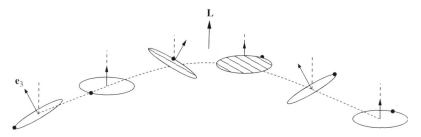

Figure 10.15 A thrown plate describes a parabolic path while wobbling. The small circle indicates the position of a mark on the rim of the plate. The direction of the $\mathbf{e}_3$ axis is indicated. The underside of the plate is shaded and each image represents the orientation of the plate at successive time intervals of $\pi/(2\omega_p)$. Note that when the $\mathbf{e}_3$ axis has precessed by 2π radians the mark has rotated by π radians.

We are finally at the point where we can give a complete description of the combined rotational and translational motion of a plate in flight[7]. Suppose that the plate is thrown like a Frisbee, then $\mathbf{L}$ will be approximately vertical as shown in Figure 10.15. The centre of mass of the plate will obey Newton's Second Law (Eq. (2.28)) and so will follow a parabolic path typical of motion of a particle in a uniform gravitational field. The wobbling motion that we have described in this section then takes place in a frame of reference with its origin at the position of the centre of mass. This separation of the motion into that *of* the centre of mass, and rotation *about* the centre of mass is valid even though the centre of mass is uniformly accelerating, as we discussed in relation to the derivation of Eq. (4.13).

10.6.3 The wobbling Earth

Does the Earth wobble as it travels spinning through space? Provided that we ignore the variation in the gravitational field of the Sun and Moon over the volume of the Earth, then there is no gravitational torque and we might well consider the Earth to be a free symmetric top. The shape of the Earth is oblate (squashed at the poles) so that I_3 is a little larger than I. The numerical values are such that

$$\frac{I_3 - I}{I} \approx 0.0033,$$

which we can substitute into Eq. (10.59) to obtain the precession frequency of $\boldsymbol{\omega}$ about $\mathbf{e}_3$ in the body-fixed frame of an Earth-bound observer:

$$\Omega = \frac{I_3 - I}{I}\,\omega_t = 3.3 \times 10^{-3} \times 2\pi/\text{day}. \tag{10.84}$$

This gives a period of about 300 days. We therefore expect that the axis of rotation of the Earth, as observed from a frame of reference fixed to the Earth, should precess about the North Pole with a period of about 10 months. This motion

---

[7] We ignore air resistance.

would be expected give rise to a periodic change in the apparent latitude (the latitude deduced from observations of the stars) of any given point on the Earth's surface. Detailed observations of the apparent latitude at many locations around the globe over the last century or so have produced data which suggest that the situation is more complicated than the above analysis suggests. The data show many irregularities, but have an underlying periodicity of one year which is thought to be due to seasonal atmospheric effects. However, in addition to the annual periodicity, there is a component with a period of 420 days known as the 'Chandler wobble'. It is this component that is thought to be the effect of the precession of ω around the polar axis. That the observed period is longer than the predicted value may be a result of the Earth not being a perfectly rigid body. In particular, the Earth's mantle is thought to be a viscous fluid, the flow of which effectively reduces the moment of inertia difference $I_3 - I$ and extends the period of the wobble. You can demonstrate an effect that a fluid interior has on the rotational properties of a body with an experiment with two eggs; simply compare the effort it takes to spin a raw egg on a flat surface as opposed to that required for a hardboiled egg. You will observe that it is more difficult to get the raw egg to spin at a given rate. This is because the fluid interior drags on the shell and dissipates energy through non-conservative viscous forces. A raw egg can not be usefully described by a moment of inertia tensor since different parts of the egg generally rotate with different angular velocities. A proper representation of the rotational motion therefore requires the use of a function of space to define the local velocity of an element of matter at any point within the egg, as well as the forces acting on it. In this chapter we shall not delve any deeper into the physics of rotating fluids, which is really the domain of fluid dynamics, but shall instead continue to explore the rich physics of rigid bodies as governed by Euler's equations.

10.7   THE STABILITY OF FREE ROTATION

We determined in Section 10.4 and Section 10.5 that it is theoretically possible to obtain fixed-axis rotation about any of the three principal axes of a free rigid body. However, we did not ask a somewhat more advanced question as to whether such rotation could be maintained for a finite time in a realistic system. Surprisingly, as we shall discover in this section, the answer is that sustained fixed-axis rotation about only two of the principal axes is achievable in practice.

For rotation about a principal axis to be stable we require that small deviations in the alignment of the angular velocity with the principal axis do not produce large effects with time. Such deviations will inevitably occur no matter how carefully we try to set a body spinning about a principal axis, but if the effects remain small we can consider them as perturbations to the motion and we will be able to ignore them at some level. However, if the perturbations come to dominate the motion we consider it to be unstable.

To see the effect that perturbations have on a body rotating about a principal axis, we look directly at the solutions to Euler's equations. Suppose that the body is rotating with angular velocity ω, which is nearly parallel to e_3. We can then make the approximation

$$\omega_1, \omega_2 \ll \omega_3,$$

which allows us to write Euler's equations, Eq. (10.53) with $\tau = \mathbf{0}$ as:

$$I_1\dot{\omega}_1 = (I_2 - I_3)\omega_2\,\omega_3, \tag{10.85}$$

$$I_2\dot{\omega}_2 = (I_3 - I_1)\omega_1\,\omega_3, \tag{10.86}$$

$$I_3\dot{\omega}_3 = (I_1 - I_2)\omega_1\,\omega_2 \approx 0. \tag{10.87}$$

Since Eq. (10.87) implies that ω_3 is approximately constant we can set $\omega_3 = \omega_t$ where ω_t is a constant, and write

$$I_1\dot{\omega}_1 = (I_2 - I_3)\omega_2\,\omega_t, \tag{10.88}$$

$$I_2\dot{\omega}_2 = (I_3 - I_1)\omega_1\,\omega_t. \tag{10.89}$$

Differentiating Eq. (10.88) with respect to time and substituting for $\dot{\omega}_2$ from Eq. (10.89) gives us

$$\ddot{\omega}_1 = -\frac{(I_2 - I_3)(I_3 - I_1)}{I_1 I_2}\,\omega_t^2\,\omega_1. \tag{10.90}$$

Eq. (10.90) is an equation describing a harmonic oscillation in ω_1 with frequency Ω, where

$$\Omega^2 = \frac{(I_2 - I_3)(I_3 - I_1)}{I_1 I_2}\,\omega_t^2. \tag{10.91}$$

Provided that $\Omega^2 > 0$, we can obtain a real frequency and the solution will be of the form

$$\omega_1 = A\,\cos(\Omega t + \delta), \tag{10.92}$$

where A and δ are constants that are fixed by the orientation of the body at $t = 0$. The oscillation will remain of small amplitude if it begins with small amplitude. However if $\Omega^2 < 0$, then Ω is imaginary and there is no oscillation of ω_1 about zero. Rather, the general solution to Eq. (10.90) becomes

$$\omega_1 = A\,e^{\kappa t} + B\,e^{-\kappa t}, \tag{10.93}$$

where $\kappa = i\Omega$ is real and A and B are constants. This solution is unstable: even a tiny ω_1 at $t = 0$ will blow up as t increases. We conclude that for a stable rotation we must therefore have

$$\Omega^2 = \frac{(I_2 - I_3)(I_3 - I_1)}{I_1 I_2}\,\omega_t^2 > 0, \tag{10.94}$$

which occurs if I_3 is the largest, or the smallest, of the three principal moments. However, if I_3 is the intermediate moment of inertia, i.e.

$$I_1 < I_3 < I_2 \text{ or } I_2 < I_3 < I_1 \tag{10.95}$$

then $\Omega^2 < 0$ and the rotation is unstable.

You can easily demonstrate the relative stability of rotations about the principal axes of a rigid body yourself. Choose an object with some clear symmetries so that the principal axes are easily identifiable: a tennis racquet makes a good choice if you can manage to spin it and catch it without injuring yourself, otherwise use a book that is secured with an elastic band so that it doesn't open. Figure 10.16 shows the principal axes of a tennis racquet. The matter distribution in a tennis racquet is such that $I_3 < I_1 < I_2$. When the racquet is spun about the long (x_3) axis stable rotation is clearly possible, as is the case when the racquet is spun in the air in a direction perpendicular to the plane of the strings. However if you attempt to spin the racquet about the x_1 axis you will see a much more erratic behaviour that makes it very difficult to catch.

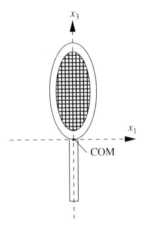

Figure 10.16   A tennis racquet with the x_1 and x_3 principal axes illustrated. The x_2 axis is through the centre-of-mass (COM) and into the page. Stable rotations are possible about the x_2 and x_3 axes but rotation about x_1 is unstable.

10.8   GYROSCOPES

10.8.1   Gyroscopic precession

We have worked hard to gain a thorough understanding of the motion of free rigid bodies through the use of Euler's equations with the torque set to zero. In this section we will 'raise the bar' just a little and explore the effect of a torque on a rapidly spinning body. This is the physics of gyroscopes. Before we get to this, let us just say a few words on the physical characteristics of the system. A gyroscope (see Figure 10.17) typically consists of a flywheel on an axle which is fixed to a supporting cage. Most of the mass of the gyroscope is contained within the flywheel which is free to rotate on the axle. There is always some mechanism for setting the flywheel spinning at high angular speed. This is typically a string wound about the axle that can be tugged sharply, or there may be an electric motor that drives the flywheel. The motion of the gyroscope is fascinating. It almost seems to defy

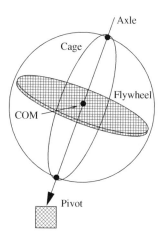

Figure 10.17   The parts of a simple gyroscope.

gravity. Instead of toppling over on the pivot when released, it precesses serenely about the vertical direction. In more sophisticated incarnations, a gyroscope may be mounted on a gimbal within a frame that allows it to take up any orientation to the frame. In such a configuration the gyroscope's axle maintains its direction as the frame tilts, and may be used to measure the orientation of the frame, a feature that leads to applications in the navigation systems for aircraft and ships. We will explore the simple gyroscope, treating it as an axially symmetric top, pivoted at its base, and subject to a gravitational torque.

   To gain a preliminary understanding of gyroscopic motion it is not necessary to work in the body-fixed frame, so we will use the lab frame to start with, but will switch to a rotating frame when we look at a more complex type of motion called nutation. Furthermore, in the first instance only, we will make things easy for ourselves by making the approximation that ω lies parallel to the symmetry axis. This is a good approximation as long as the flywheel rotates much more rapidly on the axle than the whole gyroscope precesses about the vertical direction. In this case, almost all of the angular momentum of the gyroscope comes from the rotation of the flywheel about the axle. Figure 10.18 shows the gyroscope at an angle θ to

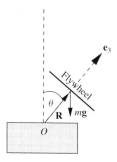

Figure 10.18   A gyroscope at angle θ to the vertical.

the vertical in a co-ordinate system that has its origin at the pivot and in which $\mathbf{e}_3$ is parallel to the instantaneous direction of the axle. If we consider only the angular momentum of the flywheel then $\mathbf{L}$ is parallel to $\mathbf{e}_3$. $\mathbf{R}$ represents the position vector of the centre-of-mass, and so the gyroscope experiences a gravitational torque about the pivot given by:

$$\boldsymbol{\tau} = \mathbf{R} \times m\mathbf{g}. \qquad (10.96)$$

The torque is directed into the page in Figure 10.18. In the inertial lab frame this torque must produce an instantaneous change in the angular momentum which is also into the page. This is achieved by rotating the direction of the $\mathbf{e}_3$ axis and hence also the direction of $\mathbf{L}$. The instantaneous change in $\mathbf{L}$ is represented in the horizontal plane in Figure 10.19. Using Eq. (4.18) we have

$$\boldsymbol{\tau}\,\delta t \approx \delta L \qquad (10.97)$$

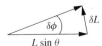

Figure 10.19   The changing horizontal component of $\mathbf{L}$ induced by the gravitational torque on the gyroscope.

and

$$L\,\sin\theta\,\delta\phi \approx \delta L, \qquad (10.98)$$

from which we obtain

$$\omega_p \equiv \dot{\phi} = \frac{\tau}{L\,\sin\theta}. \qquad (10.99)$$

As long as θ doesn't change, τ is constant and we obtain a solution in the lab frame with ω_p also constant, which represents the uniform precession of the gyroscope axle about the vertical.

The above treatment is straightforward and gets us quickly to a result that tells us how the gyroscope is able to precess, i.e. $\mathbf{L}$ rotates at exactly the correct rate so as to compensate for the gravitational torque about the pivot, so there is no torque 'left over' to cause the gyroscope to topple. However, the simple approach is unsatisfying in a couple of ways. Firstly it assumes that all the angular momentum of the system is generated by the rotation of the flywheel about the $\mathbf{e}_3$ axis, and ignores the angular momentum associated with the rotation of the whole system about the vertical direction. Secondly, it assumes that the gyroscope executes its precession at a fixed angle to the vertical and therefore says nothing about the possibility of θ changing with time, as would happen if we were not able to release the gyroscope at the correct angle. We will address these deficiencies in turn: the first in the lab frame by including the angular momentum associated with the precession; the second by using Euler's equations in a rotating frame of reference.

Let us now get rid of the approximation that ω is parallel to the body-fixed e_3 axis. As the gyroscope precesses with angular velocity ω_p about the vertical it gives rise to another contribution to the rotation about e_3, as measured in the lab frame. At any time we can project the total ω on to the symmetry axis of the gyroscope to obtain

$$\omega_3 = \omega_t + \omega_p \cos\theta, \tag{10.100}$$

where, as before, θ is the angle between the e_3 axis and the vertical direction. Previously, we made the approximation that $\omega_3 \approx \omega_t$ which is valid only as long as the top frequency is much higher than the precession frequency. Gyroscopes are constructed so that there is very little friction in the rotation of the flywheel so we shall treat ω_t as a constant. The fact that the torque is always perpendicular to the e_3 direction implies that the projection of the angular momentum onto the symmetry axis of the gyroscope is a conserved quantity, i.e.

$$\frac{dL_3}{dt} = \tau_3 = 0, \tag{10.101}$$

where

$$L_3 = I_3\,\omega_3. \tag{10.102}$$

The angular momentum will also have a contribution from the precession in a direction perpendicular to e_3 which we denote as $L_\perp$ (see Figure 10.20). Assuming the gyroscope to be a symmetric top we have $I_1 = I_2 = I$ and can write

$$L_\perp = I\omega_p \, \sin\theta. \tag{10.103}$$

The various contributions to $\mathbf{L}$ are shown in Figure 10.20 in the plane instantaneously containing the vertical and e_3. All of the vectors in Figure 10.20 are in a common plane that is precessing about the vertical direction. As such, the component of $\mathbf{L}$ in the horizontal plane describes a circle in the lab. Changes in direction of this component are a result of the action of the gravitational torque. We can now write an equation analogous to Eq. (10.98) that now includes the contribution of the precession to $\mathbf{L}$:

$$(L_3 \sin\theta - L_\perp \cos\theta)\,\delta\phi = \tau\,\delta t.$$

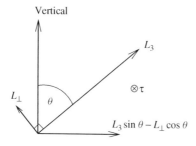

Figure 10.20   Components of $\mathbf{L}$ for a gyroscope. The torque is directed into the page.

Setting $\omega_p = \dot{\phi}$ and using Eq. (10.103) gives us

$$\omega_p(L_3 \sin\theta - I\omega_p \sin\theta \cos\theta) = \tau. \tag{10.104}$$

Eq. (10.104) is a quadratic equation in ω_p with the solution

$$\omega_p = \frac{L_3 \pm \sqrt{L_3^2 - 4I\tau/\tan\theta}}{2I \cos\theta}. \tag{10.105}$$

Eq. (10.96) gives us the magnitude of the torque

$$\tau = mgR \sin\theta, \tag{10.106}$$

which we can use together with Eq. (10.102) to rewrite Eq. (10.105) as

$$\omega_p = \frac{I_3\omega_3 \pm \sqrt{L_3^2 - 4ImgR \cos\theta}}{2I \cos\theta}. \tag{10.107}$$

What is the meaning of the two solutions given by Eq. (10.107)? The higher value of ω_p corresponds to taking the plus sign in the numerator and will give us a frequency $\omega_p \sim \omega_3$, which is fast precession, given that ω_t is large. The second solution corresponds to taking the minus sign in the numerator. We shall soon show that this 'slow' solution corresponds to the ω_p we found previously, i.e. in Eq. (10.99).

That there are two solutions to the equation of motion is a feature of gyroscopic motion that we missed with our simple analysis. To gain a deeper insight, let us look at both solutions in the limit that the torque is very small, as would be the case for a gyroscope of low mass. In this case

$$\frac{4I\tau}{\tan\theta} \ll L_3^2 \tag{10.108}$$

and the high frequency solution to Eq. (10.105) is

$$\omega_p \approx \frac{I_3\omega_3}{I \cos\theta}, \tag{10.109}$$

which is independent of the torque and represents the precession of a free symmetric top (cf. Eq. (10.83)). The slow solution may be examined, in the limit of small torque, by a binomial expansion to first order of the form:

$$(1+x)^{1/2} \simeq 1 + \frac{1}{2}x,$$

giving, (to first order in τ)

$$\sqrt{L_3^2 - 4I\tau/\tan\theta} \approx L_3 \left(1 - \frac{4I\tau}{2L_3^2 \tan\theta}\right). \tag{10.110}$$

Using this expansion in Eq. (10.105), we finally obtain:

$$\omega_p \approx \frac{\tau}{L_3 \sin\theta}. \tag{10.111}$$

As anticipated, this the result that we obtained in our simple approach (Eq. (10.99)), in which we ignored the angular momentum due to the precession.

Note that in order for the precession to be stable we require that the solution to ω_p from Eq. (10.107) is a real number. This gives us the condition that

$$I_3^2 \omega_3^2 \geq 4ImgR \cos\theta \tag{10.112}$$

or

$$\omega_3^2 \geq \frac{4ImgR \cos\theta}{I_3^2}. \tag{10.113}$$

This something new that has come out of our more detailed analysis. There is a minimum spin needed to produce gyroscopic motion.

Example 10.8.1 *A pencil spinning on its tip will not fall over if ω_3 is large enough. Determine the minimum value of ω_3.*

Solution 10.8.1 *To solve this problem we will need the moment of inertia tensor for rotations about the tip of the pencil. This requires the calculation of both the principal moment of inertia about the symmetry axis (I_3), as well as that about an axis perpendicular to the symmetry axis (I). To determine these we will assume that the pencil is cylindrical with uniform density. We have already shown that the moment of inertia of a cylinder is*

$$I_3 = \frac{1}{2}mr^2,$$

where r is the radius of the cylinder and m the mass of the pencil. The other principal moment of inertia is worked out by representing the pencil as a thin rod (see Eq. (4.28)), i.e.

$$I = \frac{mh^2}{3},$$

where h is the length of the pencil. Putting the moments of inertia into Eq. (10.113), with $R = h/2$ and $\cos\theta \approx 1$ gives

$$\omega_3^2 \geq \frac{8}{3}\frac{h^3 g}{r^4} \sim \frac{8 \times (0.15)^3 \times 10}{3 \times (0.5 \times 10^{-2})^4},$$

where we have taken the pencil to have a length of 15 cm and a radius of 5 mm, which gives

$$\omega_3 \gtrsim 12000 \, \text{rad s}^{-1}.$$

This is a large rotational frequency, of about 2000 revolutions per second.

As the previous example shows, while it is theoretically possible to make a pencil behave like a gyroscope, it is extremely difficult in practice due to the very high minimum spin needed. This is a result of the pencil having $I \gg I_3$. Gyroscopes are constructed so that $I_3 > I$ in order that the minimum value of ω_3 is not too large. To achieve this the flywheel is made to be more massive than the frame that supports it, and has most of its mass close to its outer radius.

10.8.2 Nutation of a gyroscope

We have seen that if the gyroscope is spinning at the correct frequency for the angle of tilt then one observes uniform precession. In practice we do not usually know the correct angle and tend to release at too high or low a value of θ. Moreover, we release the gyroscope from rest, rather than at the correct precession frequency. The result of these starting conditions is that the gyroscope bounces a little before settling down into a precession at constant θ. To understand this aspect of a gyroscope's behaviour it is most convenient to look at the motion in a rotating frame of reference and to use Euler's equations. There is a subtlety here; the rotating frame that we will use is one that precesses uniformly with angular velocity $\omega_p \mathbf{k}$, where $\mathbf{k}$ represents the vertical direction in the lab frame. We will call this frame the precessing frame. It is clearly non-inertial, but it is *not* a body-fixed frame since the flywheel still spins in it. Essentially, we will use the precessing frame as a temporary replacement for the lab frame, which we are allowed to do as long as we introduce the correct fictitious forces. The beauty of using the precessing frame is that the torque is necessarily zero since the vector $\mathbf{L}$ stands still in this frame. It is the cancellation of the real torque and the torque due to the ficticious forces that ensures that this is just so. Thus, precession at constant angle θ to $\mathbf{k}$ in the lab is represented by the gyroscope simply spinning about the $\mathbf{e}_3$ axis with angular speed ω_t in the precessing frame, just like a free body spinning about a principal axis. However, if we assume that the gyroscope is released with its centre-of-mass at rest in the lab frame, the initial angular velocity in the precessing frame will be $-\omega_p \mathbf{k} + \omega_t \mathbf{e}_3$. This has small[8] components in the plane perpendicular to $\mathbf{e}_3$ and so the gyroscope does not start off with a simple precession about $\mathbf{k}$ in the lab frame. Since in the precessing frame there is no torque, we can use our solution for the free symmetric top, i.e. Eq. (10.83). This gives us the frequency at which the $\mathbf{e}_3$ axis revolves about $\mathbf{L}$:

$$\Omega' \approx \frac{I_3 \omega_3}{I}, \tag{10.114}$$

where we have made the approximation that $\cos \Theta \approx 1$.

So, in the precessing frame the symmetry axis revolves around $\mathbf{L}$. To understand how things appear in the lab we have to superimpose this motion of the symmetry axis upon the precession of $\mathbf{L}$ about the vertical direction in the lab, which occurs at a frequency given by Eq. (10.99):

$$\omega_p = \frac{\tau}{I_3 \omega_3 \sin \theta} \ll \omega_t.$$

---

[8] Since ω_t is much larger than ω_p.

The net result is a type of motion known as nutation (which means 'nodding') in which the gyroscope precesses slowly about the lab **k** axis, but not at fixed θ. Instead, the axle of the gyroscope weaves an oscillatory pattern (with period $2\pi/\Omega'$) about an average angle of tilt while precessing about the vertical direction (see Figure 10.21). With a simple gyroscope nutation is often heavily damped due to friction in the pivot and is then observed as a quickly-decaying, fast oscillation in the angle of the gyroscope.

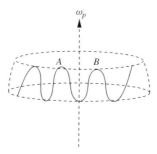

Figure 10.21   An example of a nutation pattern of a gyroscope as observed in the lab frame. The time for the gyroscope to travel from A to B is $2\pi/\Omega'$.

PROBLEMS 10

10.1 A child's hoop of mass m and radius r rolls without slipping in a straight line with speed v. To change the direction of the hoop, the child taps it with a stick, applying an impulse $\Delta\mathbf{p}$ that is perpendicular to the plane of the hoop. Where on the hoop should the child apply the blow?
Show that the hoop is deflected by an angle

$$\theta \approx \frac{\Delta p}{mv}.$$

You should assume that the plane of the hoop remains vertical throughout.
10.2 A coin may be rolled without slipping in a circular path on a horizontal surface, as long as it leans slightly towards the centre of the circle. Show that the angle of tilt ϕ of the coin is given by

$$\tan\phi \approx \frac{3v^2}{2gR},$$

where v is the constant speed of the coin, R is the radius of the circle and g is the gravitational acceleration. What happens if v decreases slowly with time due to rolling friction?
10.3 A spinning ball of diameter 15 cm may be balanced on a fingertip if the rotational speed is sufficient. Estimate the spin required for: (a) a uniform solid ball; (b) a uniform hollow thin-walled ball.

10.4 A rectangular plate of mass M has length $2a$ and width a. Determine the moment of inertia tensor for rotations about a corner. Use a Cartesian co-ordinate system in which the x_1 axis is aligned with the long axis of the plate, the x_3 axis is perpendicular to the plane of the plate and the x_2 axis is parallel to the edge that has length a.

In the same co-ordinate system the plate rotates at constant angular velocity

$$\boldsymbol{\omega} = \frac{\omega}{\sqrt{2}}(1, 1, 0).$$

Determine the magnitude of the angular momentum of the plate, and hence the torque needed to maintain the rotation. Calculate the rotational kinetic energy.

10.5 Determine the principal axes and principal moments of inertia for rotations about a corner of the solid rectangular plate in the previous problem.

10.6 A solid uniform cylinder of density ρ has radius R and height h. Use symmetry to deduce the principal axes for rotations about the centre-of-mass. Calculate the corresponding principal moments of inertia.

A cylindrical artillery shell of radius $0.1\,\text{m}$ and length $0.4\,\text{m}$ is fired from a gun into the air. Estimate the rate of precession of the symmetry axis if the barrel of the gun imparts a spin of 50 revolutions per second about the symmetry axis of the shell.

10.7 A rotating thin circular disc, moving through a fluid, is subject to a damping torque about its centre of mass that is given by

$$\tau_1 = -\kappa\omega_1,$$

$$\tau_2 = -\kappa\omega_2,$$

$$\tau_3 = 0,$$

where $\mathbf{e}_1$ and $\mathbf{e}_2$ are the principal axes that lie in the plane of the disc, $\mathbf{e}_3$ is the symmetry axis and $\boldsymbol{\omega}$ is the angular velocity. Use Euler's equations with the substitution $\eta = \omega_1 + i\omega_2$ to determine the time-dependence of $\omega_1^2 + \omega_2^2$. Describe the motion of the disc in the lab frame.

10.8 Show that for a general rigid body the rate of change of rotational kinetic energy can be expressed as

$$\frac{\mathrm{d}T}{\mathrm{d}t} = \boldsymbol{\omega} \cdot \boldsymbol{\tau}.$$

10.9 Show that the total angular momentum $\mathbf{L}$ of a system of particles may be written as

$$\mathbf{L} = M\mathbf{R} \times \dot{\mathbf{R}} + \mathbf{L}_c,$$

where $\mathbf{R}$ is the position vector of the centre-of-mass, M is the total mass of the system and $\mathbf{L}_c$ is the angular momentum of the system relative to the centre-of-mass.

10.10 (a) Show that the moment of inertia tensor of a uniform solid right circular
cone about its apex is given by

$$
\mathbf{I} = \begin{pmatrix} \frac{3}{5}m\left[\frac{R^2}{4}+h^2\right] & 0 & 0 \\ 0 & \frac{3}{5}m\left[\frac{R^2}{4}+h^2\right] & 0 \\ 0 & 0 & \frac{3}{10}mR^2 \end{pmatrix}
$$

where m is the mass of the cone, R is the radius of the base, h is the
height and the symmetry axis lies along $\mathbf{e}_3$.

(b) A solid right circular cone rolls on its side without slipping on a horizontal surface. The cone returns periodically to its starting position with
constant angular speed ω. Show that the kinetic energy of the cone is
given by

$$
T = \frac{3m\omega^2 h^2}{40}\left(\frac{6h^2+R^2}{h^2+R^2}\right).
$$

Part IV

Advanced Special Relativity

11

The Symmetries of Space and Time

11.1 SYMMETRY IN PHYSICS

Although Part II of this book succeeded in presenting all of the key ideas in Special Relativity it was weak in one crucial respect. Namely, it did not place proper emphasis on an underlying symmetry of Nature which, once appreciated, throws a whole new light on the subject and on the very way we think of space and time. It is the purpose of this part of the book to remedy that deficiency and in so doing provide the grounding for a much deep understanding of the subject which ultimately paves the way for Einstein's Theory of Gravitation: General Relativity.

Symmetry is abundant in Nature, is often intuitive and frequently perceived as beautiful, for example a snowflake or a sphere. There is a whole mathematical apparatus, called Group Theory, which exists in order to handle the mathematics of symmetry. Fortunately, we don't need to learn Group Theory to make a good deal of progress, instead we shall develop the maths as and when we need it. Generally speaking, a system possesses a symmetry if it can be transformed in some way such that the result of the transformation is to leave the system unchanged. The most trivial type of symmetry arises if we take a system and do nothing to it. The act of 'doing nothing' is a symmetry, but not a very interesting one. More interesting would be to take a circle and rotate it by any angle about an axis through its centre and perpendicular to its plane. We say that the circle is invariant under such rotations. It is also invariant under reflections about any diameter. Similarly, a square is invariant under reflections about either diagonal, or about the perpendicular bisectors of its sides.[1] The word 'invariant' is used to indicate that the object in question remains unchanged.

[1] There are more symmetries of a square which you may like to try and figure out.

Dynamics and Relativity   Jeffrey R. Forshaw and A. Gavin Smith
© 2009 John Wiley & Sons, Ltd

Before we press ahead and begin to present some specific examples of symmetry in action it is perhaps worth re-emphasising that symmetries play a very fundamental role in modern physics. Time translational invariance is the symmetry which says that the laws of physics do not change over time and it embodies the idea that an experiment performed today should yield the same result as the same experiment performed tomorrow, all other things being equal. Remarkably, the law of conservation of energy arises as a direct consequence of this symmetry.[2] Similarly, the law of conservation of momentum can be derived if we insist that the laws of physics should be invariant under translations in space (loosely speaking we might say that it does not matter where an experiment is performed) and the law of conservation of angular momentum can be derived by insisting on invariance under rotations in space (i.e. it does not matter what the orientation of an experiment is). These three symmetries are very intuitive symmetries of space and time. They embody the idea that there is no fundamentally special place, time or direction in the Universe. We have seen in Part II that Einstein added a new symmetry of space and time to this list, namely Lorentz invariance, and it is the purpose of this part of the book to emphasise the central role of that symmetry in his theory.

11.1.1 Rotations and translations

Vectors and scalars: a recap

Laws of physics, such as Coulomb's Law or Newton's laws, are built using only scalar (such as mass and electric charge), vector (such as force and acceleration) and occasionally tensor (such as the moment of inertia) quantities. By their very definition, these objects do not change if we decide to use a different system of co-ordinates. Of course the components of a vector (or tensor) do depend upon the choice of co-ordinates but the vector is still the same old vector. Objects which do depend upon the details of our co-ordinate system are not of interest to physicists, since the way we choose to parameterise points in space and time should not be important.

In this and the next subsection we explore this idea in a little more detail. Let us begin by considering two frames of reference T and T', which differ in some way that does not depend upon time. A general vector V does not care about the change of co-ordinates, although its components do generally change:

$$V = \sum_i V_i \mathbf{e}_i = \sum_i V_i' \mathbf{e}_i'. \tag{11.1}$$

The summation is over the three spatial components and the $\mathbf{e}_i$ are the unit basis vectors in T whilst the $\mathbf{e}_i'$ are the unit basis vectors in T'. So although the components of the vector do change ($V_i \neq V_i'$) the vector remains the same. We say that the two frames of reference lead to different representations of the same vector. Scalars are even simpler, for their numerical value is independent of reference frame.

---

[2] It is outside of our remit to provide the proof of the link between symmetry and conservation laws in this book.

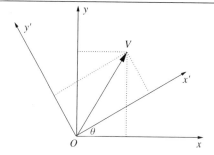

Figure 11.1   Two different frames of reference, T and T', which differ by a rotation through an angle θ.

Rotations

Let us be even more specific and consider two frames of reference that differ by a rotation through an angle θ about the z-axis, as illustrated in Figure 11.1. The components of a general vector V in the two frames are related to each other by

$$V_1' = V_1 \cos \theta + V_2 \sin \theta, \tag{11.2}$$

$$V_2' = -V_1 \sin \theta + V_2 \cos \theta, \tag{11.3}$$

$$V_3' = V_3. \tag{11.4}$$

Equivalently we may write (using the summation convention introduced in Eq. (8.7))

$$V_i' = R_{ij} V_j, \tag{11.5}$$

where the entries R_{ij} can be expressed via the matrix

$$\mathbf{R} = \begin{pmatrix} \cos \theta & \sin \theta & 0 \\ -\sin \theta & \cos \theta & 0 \\ 0 & 0 & 1 \end{pmatrix}. \tag{11.6}$$

As we shall soon see, a particularly important property of this matrix is that it is orthogonal, which means that

$$\mathbf{R}^T \mathbf{R} = \mathbf{1} \tag{11.7}$$

or, in component form,

$$R_{ji} R_{jk} = \delta_{ik}. \tag{11.8}$$

It is therefore a trivial exercise to obtain the inverse of an orthogonal matrix: one just takes the transpose: $\mathbf{R}^{-1} = \mathbf{R}^T$, i.e. $(R^{-1})_{ij} = R_{ji}$. As an aside it is very common to see Eq. (11.5) written as

$$\mathbf{V}' = \mathbf{R} \mathbf{V}. \tag{11.9}$$

This is fine so long as one is clear on the distinction between the column vector $\mathbf{V}$ and the actual physical vector V. The former is merely an ordered list of three numbers which tell us the components of V in a particular frame of reference. As such it is not equal to $\mathbf{V'}$ which is a different ordered list of numbers. This is of course not in conflict with the statement that the vector V is identical in the two frames, i.e. $V' = V$. Notice the subtle notation: we use upright boldface to denote column vectors and italic boldface to denote actual physical vectors.

Let us see the utility of vectors in action by considering a particular example. The gravitational attraction between two massive particles of mass m_1 and m_2 located at position vectors x_1 and x_2 relative to an origin O leads to the following equation of motion for particle 1:

$$m_1 \frac{d^2 x_1}{dt^2} = Gm_1 m_2 \frac{x_2 - x_1}{|x_2 - x_1|^3}. \tag{11.10}$$

Under rotations of the co-ordinate system, none of the quantities in this equation change since they are vector or scalar quantities and so the equation holds true in all frames related to each other by a rotation. Put another way, since we built the equation using vector and scalar quantities it follows that the equation does not change its form even if we change reference frame. We say that the laws of physics are invariant under a (global) change of co-ordinates.

Example 11.1.1 *When written in component form, Eq. (11.10) can be written*

$$m_1 \frac{d^2 x_{1i}}{dt^2} = Gm_1 m_2 \frac{x_{2i} - x_{1i}}{[(x_{2j} - x_{1j})(x_{2j} - x_{1j})]^{3/2}}, \tag{11.11}$$

where x_{1i} are the components of x_1 in T etc. We have again used the convention (introduced first in Section 8.2) which says that repeated indices are summed over, i.e. there is a sum over j implied in the denominator. Prove that this equation does not change its form when expressed in terms of components in T' given that T and T' are related by the rotation specified by Eq. (11.6).

Solution 11.1.1 *We can substitute for $x_{1i} = R_{ji} x'_{1j}$ etc. in the numerator of each side of Eq. (11.11). The denominator on the right hand side needs special consideration. Clearly it represents the distance between the two particles (raised to the third power) and this ought to be a scalar quantity. Let us check this. If we define $r_j \equiv x_{2j} - x_{1j}$ and*

$$d^2 \equiv r_j r_j$$

then we can write

$$\begin{aligned} d^2 &= R_{kj} r'_k R_{lj} r'_l \\ &= R_{kj} R_{lj} r'_k r'_l \\ &= R_{kj} (R^{-1})_{jl} r'_k r'_l \\ &= \delta_{kl} r'_k r'_l \\ &= d'^2. \end{aligned} \tag{11.12}$$

Thus the distance is indeed the same in both frames and we can write Eq. (11.11) as

$$m_1 R_{ji} \frac{d^2 x'_{1j}}{dt^2} = G m_1 m_2 \frac{R_{ji}(x'_{2j} - x'_{1j})}{[(x'_{2k} - x'_{1k})(x'_{2k} - x'_{1k})]^{3/2}}.$$

We are almost done, all that remains is to multiply either side by R_{li} with the implied summation over i whereupon we can use the fact that $R_{li} R_{ji} = \delta_{jl}$, i.e.

$$m_1 \frac{d^2 x'_{1l}}{dt^2} = G m_1 m_2 \frac{(x'_{2l} - x'_{1l})}{[(x'_{2k} - x'_{1k})(x'_{2k} - x'_{1k})]^{3/2}}.$$

And we have proven that the equation does not change its form under a rotation of the co-ordinate system. We chose to perform this calculation explicitly in component notation and hopefully you managed to thread your way through the maze of indices. We could have worked in terms of matrices and column vectors, in which case we write

$$d^2 = (\mathbf{R}^{-1} \mathbf{r}')^T (\mathbf{R}^{-1} \mathbf{r}')$$

$$= \mathbf{r}'^T (\mathbf{R}^{-1})^T (\mathbf{R}^{-1}) \mathbf{r}'$$

$$= \mathbf{r}'^T \mathbf{R} \mathbf{R}^T \mathbf{r}'$$

$$= \mathbf{r}'^T \mathbf{r}'$$

$$= d'^2$$

and Eq. (11.11) becomes

$$m_1 \mathbf{R}^{-1} \frac{d^2 \mathbf{x}'_1}{dt^2} = G m_1 m_2 \frac{\mathbf{R}^{-1}(\mathbf{x}'_2 - \mathbf{x}'_1)}{[(\mathbf{x}'_2 - \mathbf{x}'_1)^T (\mathbf{x}'_2 - \mathbf{x}'_1)]^{3/2}},$$

which reduces to the required form after multiplying both sides by $\mathbf{R}$.

The previous example illustrates the usefulness of the scalar product between two vectors since the proof of Eq. (11.12) can easily be broadened to show that

$$a_i b_i = a'_i b'_i \tag{11.13}$$

for any two vectors $\mathbf{a}$ and $\mathbf{b}$. It should be stressed that this is not inevitable. There are an infinity of ways in which two vectors can be combined to give a pure number but only one way yields a pure number that is also scalar. For example, given our two vectors we might combine them as $a_1 b_1 - a_2 b_2 + a_3 b_3$. The resultant number is not a scalar quantity for its value does depend upon whether we are in frame T or T'. So the scalar product is the only possible way to combine two vectors in order to produce a scalar. There is likewise only one way to produce a vector quantity from two vectors and that is the vector product. Specifically this means that under the rotation $\mathbf{R}$ the vector product must necessarily satisfy

$$(\mathbf{R}V) \times (\mathbf{R}W) = \mathbf{R}(V \times W), \tag{11.14}$$

where V and W are any two vectors.

In this subsection we have been investigating how vectors make manifest the necessary invariance in the form of the equations of physics as we move between frames of reference which differ by a rotation. In the language of many textbooks, we have been considering 'passive' rotations of the co-ordinate axes. We have not however made any statement as to whether or not the physical world possesses a rotational symmetry. To explore this question requires a somewhat different approach: we need to ask what happens if we rotate the actual position vectors corresponding to all parts of our experiment? If we want to insist that the physical world is rotationally symmetric then performing such a rotation should not alter the form of the equations of motion for this is the mathematical expression of the statement that the results of an experiment do not depend upon the orientation of the experiment. Again things should become clearer if we pick a specific example.

Let us consider a particular experiment in which a charged particle is moving in a magnetic field. The particle moves according to

$$m\ddot{\boldsymbol{x}} = q\dot{\boldsymbol{x}} \times \boldsymbol{B}. \tag{11.15}$$

Now suppose that the magnetic field is generated by the apparatus of our experiment (for example by a solenoid). We can ask what happens if we rotate all elements of our experiment, including the solenoid, by the same amount? The new vectors can all be obtained from the old vectors through the action of some rotation matrix $\boldsymbol{R}$, and in particular

$$\boldsymbol{x}' = \mathbf{R}\boldsymbol{x}, \tag{11.16}$$

$$\boldsymbol{B}' = \mathbf{R}\boldsymbol{B}. \tag{11.17}$$

Using Eq. (11.15) we thus have

$$m\mathbf{R}^{-1}\ddot{\boldsymbol{x}}' = q(\mathbf{R}^{-1}\dot{\boldsymbol{x}}') \times (\mathbf{R}^{-1}\boldsymbol{B}') \tag{11.18}$$

$$= q\mathbf{R}^{-1}(\dot{\boldsymbol{x}}' \times \boldsymbol{B}') \tag{11.19}$$

where the second line is necessary if the vector product is to be a vector quantity. We can pre-multiply each side by $\mathbf{R}$ to get

$$m\ddot{\boldsymbol{x}}' = q\dot{\boldsymbol{x}}' \times \boldsymbol{B}'. \tag{11.20}$$

Thus, we see that the Lorentz force law is invariant under 'active' rotations of all parts of the system. Now we change the situation somewhat and move to a fictitious universe in which there exists a universal uniform magnetic field $\boldsymbol{B}$ which permeates the whole of space. Clearly this universe is not rotationally symmetric since the magnetic field picks out a very special direction. Charged particles travelling parallel to this direction would feel no force whereas those travelling in any other direction would be deflected. Clearly, the results of experiments will now depend upon orientation. We expect that this feature should express itself in a non-invariance of the form of the laws of physics and this is indeed the case as we can easily see. Eq. (11.18) is now replaced by

$$m\mathbf{R}^{-1}\ddot{\boldsymbol{x}}' = q(\mathbf{R}^{-1}\dot{\boldsymbol{x}}') \times \boldsymbol{B}' \qquad (11.21)$$

since in this universe we cannot actively rotate the background magnetic field and so $\boldsymbol{B}' = \boldsymbol{B}$. This equation can be simplified to

$$m\ddot{\boldsymbol{x}}' = q\dot{\boldsymbol{x}}' \times (\mathbf{R}\boldsymbol{B}'). \qquad (11.22)$$

Thus we see that the equation of motion for a charged particle varies depending upon the orientation of our apparatus. In physical terms, the effective magnetic field which appears in the Lorentz force law varies with orientation.

11.1.2 Translational symmetry

Having dealt with pure rotations, let us now focus upon the consequences of shifting origin. Again we shall speak of two frames of reference, T and T', but this time T' differs from T in that the origin in T' lies at position $\boldsymbol{R}$ relative to the origin in T, as illustrated in Figure 11.2. Clearly all vectors are once again blind to this change of frame. In fact, as we move from T to T' not only does a general vector V remain unchanged its components are also unchanged:

$$V_i' = V_i. \qquad (11.23)$$

There is however a subtlety we ought to be sensitive to. When we speak of position vectors we are stating a position relative to some origin. Thus when we change frames, we should remember that we have also changed the point of reference for position vectors.

Let us return again to the example of two massive particles acting under gravity and in particular let us recast Eq. (11.10) in terms of position vectors measured

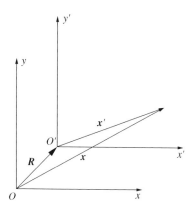

Figure 11.2   Two different frames of reference, T and T', which differ by a translation.

relative to the origin in T', i.e. we must write $x' = x - R$. This may look odd, for we have previously been stressing that vectors remain unchanged as we move from frame to frame. However, a position vector relative to one origin is not the same as the position vector representing the same point but relative to a different origin, that is why $x' \neq x$. The equation of motion for particle 1 now becomes

$$m_1 \frac{\mathrm{d}^2(x_1' - R)}{\mathrm{d}t^2} = Gm_1m_2 \frac{(x_2' - R) - (x_1' - R)}{\left|(x_2' - R) - (x_1' - R)\right|^3} \qquad (11.24)$$

and since R is constant this reduces to

$$m_1 \frac{\mathrm{d}^2 x_1'}{\mathrm{d}t^2} = Gm_1m_2 \frac{x_2' - x_1'}{\left|x_2' - x_1'\right|^3}. \qquad (11.25)$$

Thus the form of the equation is once again unchanged. It was not automatic that this form invariance should occur, in particular it was important that we had the opportunity to differentiate the vector R on the left hand side. If translational symmetry is a good symmetry of Nature then we should require all the laws of physics to possess the same form invariance as we have just discovered for the law of gravitation.

11.1.3 Galilean symmetry

In the previous subsection we showed how vectors and scalars are the building blocks which ensure that the mathematical expression of the laws of physics accord with the fact that Nature does not care how we choose to set up our system of co-ordinates. It is very natural to ask if there are any other symmetries of Nature which constrain the form of physical laws in analogy to the way that co-ordinate invariance constrains us to build the laws of physics using vectors and scalars. Of course we immediately know of one such symmetry from Part II: the principle of special relativity which states that physics looks the same in all inertial frames. It was Einstein who elevated Galileo's observation that there appears to be no experiment able to ascertain whether an object is at rest or moving with uniform velocity into a fundamental symmetry of Nature. In its Galilean form the principle of relativity would say that the laws of physics should take the same form in inertial frames S and S' where

$$x' = x - Vt. \qquad (11.26)$$

We might think of the situation as a translation (Figure 11.2) but where the translational vector R depends linearly on time, i.e. $R = Vt$. Clearly it is a significant additional restriction on any physical law that it should be in accord with the relativity principle.

As a specific example, let us return once again to the two masses interacting gravitationally. Eq. (11.24) still holds true but now R is not a constant vector.

Fortunately, this does not prevent Eq. (11.25) from remaining true since the dependence upon $\boldsymbol{R}$ always cancels in the right hand side, regardless of the dependence of $\boldsymbol{R}$ upon the time, whilst it also disappears from the left hand side after differentating twice with respect to time. Thus the law of gravity is invariant under Galilean transformations. Notice that it would not be invariant if $\boldsymbol{R}$ were to depend upon some higher power of t.

11.2   LORENTZ SYMMETRY

At the end of the last section we discussed invariance under Galilean transformations. However we know from Part II that although Galilean transformations are a good approximation at low velocities they ought really to be replaced by the Lorentz transformations if physics is to accord with both of Einstein's postulates. Since it is our intention that all laws of physics should be consistent with Einstein's theory it would be to our advantage to find a way of representing physical objects such that Lorentz invariance is explicit from the outset, in much the same way that the use of vectors makes explicit invariance under co-ordinate transformations (rotations and translations).

Let us state our intention. We would like to build all of the equations in physics using only mathematical objects which do not change as one alters the inertial frame of reference. Ordinary vectors and scalars provide the paradigm since equations built out of them do not change under a change of co-ordinates. Ordinary scalars and vectors will not suffice however, since transformations between inertial frames mix up the spatial and temporal co-ordinates of an event. Now physics is concerned entirely with the relationships between *events* in space and time. For every event we can represent it, in any given inertial frame, by a list of four numbers (t, x, y, z). Now these numbers may change as we move from inertial frame to inertial frame but the event remains the same. The invariant idea of 'an event' suggests immediately that we might try to represent events by vectors in a four-dimensional space. At this stage in our development this is little more than an idea but it is an idea that will soon gain in stature.

Let us begin by recapping the Lorentz transformations. As in Part II, when it is useful to focus on two particular inertial frames we shall always pick the frames S and S' related in the usual way, i.e. the axes are aligned, the origins coincide at $t = t' = 0$ and S' moves in the positive x direction with speed u. Accordingly we can write the Lorentz transformations written in Eq. (6.28) in a particularly suggestive manner:

$$ct' = ct \cosh\theta - x \sinh\theta, \tag{11.27a}$$

$$x' = -ct \sinh\theta + x \cosh\theta, \tag{11.27b}$$

$$y' = y, \tag{11.27c}$$

$$z' = z, \tag{11.27d}$$

where

$$\cosh\theta = \gamma(u) = \frac{1}{\sqrt{1 - u^2/c^2}},$$

$$\sinh\theta = \gamma(u)\frac{u}{c}, \tag{11.28}$$

i.e. $\tanh\theta = u/c$. In matrix form we can equivalently write

$$\begin{pmatrix} ct' \\ x' \\ y' \\ z' \end{pmatrix} = \begin{pmatrix} \cosh\theta & -\sinh\theta & 0 & 0 \\ -\sinh\theta & \cosh\theta & 0 & 0 \\ 0 & 0 & 1 & 0 \\ 0 & 0 & 0 & 1 \end{pmatrix} \begin{pmatrix} ct \\ x \\ y \\ z \end{pmatrix}. \tag{11.29}$$

We have done nothing except to write the Lorentz transformations of Part II in a different way, however the similarity to the formalism for rotations, embodied in Eq. (11.5) and Eq. (11.6), is clearly striking. Roughly speaking, it seems that the Lorentz transformations are something akin to rotating a vector in a four-dimensional space through an imaginary angle. Furthermore, we also know from Eqs. (7.33) and (7.34) in Part II that the energy and momentum of a particle transform in precisely the same way:

$$\begin{pmatrix} E'/c \\ p'_x \\ p'_y \\ p'_z \end{pmatrix} = \begin{pmatrix} \cosh\theta & -\sinh\theta & 0 & 0 \\ -\sinh\theta & \cosh\theta & 0 & 0 \\ 0 & 0 & 1 & 0 \\ 0 & 0 & 0 & 1 \end{pmatrix} \begin{pmatrix} E/c \\ p_x \\ p_y \\ p_z \end{pmatrix}. \tag{11.30}$$

Having seen these results, we are encouraged to follow Minkowski[3] in supposing that we really should think of space and time not as seperate entities but rather as forming a unified four dimensional 'space-time' (often called 'Minkowski space') and that the equations in physics should be built using vectors and scalars in this space-time for they are the objects which do not vary as we move from one inertial frame to another. Actually we should pause for a moment and admit that there may be other types of object available to us, such as four-tensors or perhaps even more exotic objects[4] but to admit such a possibility does not undermine the potential value of four-vectors and four-scalars.

For example, an event in space-time would then have a position 'four-vector' $\mathbf{X} = (ct, x, y, z)$, and instead of seperately speaking of the energy and momentum of a particle we should speak of its momentum four-vector $\mathbf{P} = (E/c, p_x, p_y, p_z)$. Other directional quantities should likewise be described by an appropriate four-vector[5]. We are laying claim to the idea that space and time form a four dimensional space which supports the existence of scalars and vectors. However if this space is to be useful to us it should possess a well defined scalar product.

---

[3] Hermann Minkowski (1864-1909).

[4] Such objects do actually exist. For example, to describe the relativistic motion of electrons we should use objects known as 'spinors'.

[5] From this point onwards we use upper case boldface characters to represent four-vectors.

Let us take two four-vectors, $\mathbf{A} = (A_0, A_x, A_y, A_z)$ and $\mathbf{B} = (B_0, B_x, B_y, B_z)$. The question is 'can we combine these two four-vectors to produce a pure number in such a way that the result does not depend upon the choice of inertial frame?'. The answer is in the affirmative for we can define the scalar product as follows:

$$\mathbf{A} \cdot \mathbf{B} = A_0 B_0 - A_x B_x - A_y B_y - A_z B_z. \tag{11.31}$$

This is very similar to the way we multiply vectors in three dimensional space except for the fact that one of the terms has opposite sign to all of the others (it is the term multiplying the components of the two vectors in the time direction).[6] We can easily check that this definition does indeed yield a result which is the same in S and S' since

$$
\begin{aligned}
\mathbf{A'} \cdot \mathbf{B'} &= A_0' B_0' - A_x' B_x' - A_y' B_y' - A_z' B_z' \\
&= (A_0 \cosh\theta - A_x \sinh\theta)(B_0 \cosh\theta - B_x \sinh\theta) \\
&\quad - (A_x \cosh\theta - A_0 \sinh\theta)(B_x \cosh\theta - B_0 \sinh\theta) \\
&\quad - A_y B_y - A_z B_z \\
&= A_0 B_0 - A_x B_x - A_y B_y - A_z B_z \\
&= \mathbf{A} \cdot \mathbf{B}
\end{aligned}
\tag{11.32}
$$

and we have made use of $\cosh^2\theta - \sinh^2\theta = 1$. Thus we have a recipe for combining two four-vectors into a four-scalar. If we take the scalar product of the position four-vector of an event with itself we obtain

$$\mathbf{X} \cdot \mathbf{X} = c^2 t^2 - x^2 - y^2 - z^2. \tag{11.33}$$

In space-time language, this is the squared distance of the event from the origin. Similarly, the squared length of the momentum four-vector in space-time is given by

$$\mathbf{P} \cdot \mathbf{P} = E^2/c^2 - p_x^2 - p_y^2 - p_z^2. \tag{11.34}$$

Since all inertial observers must agree upon the value of this quantity, we can evaluate it in the inertial frame where the momentum of the particle is zero in which case $E = mc^2$ and thus we know that $\mathbf{P} \cdot \mathbf{P} = m^2 c^2$, i.e.

$$m^2 c^2 = E^2/c^2 - p^2$$
$$\Rightarrow E^2 - c^2 p^2 = m^2 c^4. \tag{11.35}$$

This is none other than the result we presented first in Eq. (7.36). Viewed this way, the mass of a particle is simply the length of the particle's momentum four-vector (divided by c).

---

[6] The overall sign is a matter of convention, as in fact it is when we define the scalar product in three dimensions.

Let us close this section by reflecting upon the necessity of space-time. Clearly we can always represent events by a string of four numbers whose values depend upon the inertial frame of reference and clearly an event is independent of inertial frame. What does not follow however is that the string of four numbers should constitute a vector in a space which possesses a well defined scalar product. To illustrate the point, we could have attempted to seek a way to write down the laws of physics such that they are invariant under Galilean transformations. However our attempt to introduce vectors in four dimensional space-time would be plagued by the fact that we cannot define a scalar product. This failure implies that the invariant distance between any two points in Galilean space-time is not defined. Thus Galilean four-vectors are not particularly useful objects in drawing up the laws of physics. It is perhaps worth going into a little more detail. In Galilean relativity the space-time four-vector of an event transforms according to

$$
\begin{pmatrix} ct' \\ x' \\ y' \\ z' \end{pmatrix} = \begin{pmatrix} 1 & 0 & 0 & 0 \\ -u/c & 1 & 0 & 0 \\ 0 & 0 & 1 & 0 \\ 0 & 0 & 0 & 1 \end{pmatrix} \begin{pmatrix} ct \\ x \\ y \\ z \end{pmatrix}
\tag{11.36}
$$

as one moves between S and S'. The time t of an event is a four-scalar but the spatial interval between two points is not and there is no way to combine both intervals into an invariant distance in space-time. However, we can still proceed, even without the existence of a distance measure. The four-vector corresponding to the velocity of a particle can be written

$$
\mathbf{U} = \frac{d\mathbf{X}}{dt} = (c, \dot{x}).
\tag{11.37}
$$

This is a four-vector since $\mathbf{X}$ is a four-vector and t is a four-scalar, so the ratio in the difference of two such quantites must itself be a four-vector. Similarly the four-acceleration can be written

$$
\mathbf{A} = \frac{d\mathbf{U}}{dt} = (0, \ddot{x}).
\tag{11.38}
$$

Clearly there is little advantage in using the four-vector formalism since the time component of these four vectors is either constant or zero. For example, Eq. (11.10) gains nothing from being rewritten in terms of four-vectors. Even so, there is one interesting insight we can gain using Galilean four-vectors. The conservation of four-momentum implies that

$$
\sum_i m_i(c, \dot{x}_i) = \sum_f m_f(c, \dot{x}_f),
\tag{11.39}
$$

where the indices i and f label the particles in a closed system at two different times. Apart from informing us that the momentum in three dimensions is conserved this equation also informs us that the total mass of the system is conserved, i.e.

$$
\sum_i m_i = \sum_f m_f.
\tag{11.40}
$$

PROBLEMS 11

11.1 If a particle moves with speed u along the x-axis, show that if $\cosh\eta = \gamma(u)$ and $\tanh\eta = u/c$ then

$$\eta = \frac{1}{2}\ln\left(\frac{E+cp}{E-cp}\right).$$

The variable η is known as the 'rapidity' of the particle.

11.2 Represent the Galilean transformation

$$t' = t$$
$$x' = x - ut \tag{11.41}$$

as a 2×2 matrix equation, i.e. $\boldsymbol{x}' = \boldsymbol{G}(u)\,\boldsymbol{x}$. Now consider a second transformation represented by $\boldsymbol{G}(v)$. Compute the matrix $\boldsymbol{G}(u)\,\boldsymbol{G}(v)$ and show that it is also a Galilean transformation.

In the same co-ordinate basis, Lorentz transformations can be generated by the 2×2 matrix (see Eq. (11.29)):

$$\boldsymbol{L}(\eta) = \begin{pmatrix} \cosh\eta & -\sinh\eta \\ -\sinh\eta & \cosh\eta \end{pmatrix}.$$

Show that $\boldsymbol{L}(\eta_1)\boldsymbol{L}(\eta_2) = \boldsymbol{L}(\eta_1 + \eta_2)$.

11.3 Prove Eq. (11.14). You may find the following identity useful:

$$\varepsilon_{ijk} = R_{ia}R_{jb}R_{kc}\varepsilon_{abc}.$$

12

Four-vectors and Lorentz Invariants

In the last chapter we arrived at the conclusion that physical laws should utilise vectors in the four dimensional space-time of Minkowski. In this chapter we explore more fully the four-vector formalism and in so doing it should become clear that this is indeed the language of Special Relativity.

The prototype four-vector is the one which specifies the displacement between two events located at positions $\mathbf{X}_1$ and $\mathbf{X}_2$, namely

$$\Delta \mathbf{X} = \mathbf{X}_2 - \mathbf{X}_1 \tag{12.1}$$

and the square of the invariant distance between the two events is

$$\begin{aligned} \Delta \mathbf{X} \cdot \Delta \mathbf{X} &= c^2 (\Delta t)^2 - (\Delta x)^2 - (\Delta y)^2 - (\Delta z)^2, \\ &= c^2 (\Delta \tau)^2. \end{aligned} \tag{12.2}$$

The second line defines what is called the 'proper time' interval between the two events $\Delta \tau$. We will usually speak of the proper time rather than the 'invariant distance' although the two are the same thing up to a factor of c. If it is possible to find a frame in which the two events occur at the same point then the proper time is simply the time interval between the two events in that inertial frame. For example, in a frame of reference attached to your body the proper time interval between any two events in your life is simply the time difference measured by the watch on your wrist. Notice that $(\Delta \tau)^2$ can in principle take on the value of any real number; positive, negative or zero. This is quite different from distances in ordinary Euclidean space, and we shall explore the consequences in the next chapter. Suffice to say here that if $(\Delta \tau)^2 < 0$ then the proper time interval is

Dynamics and Relativity   Jeffrey R. Forshaw and A. Gavin Smith
© 2009 John Wiley & Sons, Ltd

imaginary, which means that it is not possible for an observer in any inertial frame to be present at both events. Starting from this displacement four-vector we can construct a number of other very useful four-vectors without too much hard work.

12.1   THE VELOCITY FOUR-VECTOR

The velocity four-vector is defined by

$$\mathbf{U} = \frac{d\mathbf{X}}{d\tau},\tag{12.3}$$

$$= \left(c\frac{dt}{d\tau}, \frac{d\mathbf{x}}{d\tau} \right).\tag{12.4}$$

It is a four-vector since the derivative is defined as $\Delta\mathbf{X}/\Delta\tau$ in the limit of vanishingly small $\Delta\tau$. The numerator is our prototypical four-vector and the denominator is our prototypical four-scalar so the ratio must be a four-vector[1]. We can rewrite the differential element of proper time using

$$(d\tau)^2 = (dt)^2 \left(1 - \frac{1}{c^2}\left(\frac{d\mathbf{x}}{dt}\right)^2 \right)\tag{12.5}$$

which implies that

$$\frac{d\tau}{dt} = \frac{1}{\gamma(u)}\tag{12.6}$$

and therefore that

$$\mathbf{U} = \frac{dt}{d\tau}\left(c, \frac{d\mathbf{x}}{dt} \right),$$

$$= \gamma(u)(c, \mathbf{u}).\tag{12.7}$$

Thus the velocity four-vector can be simply related to the velocity three-vector $\mathbf{u}$. Notice that the magnitude of the four-velocity is given by

$$\mathbf{U} \cdot \mathbf{U} = c^2\tag{12.8}$$

and so everything moves through space-time with the same four-speed.

We can use the four-velocity to re-derive the formulae which relate the velocity of a particle in one inertial frame of reference to that in another. As usual we focus upon the particular case of inertial frames S and S'. This is a very straightforward task, for if we suppose that we know a four-velocity in S we can obtain the corresponding four-velocity in S' by applying the transformation specified by the matrix in Eq. (11.29). In order to make precise contact with the results in Part II we shall suppose that we are given a velocity four-vector in S' and asked for its

---

[1] This is a particular example of what is sometimes called the quotient theorem.

components in S. In this case we must use the inverse transformation matrix, which you can easily check is equal to

$$\Lambda^{-1} = \begin{pmatrix} \cosh\theta & \sinh\theta & 0 & 0 \\ \sinh\theta & \cosh\theta & 0 & 0 \\ 0 & 0 & 1 & 0 \\ 0 & 0 & 0 & 1 \end{pmatrix}. \tag{12.9}$$

Thus we can write

$$\gamma(u) \begin{pmatrix} c \\ u_x \\ u_y \\ u_z \end{pmatrix} = \gamma(u') \begin{pmatrix} \cosh\theta & \sinh\theta & 0 & 0 \\ \sinh\theta & \cosh\theta & 0 & 0 \\ 0 & 0 & 1 & 0 \\ 0 & 0 & 0 & 1 \end{pmatrix} \begin{pmatrix} c \\ u'_x \\ u'_y \\ u'_z \end{pmatrix}, \tag{12.10}$$

where $\tanh\theta = V/c$ and V is the relative speed between S and S'. These are the equations which explain how to transform velocities from S' to S but to make the link with Eqs. (6.33) and (6.34) explicit we would like to write down formulae for u_x, u_y and u_z in terms of V and the velocity in S'. The factor of $\gamma(u)$ on the left hand side prevents us from writing the result immediately however the first of the four equations encoded in Eq. (12.10) tells us that

$$\gamma(u) = \gamma(u')(\cosh\theta + \sinh\theta\, u'_x/c) \tag{12.11}$$

and we can use this in the remaining three equations. Doing so gives

$$u_x = \frac{1}{(\cosh\theta + \sinh\theta u'_x/c)}(c\sinh\theta + u'_x \cosh\theta), \tag{12.12}$$

$$u_y = \frac{1}{(\cosh\theta + \sinh\theta u'_x/c)}u'_y, \tag{12.13}$$

and a similar equation for the z component. Using $\tanh\theta = V/c$ and $\cosh\theta = \gamma(V)$ gives the final answer:

$$u_x = \frac{1}{(1 + Vu'_x/c^2)}(V + u'_x) \text{ and} \tag{12.14}$$

$$u_y = \frac{1}{\gamma(V)} \frac{1}{(1 + Vu'_x/c)}u'_y, \tag{12.15}$$

which are identical to Eqs. (6.33) and (6.34).

12.2 THE WAVE FOUR-VECTOR

Let us now consider a travelling wave whose equation in S is written

$$f(x, t) = A\cos(k \cdot x - \omega t). \tag{12.16}$$

It might correspond to a disturbance in some medium or it might describe the propagation of a plane polarised light wave. Before asking how such a wave looks

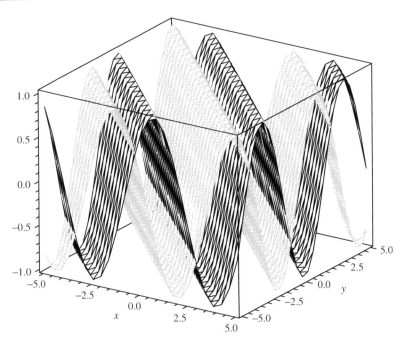

Figure 12.1 The plane wave $f(x, t)$ at two different times. The lighter shaded wave is at the earlier time and the wave travels in the $(1,1)$ direction.

from the viewpoint of an observer in S' let us first be clear on what kind of wave Eq. (12.16) describes. Figure 12.1 shows a plot of $f(x, t)$ at two different times for a particular choice of wavevector k. So that we could draw the picture in the page we picked a two-dimensional wave. The figure illustrates that Eq. (12.16) describes a wave travelling in the direction indicated by k (to produce the figure we picked $k = (1, 1)$). In addition, the displacement of the wave is constant along lines in the xy-plane which lie perpendicular to the direction of propagation k. In three-dimensions the situation is much the same except that the displacement of the wave is now constant on planes that lie perpendicular to the wavevector k. For this reason such waves are often referred to as 'plane waves'. The speed of propagation in S is given by $v = \omega/k$ (where $k = |k|$).

Now we return to the task in hand. What does this wave look like from the viewpoint of an observer in S'? This question is explored in detail in one of the problems at the end of this chapter but for now we need only the result that the phase of the wave must be a four-scalar, i.e.

$$\phi = k \cdot x - \omega t \qquad (12.17)$$

must be the same in all inertial frames. This follows since a particular value of ϕ corresponds to a particular state of the wave and this cannot depend upon inertial frame. For example, the phase difference between a maximum of displacement and the adjacent minima should equal $\pm\pi/2$ in all inertial frames. Now we notice that the phase can be written as a Minkowski scalar product:

$$\phi = -\mathbf{K} \cdot \mathbf{X}, \tag{12.18}$$

where $\mathbf{X}$ is a position four-vector and $\mathbf{K} = (\omega/c, \mathbf{k})$. Now since ϕ is a four-scalar and $\mathbf{X}$ a four-vector it follows that $\mathbf{K}$ must also be a four-vector.

We are now in a position to re-derive the Doppler effect for light. Consider the situation illustrated in Figure 6.4, i.e. a light source is at rest in S' such that it radiates plane waves in the direction of an observer in S. If in S' the light wave has a wave four-vector given by

$$\mathbf{K}' = (k', -k', 0, 0) \tag{12.19}$$

then it will describe plane light waves travelling in the negative x' direction and we have used the fact that the speed of the wave is c hence $\omega' = ck'$. As an aside we ought to comment on our notation, which is rather standard but also potentially rather confusing. We have used a prime on the four-vector itself ($\mathbf{K}'$) but four-vectors are frame independent objects so strictly speaking $\mathbf{K}' = \mathbf{K}$. What the prime really indicates is that the explicit representation of $\mathbf{K}$ on the right hand side is understood to be in S'. Perhaps a better notation would be to write something like $\mathbf{K} = (k', -k', 0, 0)_{S'}$ but one rarely sees this in the literature and so we stick to the slightly imprecise notation of Eq. (12.19). Returning to the task in hand, given Eq. (12.19) we can immediately determine the corresponding wave four-vector in S. It is

$$\begin{pmatrix} \omega/c \\ k_x \\ k_y \\ k_z \end{pmatrix} = \begin{pmatrix} \cosh\theta & \sinh\theta & 0 & 0 \\ \sinh\theta & \cosh\theta & 0 & 0 \\ 0 & 0 & 1 & 0 \\ 0 & 0 & 0 & 1 \end{pmatrix} \begin{pmatrix} k' \\ -k' \\ 0 \\ 0 \end{pmatrix}. \tag{12.20}$$

Three of these equations imply that $k_y = k_z = 0$ and $k_x = -\omega/c = -k$, as it indeed should be for a light wave, whilst the fourth implies that

$$k = k'(\cosh\theta - \sinh\theta)$$
$$= \gamma(v)k'(1 - v/c). \tag{12.21}$$

Putting $f = \omega/2\pi$ and $k = \omega/c$:

$$f = \gamma(v)f'(1 - v/c)$$
$$= f'\sqrt{\frac{1 - v/c}{1 + v/c}} \tag{12.22}$$

which is Eq. (6.9).

12.3   THE ENERGY-MOMENTUM FOUR-VECTOR

Now we turn our attention to energy and momentum. In Part II we had to work rather hard in order to motivate the relativistic equations for momentum and energy. Recall that we considered a particular scattering process and viewed it in two different inertial frames with the goal of finding a definition of momentum which was consistent with a universal law for the conservation of momentum. Energy conservation then arose almost as if by accident. Finally we can present a much more transparent account of the underlying physics.

Given the velocity four-vector presented in Eq. (12.7) we can define what we shall call the momentum four-vector:

$$\mathbf{P} = m\mathbf{U}$$
$$= \gamma(u)m(c, \boldsymbol{u}). \tag{12.23}$$

This is evidently a four-vector since $\mathbf{U}$ is a four-vector and m is a four-scalar. If we should seek to generalise the law of conservation of momentum so that it holds in all inertial frames then we need look no further than Eq. (12.23) and invoke the law of conservation of four-momentum. The new conservation law states that for an isolated system of particles the quantity $\sum_i \mathbf{P}_i$ is fixed where the summation is over all particles in the system. Given Eq. (12.23) it follows that

$$\sum_i \gamma(u_i)m_i c \tag{12.24}$$

and

$$\sum_i \gamma(u_i)m_i \boldsymbol{u} \tag{12.25}$$

are seperately conserved. These are none other than the new laws for energy and momentum conservation which we worked so hard to determine in Part II (see Eq. (7.15) and Eq. (7.25)). Defining the energy and momentum of a particle of mass m moving with velocity $\boldsymbol{u}$ to be

$$E = \gamma(u)mc^2 \text{ and} \tag{12.26}$$

$$\boldsymbol{p} = \gamma(u)m\boldsymbol{u} \tag{12.27}$$

it follows that the momentum four-vector can be written

$$\mathbf{P} = (E/c, \boldsymbol{p}). \tag{12.28}$$

Now since $\mathbf{P}$ is a four-vector the quantity $\mathbf{P} \cdot \mathbf{P}$ is a four-scalar. As such it can be evaluated in any inertial frame. If we evaluate it in the inertial frame in which the particle is a rest we find

$$\mathbf{P} \cdot \mathbf{P} = m^2 c^2 \tag{12.29}$$

and thus in a general inertial frame

$$E^2/c^2 - p^2 = m^2c^2, \qquad (12.30)$$

which is none other than Eq. (7.36). Also, since **P** is a four-vector we now understand why it transforms just like the position four-vector under Lorentz transformations, as we noted in Eq. (11.30).

12.3.1 Further examples in relativistic kinematics

The conservation of four-momentum is of great utility in studying relativistic particle collisions, as is illustrated in the following examples.

Example 12.3.1 *In Section 7.2.3 we explored the Compton scattering process $\gamma + e^- \rightarrow \gamma + e^-$. Re-derive Eq. (7.44) making use of the four-vector formalism.*

Solution 12.3.1 *It is a good idea in problems like this to first write down the four-momenta of the various particles before and after the collision, i.e.*

$$\mathbf{P}_\gamma = \frac{1}{c}(E, E, 0, 0),$$

$$\mathbf{P}_e = (m_e c, 0, 0, 0),$$

$$\mathbf{P}'_\gamma = \frac{1}{c}(E', E'\cos\theta, E'\sin\theta, 0).$$

We would rather not write down the explicit representation for the four-momentum of the scattered electron since we are aiming to express the scattered photon energy purely in terms of the incoming photon energy and the photon scattering angle θ. Four-momentum conservation informs us that

$$\mathbf{P}_\gamma + \mathbf{P}_e = \mathbf{P}'_\gamma + \mathbf{P}'_e.$$

It is now clear how we should proceed if we would like to find a relationship between E, E' and θ: we should exploit the fact that $\mathbf{P}'_e \cdot \mathbf{P}'_e$ is Lorentz invariant and equal to $m_e^2 c^2$. Thus we write

$$\mathbf{P}_\gamma + \mathbf{P}_e - \mathbf{P}'_\gamma = \mathbf{P}'_e$$

and then 'square' each side (i.e. take the scalar product of each side with itself):

$$(\mathbf{P}_\gamma + \mathbf{P}_e - \mathbf{P}'_\gamma)^2 = m_e^2 c^2.$$

The problem is essentially solved now and all that remains is for us to expand out the left hand side. We could combine the three four-vectors into one big four-vector and square that or we could stay in four-vector formalism for as long as possible. The latter has the advantage that we can make direct use of such results as

$\mathbf{P}_e \cdot \mathbf{P}_e = m_e^2 c^2$ and $\mathbf{P}_\gamma \cdot \mathbf{P}_\gamma = 0$, *i.e.*

$$\mathbf{P}_\gamma^2 + \mathbf{P}_e^2 + \mathbf{P}_\gamma'^2 + 2\mathbf{P}_\gamma \cdot \mathbf{P}_e - 2\mathbf{P}_\gamma' \cdot (\mathbf{P}_\gamma + \mathbf{P}_e)$$
$$= m_e^2 c^2.$$

Now is a good time to expand out the four-vectors:

$$m_e^2 c^2 + 2E m_e - 2E' m_e - 2E' E(1 - \cos\theta)/c^2 = m_e^2 c^2,$$

which can be rearranged to give

$$\frac{1}{E'} = \frac{1}{E} + \frac{1 - \cos\theta}{m_e c^2}$$

and this is Eq. (7.44). Clearly the four-vector formalism allows us to see our way to the answer in a much more elegant manner. Note also that the elegance would be lost if we had rushed into component form at too early a stage. As in the manipulations of ordinary vectors in three dimensions it is usually wise to try and stay in vector notation for as long as possible.

Example 12.3.2 *Consider the particle physics process* $\gamma + p \rightarrow \pi^0 + p$ *in which a photon* (γ) *collides with a proton* (p) *to produce a neutral pion* (π^0) *and a proton. If the initial proton is at rest what is the minimum energy (called the 'threshold energy') that the photon must have in order for the process to occur? [The proton has mass 938 MeV/c^2 and the pion has mass 135 MeV/c^2.]*

Solution 12.3.2 *As usual it is a good idea to start with a sketch illustrating the process, like that shown in Figure 12.2. Since the sum of the rest masses in the final state is greater than that in the initial state it is clear that the photon must deliver*

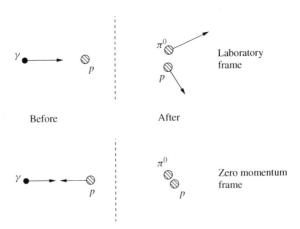

Figure 12.2   The reaction $\gamma + p \rightarrow \pi^0 + p$ viewed in two different inertial frames.

some kinetic energy which can be traded off for the additional mass. However it would be a grave error to suppose that the photon needs a kinetic energy which exactly compensates the mass difference between the initial and final state. This is wrong because momentum conservation must hold. Therefore, since the initial state has non-zero momentum so too must the final state. As a result, the final state particles must always be produced in motion and therefore they will carry some kinetic energy. How then shall we proceed?

We can neatly circumvent the problem of momentum conservation by thinking not in the laboratory frame but in an inertial frame where the total momentum of the system is zero, i.e. the incoming photon and proton have equal and opposite momentum as illustrated in the lower pane in Figure 12.2. In this 'zero momentum frame' it is possible to produce the final state pion and proton at rest and clearly this is the configuration which corresponds to the smallest possible photon energy since none of its energy is wasted on giving the final state particles some motion. The challenge is to relate quantities in the zero momentum frame to those in the laboratory frame. Jumping between frames can be a time consuming affair but not if we restrict our attention to Lorentz invariant quantities. The relevant invariant in this problem is the four-scalar associated with the total four-momentum of the system, i.e.

$$M^2 c^2 = (\mathbf{P}_\gamma + \mathbf{P}_p)^2,$$
$$= (\mathbf{P}_\pi + \mathbf{P}'_p)^2.$$

The quantity M is called the 'invariant mass' of the system and the second line follows from the first by the conservation of four-momentum, i.e.

$$(\mathbf{P}_\gamma + \mathbf{P}_p)^2 = (\mathbf{P}_\pi + \mathbf{P}'_p)^2. \tag{12.31}$$

The left hand side of this equation involves the photon energy, which we seek to find, i.e. using

$$\mathbf{P}_\gamma = \frac{1}{c}(E, E, 0, 0) \ and$$
$$\mathbf{P}_p = (m_p c, 0, 0, 0)$$

the left hand side is simply equal to

$$(\mathbf{P}_\gamma + \mathbf{P}_p)^2 = \mathbf{P}_\gamma^2 + \mathbf{P}_p^2 + 2\mathbf{P}_\gamma \cdot \mathbf{P}_p$$
$$= m_p^2 c^2 + 2E m_p. \tag{12.32}$$

We are permitted to calculate the right hand side of Eq. (12.31) in any convenient frame of reference since it is a four-scalar. At threshold it makes sense to compute it in the zero momentum frame since in this frame we know that

$$\mathbf{P}_\pi = (m_\pi c, 0, 0, 0) \ and$$
$$\mathbf{P}'_p = (m_p c, 0, 0, 0).$$

The right hand side of Eq. (12.31) can thus be evaluated to

$$(\mathbf{P}_\pi + \mathbf{P}'_p)^2 = \mathbf{P}_\pi^2 + \mathbf{P}'^2_p + 2\mathbf{P}_\pi \cdot \mathbf{P}'_p$$
$$= m_\pi^2 c^2 + m_p^2 c^2 + 2m_\pi m_p c^2. \qquad (12.33)$$

Equating Eq. (12.32) and Eq. (12.33) gives

$$m_p^2 c^2 + 2Em_p = m_\pi^2 c^2 + m_p^2 c^2 + 2m_\pi m_p c^2,$$

which can be re-arranged in order to determine the threshold energy for the incoming photon:

$$E = \frac{m_\pi^2 c^2 + 2m_\pi m_p c^2}{2m_p}$$
$$= \frac{135^2 + 2 \times 135 \times 938}{2 \times 938} MeV$$
$$= 145 \ MeV.$$

12.4   ELECTRIC AND MAGNETIC FIELDS

As a final example we turn our attention to the subject of electromagnetism wherein lies perhaps the most important application of relativity theory in everyday life.

We start with a puzzle. Consider a wire carrying a current along its length. At some instant in time a positively charged particle travels parallel to the wire and in the direction of the current. Viewed from a frame in which the wire is at rest, the charged particle is subsequently drawn towards the wire by the Lorentz force which arises as a result of the magnetic field around the wire. Now let us consider the same circumstance from the viewpoint of a frame of reference in which the charged particle is at rest. In this frame, the Lorentz force is zero since the particle's velocity is zero[2]. It therefore seems that the charged particle will remain at rest and we have a contradiction.

The resolution to this apparent paradox lies in Einstein's theory of relativity. In the rest frame of the charged particle, the electrons which carry the current in the wire are closer together as a result of Lorentz contraction and hence their charge density is greater than if they were at rest by a factor of $\gamma(u)$ where u is the speed of the electrons in the rest frame of the charged particle. The ionic lattice against which the electrons move is also Lorentz contracted but by a lesser amount (since the ions are at rest relative to the wire). Consequently, there is not a perfect cancellation of the electric field due to the ions with that due to the electrons and the positively charged particle is compelled to accelerate. Since, from the viewpoint of the charged particle, the electron density is greater than the ionic charge density, the postively charged particle is drawn towards the wire. We thus see that what is a magnetic field in one frame of reference is an electric field in another frame.

---

[2] The Lorentz force is given by $\mathbf{F} = q\mathbf{v} \times \mathbf{B}$.

This really is quite remarkable: even the most basic of phenomena in the study of electricity and magnetism requires relativity theory for a consistent interpretation. It is all the more remarkable given that the effect is sensitive to the drift speed of the electrons in a wire, which is no more than a few millimetres per second. Strictly speaking this ought not to have come as too great a surprise since we already stated that Einstein was impressed by the fact that Maxwell's equations of electromagnetism were inconsistent with Galilean relativity and that he built his theory so as to respect Maxwell's theory. Nevertheless, this is our first concrete illustration of the fact.

That one can view the occurrence of magnetic phenomena as a purely relativistic effect is further illustrated by the following example. This time let us consider a current I which flows as a result of the linear motion of an ensemble of charged particles. Using Ampère's Law we can deduce the magnetic field which arises at a distance r from the wire:

$$B = \frac{\mu_0 I}{2\pi r}.$$

In a real wire the charged particles are electrons and they are accompanied by positively charged ions such that the wire as a whole is electrically neutral but now we shall consider a current of free charges. In which case there is also an electric field at a distance r from the wire that is equal to

$$E = \frac{\rho}{2\pi r \varepsilon_0} = \frac{I}{2\pi \varepsilon_0 v r},$$

where ρ is the net charge per unit length of the charged particles and we have used $I = v\rho$ to rewrite this in terms of the current. Now we notice that the ratio of electric and magnetic fields is given by

$$\frac{cB}{E} = \frac{v}{c}$$

and we have used the fact that $\varepsilon_0 \mu_0 = 1/c^2$. Viewed this way, it is clear that the magnetic field is a small relativistic correction to the electric field. Even so, it is an effect which has huge technological and commercial relevance.

PROBLEMS 12

12.1 A rocket of initial mass m_i starts from rest and propels itself forwards by emitting photons backwards. The final mass of the rocket, after its engine has finished firing, is m_f. By considering the four-momenta of the rocket before and after it emitted the photons, and the net four-momentum of the photons, show that the final speed of the rocket, u, must satisfy

$$\frac{m_i}{m_f} = \gamma(u)\left(1 + \frac{u}{c}\right).$$

Hence deduce the final speed of the rocket.

12.2 An electron and a positron can collide and produce a proton and an antiproton, i.e. $e^- + e^+ \rightarrow p + \bar{p}$. Find the minimum kinetic energy of the positron in (a) a frame of reference in which the total momentum of the particles is zero; (b) a frame of reference in which the positron collides with a stationary electron.

[The masses of the electron and the positron are identical, and equal to $0.51\,\text{MeV}/c^2$. The masses of the proton and the antiproton are also identical, and equal to $938.3\,\text{MeV}/c^2$.]

12.3 Prove that the minimum invariant mass of an arbitrary system of particles is greater than or equal to the sum of the masses of the individual particles.

12.4 A photon with energy above 1.02 MeV has an energy greater than the rest energy of an electron-positron pair. Nevertheless the process

$$\gamma \rightarrow e^- + e^+$$

cannot occur in the absence of other matter or radiation. Why not?

12.5 This question is about the so-called transverse Doppler effect. Consider a frame S in which a transverse wave

$$y(x, t) = \sin(kx - \omega t)$$

propagates. As seen by an observer at rest in S, this wave is travelling along the $+x$ direction with wavelength $2\pi/k$ and angular frequency ω. The speed of propagation is $u = \omega/k$.

Now consider a frame S' which is moving at speed V in the $+y$ direction. The origins of S and S' coincide at $t = t' = 0$. Show that an observer in S' sees the following transverse wave:

$$y' = -Vt' + \frac{1}{\gamma}\sin(\mathbf{k}' \cdot \mathbf{x}' - \omega't').$$

Deduce $\mathbf{k}'$ and ω', and hence show that $\mathbf{K} = (\omega/c, \mathbf{k})$ transforms as a four-vector.

What is the speed of propagation in S'? Show that it reduces to the correct values in the limits $u \rightarrow c$ and $u \ll c$.

12.6 A pion of momentum 32 MeV/c decays into a muon and a neutrino. Using the conservation of four-momentum, and the fact that the neutrino is (to a good approximation) massless, show that

$$\mathbf{P}_\pi \cdot \mathbf{P}_\mu = \frac{(m_\pi^2 + m_\mu^2)c^2}{2},$$

where $\mathbf{P}_\pi$ and $\mathbf{P}_\mu$ are the momentum four-vectors of the pion and muon. If the outgoing muon travels at 90° relative to the direction of the incoming pion, use the above expression to determine the kinetic energy of the muon. At what angle does the neutrino travel relative to the incoming pion? [The mass of the pion is $140\,\text{MeV}/c^2$ and that of the muon is $106\,\text{MeV}/c^2$.]

12.7 A photon of frequency 2.0×10^{15} Hz travels at an angle of $20°$ relative to the positive x-axis. What angle does the photon make with the positive x-axis according to an observer who travels along the positive x-axis at a speed of $0.87c$? What is the frequency of the photon according to this observer?

12.8 In the laboratory frame, a ϕ particle (mass $1020\,\text{MeV}/c^2$) travels in the direction of the unit vector $\hat{\mathbf{n}}$ with a momentum of $3000\,\text{MeV}/c$. After a time it decays into two kaons, each of mass $494\,\text{MeV}/c^2$ and each also travelling in the $\hat{\mathbf{n}}$ direction.

 (a) What are the energy and momentum of each kaon in the rest frame of the ϕ particle?
 (b) The laboratory frame is moving with velocity $\mathbf{v}$ relative to the rest frame of the ϕ particle. In which direction is $\mathbf{v}$ and what is its magnitude?
 (c) Perform a Lorentz transformation on the kaon momenta obtained in part (a) in order to deduce the momentum of each kaon in the laboratory frame.

12.9 A π^+ meson (mass $140\,\text{MeV}/c^2$) collides with a neutron (mass $940\,\text{MeV}/c^2$) to produce a K^+ meson (mass $494\,\text{MeV}/c^2$) and a Λ hyperon (mass $1115\,\text{MeV}/c^2$).
 What is the minimum energy of the π^+ meson for the reaction to proceed in the frame in which the neutron is at rest?

12.10 A K meson (mass $498\,\text{MeV}/c^2$) is travelling through the laboratory when it decays into two π mesons (each of mass $140\,\text{MeV}/c^2$). One of the π mesons is produced at rest. What is the energy of the other?

12.11 In its rest frame, a π^0 meson decays isotropically into two photons. If one such meson is moving in the laboratory frame with a speed u show that the probability of a photon being emitted into the solid angle $d\Omega$ is given by

$$\frac{dP}{d\Omega} = \frac{1}{4\pi}\,\frac{1 - \left(\dfrac{u}{c}\right)^2}{\left(1 - \dfrac{u}{c}\cos\theta\right)^2},$$

where θ is the photon angle relative to the meson's direction of travel as measured in the laboratory frame.

13

Space-time Diagrams and Causality

Right back in Section 6.1.3 we stated that it is sometimes possible for two observers to disagree on the time ordering of a pair of events. As promised then, we shall now take a closer look at this extraordinary statement. Our journey will eventually lead us, at the end of this chapter, to a new way to think about Einstein's theory which places the emphasis much more on space-time and the notions of past, present and future than on the constancy of the speed of light.

We shall find it very helpful to draw 'space-time diagrams' and Figure 13.1 illustrates a particularly simple space-time diagram: an event is represented by a point in the $x - t$ plane. Of course actual events in space-time are represented by points in a four-dimensional space but the diagrams are easier to draw if we imagine there is only one spatial dimension. A slightly more interesting space-time diagram is illustrated in the left pane of Figure 13.2 which shows the history of a light-front which originates at the origin at time $t = 0$. As time progresses the light spreads out such that at some time t the light-front is located at $x = \pm ct$. This particular example is perhaps better visualised by going to two spatial dimensions, in which case the light spreads out such that at some time t the light-front is a circle of radius ct. In this case the history of the light-front is a cone[1], as illustrated in the right pane of Figure 13.2.

Our third example of a space-time diagram is shown in Figure 13.3. This diagram shows the curve in space-time which corresponds to the entire lifetime of a hypothetical person. The person was born at A and will die at B, and when they are at O their future lies in the upper half-plane whilst their past lies in the lower half plane. Such a curve through space-time corresponding to the history of some object

---

[1] The equation of the cone is $x^2 + y^2 = c^2t^2$.

Dynamics and Relativity   Jeffrey R. Forshaw and A. Gavin Smith
© 2009 John Wiley & Sons, Ltd

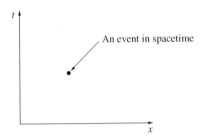

Figure 13.1   Space-time diagram illustrating the location of a single event.

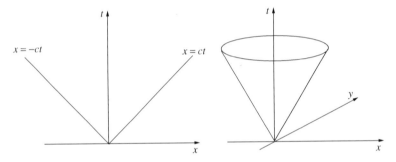

Figure 13.2   Space-time diagram illustrating the history of a light-front emanating from the origin in one spatial dimension (left) and two spatial dimensions (right).

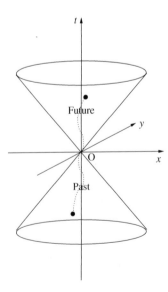

Figure 13.3   Space-time diagram illustrating the world line of a hypothetical person.

or other has a special name: it is called a 'world line'. Also shown in Figure 13.3 are the light cones $x^2 + y^2 = c^2t^2$. When our hypothetical person is at O, their entire future must lie inside the light cone with $t > 0$, since to escape from that region would require the person to travel faster than the speed of light. If this is not immediately clear then it might help to note that since the speed must always be smaller than c it follows that the gradient of the wordline must always be steeper than the slope of the light cone. This region of the space-time diagram, marked 'future' in Figure 13.3, is called the future light cone of the person at O. Similarly, their entire past must lie inside the cone with $t < 0$ (their past light cone) otherwise the person would have travelled faster than light speed at some time in their past.

We are now ready to explain what is meant by causality. Let's start with a definition. Two events A and B are said to be causally connected if event B lies in either the future or past light cone of event A. Stated slightly more succinctly, events A and B must lie within each-other's light cones. Conversely, if two events lie outside of each-other's light cones then they are said to be causally disconnected. Figure 13.4 illustrates what is going on with an example. Three events, A, B and C, are represented on a space-time diagram. B lies in the future light cone of A and is therefore causally connected to A. Another way of saying this is to say that A lies in the past light cone of B, which is illustrated by the dotted lines in Figure 13.4. C is just outside of A's future light cone and is therefore causally disconnected from A. However, C does lie within the past light cone of B and so events B and C are causally connected. All of this may sound rather academic but it is far from that. Let us see why. In principle event A can influence event B but it cannot influence event C. Indeed, event A might be the birth of our hypothetical person whilst event B could be their graduation from university. Event C however knows nothing of the person's birth since it occurs too far away from A for even a pulse of light to travel in the time available. The event at C is however causally connected to B. For example it might correspond to the launch of an extra terrestial spaceship from some distant part of the Universe such that the spaceship finally arrives on Earth just in time for graduation. If Einstein's Special Theory of Relativity is to satisfy the demands of causality then we must insist that the time order of causally connected events is agreed upon by all inertial observers, i.e. everyone agrees that birth precedes death. It is very hard to imagine a universe without causality, never mind

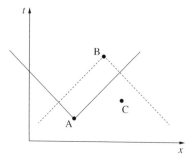

Figure 13.4   Three events A, B and C along with the future light cone of A and the past light cone of B.

to write down the laws of physics in such a universe. Notice however that there is no *a priori* reason for everyone to agree upon the time order of causally disconnected events since such events are by definition unable to influence each other.

Having introduced space-time diagrams, light cones and the definition of causality we are now ready to show that, in Einstein's theory, only causally disconnected events can have their time ordering changed.

13.1   RELATIVITY PRESERVES CAUSALITY

We are going to present two different explanations for why Special Relativity respects causality. The first of our explanations is based upon the methods utilised in Part II whilst the second makes use of space-time diagrams and Lorentz invariance.

Starting from the Lorentz transformations, we can write the time interval between any two events in S' ($\Delta t'$) in terms of the corresponding time interval in S (Δt), i.e. using Eq. (6.28b) we get

$$\Delta t' = \gamma(\Delta t - v\Delta x/c^2).   \tag{13.1}$$

Immediately we can see that the sign of $\Delta t'$ need not be the same as that for Δt. This means that the time ordering of the two events could be different in the two inertial frames. However, Eq. (13.1) also tells us that the time intervals can only be of opposite sign if, for $\Delta t > 0$,

$$\frac{v\Delta x}{c^2} > \Delta t,$$

$$\text{i.e. } \Delta x > \frac{c^2\Delta t}{v}.   \tag{13.2}$$

If $v > 0$ and since $|v| < c$ it follows therefore that the ordering of events can be switched only if

$$\Delta x > c\Delta t.   \tag{13.3}$$

If $v < 0$ then you should be able to confirm that this inequality changes to $\Delta x < -c\Delta t$. In words, we have found that the spatial separation between the two events is too great for even light to travel between the events in the time available (Δt). This is just one way of saying that the events are causally disconnected. In the language of light-cones, we have proved that it is only possible to reverse the time ordering of a pair of events if the events do not lie within their respective light-cones.

Let us now think more in terms of space-time. We start by considering three events, one located at O in S, one at a point A in S and the third event at a point B in S. Figure 13.5 illustrates the respective positions of the three events. Notice that we have chosen events A and B to lie inside and outside of the future and past light cones of the event at O. We should like to know what the co-ordinates of these three events are when viewed in some other inertial frame S'. Of course the Lorentz transformations will give us the answer, but we can make interesting

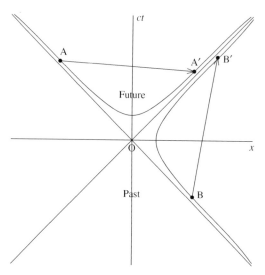

Figure 13.5   Space-time diagram illustrating lines of constant proper time. Two events are labelled by the points A and B in a particular inertial frame. In a different inertial frame the same two events have the co-ordinates labelled by the points A′ and B′.

progress without them. We know that the proper time interval[2] between any pair of events is independent of inertial frame, i.e.

$$(\Delta \tau)^2 = (\Delta t)^2 - \left(\frac{1}{c}\Delta x\right)^2 = (\Delta t')^2 - \left(\frac{1}{c}\Delta x'\right)^2. \tag{13.4}$$

Let us now imagine we are measuring the events in S'. Graphically, we'll still represent the events using Figure 13.5 but we should re-interpret the x axis as the x' axis and the t axis as the t' axis. The event at O stays at the origin, since as always we're assuming that S and S' have their origins coincident at $t = t' = 0$. But what happens to the events located at A and B in S? Figure 13.5 contains the answer. Two curves are drawn on the figure, they are such that all of the points on a given curve are the same space-time distance away from the origin, i.e. they are at a fixed value of $(\Delta \tau)^2$. In fact such curves are necessarily hyperbolae since $(\Delta \tau)^2 = (\Delta t)^2 - (\frac{1}{c}\Delta x)^2$ is none other than the equation of a hyperbola. Now it follows that in moving from S to S' the event at A can only move to another point on the hyperbola passing through A. It might for example move from A to A′. Similarly the event at B must remain on the hyperbola passing though B and in the figure we have shown it moving to the point B′. We can now immediately see why Einstein's theory does not violate causality. Events which lie in the future (or past) light cone of O can only ever be transformed to another point which is also in the future (or past) light cone of O under a Lorentz transformation. There is simply no possibility for an event which lies in the future of O in one inertial frame to lie in

---

[2] Or equivalently the distance in space-time.

the past of O in some other frame. The picture is however strikingly different for events which lie outside of O's light cones, such as the event at B. The hyperbolae of constant proper time cross the $t = 0$ axis and hence it is perfectly possible for events which lie in O's past in one frame (such as the event at B) to appear in O's future in another frame (the event is at B' in this frame). In two or three spatial dimensions none of the conclusions we have just drawn change but the hyperbolic curves of constant proper time become hyperbolic surfaces of constant proper time.

Before we finish this section we shall introduce some common terminology. Notice that all events that lie in either the future or the past light cone of the event at O necessarily satisfy $(\Delta\tau)^2 > 0$ whereas all events that lie outside of these light cones always satisfy $(\Delta\tau)^2 < 0$. If $(\Delta\tau)^2 > 0$ for a pair of events we say that the events are separated by a 'timelike' interval and if $(\Delta\tau)^2 < 0$ we say they are separated by a 'spacelike' interval. In the special case that $(\Delta\tau)^2 = 0$ the interval is said to be 'lightlike'. Events that are timelike separated (such as the events at O and at A) are always causally connected whereas events that are spacelike seperated (such as the events at O and at B) are always causally disconnected. Only something travelling at the speed of light can be present at both events if they are separated by a lightlike interval.

13.2 AN ALTERNATIVE APPROACH

We have established that Einstein's theory of Special Relativity can be understood in terms of an underlying four dimensional space-time continuum and that physical laws are built out of objects such as vectors and scalars in this four dimensional space. Equations built in such a way will automatically satisfy Einstein's postulates, just as equations built out of three dimensional vectors and scalars are automatically independent of any particular choice of co-ordinate system. Almost as if by magic, we found that Special Relativity is also a causal theory although the idea of causality was never mentioned when we originally formulated the theory in Part II. In this section we would like to promote causality to a much more central concept within the theory and at the same time we shall develop a new way of viewing the special role played by the speed of light.

Our main aim is to reformulate the theory of Special Relativity. Rather than start with Einstein's two postulates we shall start by boldly assuming that space and time form a four dimensional continuum which we shall, of course, refer to as space-time. What properties of space-time shall we assume? In the first case we shall insist that it supports the notion of distance between two points. More specifically, if we consider two neighbouring points in space-time located at co-ordinates (ct, x, y, z) and $(ct + cdt, x + dx, y + dy, z + dz)$ then the squared distance between these two points $(ds)^2$ is defined by

$$(ds)^2 = (cdt, dx, dy, dz) \begin{pmatrix} g_{00} & 0 & 0 & 0 \\ 0 & -1 & 0 & 0 \\ 0 & 0 & -1 & 0 \\ 0 & 0 & 0 & -1 \end{pmatrix} \begin{pmatrix} cdt \\ dx \\ dy \\ dz \end{pmatrix}$$

$$= g_{00}(cdt)^2 - (dx)^2 - (dy)^2 - (dz)^2. \tag{13.5}$$

The speed c has entered into the way we define the co-ordinates in space-time but at this stage we stress that it is simply an entirely arbitrary constant speed introduced purely to make g_{00} dimensionless, i.e. it is needed because we choose to measure one of the space-time co-ordinates in different units to the others. At this stage in our considerations, g_{00} is to be viewed simply as a dimensionless constant characteristic of the space-time.

What have we assumed? Certainly the matrix we have written down looks very special with all of its entries along the diagonal and it is true that the most general distance measure for what is called in mathematics a Riemannian space[3] would allow much more general 4×4 matrices. Clearly the choice of matrix is very intimately connected with the geometry of the space for it tells us how to compute the distance between neighbouring points. Since this matrix is so important, it has a name: it is called the 'metric' of the space. Perhaps before we answer the question posed at the start of this paragraph we should get better acquainted with the idea of a metric.

Example 13.2.1 *What is the metric that determines distances on the surface of a sphere of radius R? You should work in spherical polar co-ordinates.*

Solution 13.2.1 *The surface of a sphere is a two dimensional space. As illustrated in Figure 13.6, neighbouring points A and B are separated by a distance ds which satisfies*

$$(ds)^2 = (R\,d\theta)^2 + (R\sin\theta\,d\varphi)^2. \tag{13.6}$$

This result is valid in the limit of vanishing distance since in that limit the relevant portion of the sphere looks flat (i.e. Euclidean) and we can use Pythagoras' Theorem.

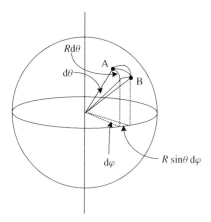

Figure 13.6   The distance between two points on the surface of a sphere of radius R.

---

[3] A space is Riemannian if the squared distance between two neighbouring points is of the form $(ds)^2 = g_{ij}\,dx_i\,dx_j$.

It is now easy enough to read off the metric for this space:

$$g = \begin{pmatrix} R^2 & 0 \\ 0 & R^2 \sin^2 \theta \end{pmatrix} \tag{13.7}$$

in the (θ, φ) basis. As an aside, notice that we compute distances on the surface of a sphere by integrating the distance measure, e.g. the circumference can be obtained by integrating ds along the curve φ =constant, $0 < \theta < 2\pi$:

$$circumference = \int ds = \int\limits_0^{2\pi} R \, d\theta \sqrt{1 + \sin^2 \theta \left(\frac{d\varphi}{d\theta}\right)} \tag{13.8}$$

$$= \int\limits_0^{2\pi} R d\theta = 2\pi R. \tag{13.9}$$

The metric of ordinary three dimensional Euclidean space when expressed in Cartesian co-ordinates is just the unit matix, whilst in spherical polar co-ordinates it is given by

$$g = \begin{pmatrix} 1 & 0 & 0 \\ 0 & R^2 & 0 \\ 0 & 0 & R^2 \sin^2 \theta \end{pmatrix} \tag{13.10}$$

in the (r, θ, φ) basis. The metric is an inherent geometrical feature of the space and it is therefore a tensor[4], so although for any given space the matrix representation of the metric depends upon the chosen co-ordinate basis the metric itself remains unchanged. Notice that whilst the metric of Euclidean space written in Eq. (13.10) describes a flat space (i.e. a space in which Pythagoras' Theorem always works) the same cannot be said of the metric written in Eq. (13.7) which describes a curved space, e.g. right-angled triangles drawn on the surface of a sphere do not satisfy Pythagoras' Theorem. Equivalently, it is not possible to identify a two-dimensional co-ordinate basis in which the metric of Eq. (13.7) is represented by the unit matrix.

We're now ready to return to space-time and the metric of Eq. (13.5). If we assume that the metric is constant, i.e. that space-time has the same geometry at all points, then its diagonal form follows rather generally since any non-singular matrix can be diagonalised by an appropriate change of basis. The presence of the diagonal entries equal to −1 arises if we insist that the space should be Euclidean if we take slices through it of constant time. That we chose −1 rather than +1 (or any other number) is a matter of convention. For example if in Euclidean space we chose a metric equal to minus the unit matrix then all distances would be multiplied by the square root of minus one, which is not very economical but otherwise wholly acceptable. In summary, we have established that Eq. (13.5) is in fact the most general metric which satisfies the constraint that slices of constant

---

[4] See Section 10.2 for a discussion of the moment of inertia tensor.

time be Euclidean. All that remains is to specify the top left hand entry, g_{00}. Since it is a dimensionless constant, we can always choose the constant c in Eq. (13.5) such that $g_{00} = \pm 1$ without altering the value of the distance ds. We have now almost completely specified the metric of space-time. All that remains is for us to settle on the sign of g_{00}. It is at this point that causality enters.

We state the result first and then prove it: the metric must have $g_{00} = +1$ if it is to be the metric of a space-time that satisfies the demands of causality. We already know that space-time with $g_{00} = +1$ is causal, because that space-time is the Minkowski space-time of Special Relativity and, following the discussion in the previous section, we know that to be a causal space-time. The core of the argument involved showing that lines of constant proper time are hyperbolae and that any hyperbola lying in either the future or the past light cone of some point O always remains inside that light cone[5]. In contrast a hyperbola that lies outside of either light cone will span times which lie both in the future and in the past of O. Since a shift from one inertial frame to another corresponds to sliding events around on their corresponding hyperbolae it follows that all observers always agree upon the time ordering of causally connected events. Notice also that this argument only works if all matter is constrained to move on timelike trajectories (which means they must always travel with speed c or less) otherwise a particle could start at O and follow a wordline outside of O's future light cone whereupon an observer in a second inertial frame could conclude that an event on the particle's trajectory which lies outside of O's future light cone could have occured in O's past; so causality is also acting to constrain the laws of dynamics as well as the structure of space-time. Our task is now to explain why $g_{00} = -1$ does not lead to a space-time that respects causality. The invariant distance between two neighbouring events in this space-time is given by

$$(\mathrm{d}s)^2 = -(c\,\mathrm{d}t)^2 - (\mathrm{d}x)^2 - (\mathrm{d}y)^2 - (\mathrm{d}z)^2 , \qquad (13.11)$$

which is just the metric of a four dimensional Euclidean space (recall the overall sign is unimportant). The locus of all points in this space-time that lie a fixed distance from the origin O is therefore, in one spatial dimension, simply a circle of radius $\sqrt{-(\Delta s)^2}$. In two spatial dimensions we have the surface of a sphere and in three spatial dimensions it is the three dimensional generalisation of a spherical surface, often called a 'three-sphere'. Now we know that the equations of physics must be the same for all co-ordinate systems that preserve the invariant distance between any two events. However, as illustrated in Figure 13.7, if in a frame S an event is located at A, which lies in the future of an event at O, then there always exists a second frame of reference S' in which that very same event occurs in O's past. The figure shows the location B in S' of the event located at A in S. The equivalent of Lorentz transformations in this Euclidean space are simple rotations, i.e.

$$\begin{pmatrix} ct' \\ x' \end{pmatrix} = \begin{pmatrix} \cos\theta & -\sin\theta \\ \sin\theta & \cos\theta \end{pmatrix} \begin{pmatrix} ct \\ x \end{pmatrix}. \qquad (13.12)$$

---

[5] Moreover, the hyperbolae never intersect each other.

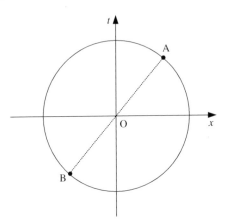

Figure 13.7   Transformations in Euclidean space-time.

In Figure 13.7, the two frames are clearly related to each by a rotation through $\theta = \pi$ radians. There is therefore no frame-independent notion of past, future or present in this space-time, which means that it does not support the idea of cause and effect in the laws of physics. We must therefore reject this space-time and are left with only one possibility: space-time must be Minkowski space-time with a metric

$$g = \begin{pmatrix} 1 & 0 & 0 & 0 \\ 0 & -1 & 0 & 0 \\ 0 & 0 & -1 & 0 \\ 0 & 0 & 0 & -1 \end{pmatrix}. \tag{13.13}$$

Under only a few rather natural assumptions we have arrived at the conclusion that Minkowski space-time is the only possible space-time. The constant c was originally introduced only to calibrate distances in the time direction, with the causal structure of the theory elevating it to the status of a limiting speed. Actually we should mention that it is still possible that c could be infinite and this would lead us to Galilean relativity. There is no purely theoretical argument to reject this possibility and it is experiment that informs us that c is in fact finite. Armed with the metric we can now go ahead and re-derive all of the familiar results we have encountered so far in Special Relativity. For example, the space-time we have introduced supports the existence of four-vectors. The displacement four-vector

$$\Delta \mathbf{X} = (c\Delta t, \Delta x, \Delta y, \Delta z) \tag{13.14}$$

is our prototypical four-vector and the metric tensor tells us how to form the scalar product, i.e.

$$(\Delta \mathbf{X})^T g (\Delta \mathbf{X}) = (c\Delta t)^2 - (\Delta x)^2 - (\Delta y)^2 - (\Delta z)^2. \tag{13.15}$$

We know from Section 11.2 that the Lorentz transformations preserve the scalar product defined in Eq. (13.15). Thus we recognise that the Minkowski space-time equivalent of Euclidean rotations involving the time dimension are the Lorentz transformations and that these correspond physically to a change of inertial frame. We are free to introduce other vectors in space-time. They may be useful in drawing up the laws of physics provided that they transform according to the Lorentz transformations and that the scalar product between any two four-vectors is determined by the metric, i.e. for four-vectors **A** and **B**

$$\mathbf{A} \cdot \mathbf{B} = (\mathbf{A})^T \mathbf{g}(\mathbf{B}) = A_0 B_0 - A_1 B_1 - A_2 B_2 - A_3 B_3. \qquad (13.16)$$

Defined this way, the scalar product is guaranteed to be the same in all frames and is thus a four-scalar. Armed with Minkowski space-time, four-vectors and four-scalars we can make progress in physics. In fact our logical development has brought us all the way to the start of Chapter 12. Still, since we have not made use of Einstein's postulate that the speed of light is a universal constant, c (which has by now also appeared in the Lorentz transformation equations) remains nothing other than the constant that calibrates space-time distances in the time direction. The causal structure of the theory dictates that c must be a limiting speed, i.e. we require that particles always follow timelike trajectories through space-time. Only after we have introduced the energy-momentum four-vector does the more familiar interpretation of c emerge: for there can exist particles for which $m = 0$ and $E = cp$ but only if such particles travel at a speed equal to c in all inertial frames. Thus massless particles may exist in a Minkowski space-time provided they always travel with speed c. Since light is made of massless photons, we may go ahead and refer to c as the speed of light. From the space-time view there is clearly nothing very special about light. Indeed, the four-speed (defined as $\sqrt{U \cdot U}$) of any particle (including those with mass) is, from Eq. (12.8), always equal to c, which means that everything travels through space-time with the same speed. In our three dimensional world, massless particles appear special since only they travel with the same speed in all inertial frames.

It is fair to say that one of the main goals of this section of the book is to present the reader with a new way of viewing Einstein's statement that the speed of light is the same in all inertial frames. In particular, we have traced the roots of this statement all the way back to space-time and causality. Incidentally, Einstein's first postulate, that the laws of physics are the same in all inertial frames follows automatically once we have specified that we should work in Minkowski space-time, for we know that moving between inertial frames is just a co-ordinate change in space-time and the laws of physics should be trivially independent of co-ordinates.

14

Acceleration and General Relativity

14.1 ACCELERATION IN SPECIAL RELATIVITY

There is nothing to stop us from describing accelerated motion in Special Relativity. Perhaps the most natural question to ask is: what are the components of an acceleration in S given the corresponding components in S' (where S and S' are the usual two inertial frames)? Starting from the velocity addition formula, Eq. (6.33), we have that

$$dv_x = dv_x' \left(\frac{1}{1 + uv_x'/c^2} - \frac{(u + v_x')u/c^2}{(1 + uv_x'/c^2)^2} \right). \tag{14.1}$$

In conjunction with $dt = \gamma(u)(1 + uv_x'/c^2)\, dt'$ this equation implies that

$$a_x = a_x' \left(\frac{1 - u^2/c^2}{(1 + uv_x'/c^2)^2} \right) \frac{1}{\gamma(u)} \frac{1}{1 + uv_x'/c^2},$$

$$\text{i.e. } a_x = \left(\frac{1}{\gamma(u)(1 + uv_x'/c^2)} \right)^3 a_x', \tag{14.2}$$

where $a_x = dv_x/dt$ and $a_x' = dv_x'/dt'$ are the accelerations in the x-direction in S and S'. Simlarly we can use Eq. (6.34) to establish that

$$a_y = \left(\frac{1}{\gamma(u)(1 + uv_x'/c^2)} \right)^2 \left(a_y' - a_x' \frac{uv_y'}{c^2 + uv_x'} \right). \tag{14.3}$$

Dynamics and Relativity Jeffrey R. Forshaw and A. Gavin Smith
© 2009 John Wiley & Sons, Ltd

These are not particularly elegant formulae. Not surprisingly, unlike Galilean relativity, the components of an acceleration are not the same in all inertial frames. In particular, a constant acceleration in one inertial frame is not a constant acceleration in a different inertial frame.

We can however define an acceleration four-vector as simply the rate of change of the velocity four-vector with respect to the proper time, i.e.

$$\mathbf{A} \equiv \frac{d\mathbf{V}}{d\tau}. \tag{14.4}$$

Starting from Eq. (12.7) we can write $\mathbf{A}$ in terms of its components in some inertial frame, i.e.

$$\mathbf{A} = \gamma(\dot{\gamma}c, \dot{\gamma}\mathbf{u} + \gamma\mathbf{a}), \tag{14.5}$$

where the dot indicates differentiation with respect to time as measured in the inertial frame and $\mathbf{a} = \dot{\mathbf{u}}$. The length of this four-vector will turn out to be of some use to us and the quickest way to figure it out is to compute it in the inertial frame in which $\mathbf{u} = \mathbf{0}$. It does not matter that this frame is useful only for an instant in time (after which the particle may have developed a non-zero velocity); an instant in time is long enough. In this instantaneous rest frame, $\mathbf{A} = (0, \mathbf{a})$ and hence

$$\mathbf{A} \cdot \mathbf{A} = -a^2. \tag{14.6}$$

Since $\mathbf{A}$ is a four-vector, this result is valid in all other frames. Note that a is the magnitude of the three-acceleration in the inertial frame in which the particle is instantaneously at rest: it is often called the 'proper acceleration' of the particle.

14.1.1 Twins paradox

As an example, let us consider the so-called twins paradox. Suppose that one twin accelerates away from the Earth at a constant proper acceleration equal to g, leaving the other twin behind. This rate of acceleration will lead the astronaut twin to feel their weight inside the spaceship. After 10 years they switch the rockets on their spaceship such that for the next 10 years they decelerate also at a rate g. At which time they again reverse the rockets and accelerate (again at g) back towards the Earth for 10 further years before finally reversing the rockets one last time for their arrival back on Earth some 10 years later. According to the twin who travelled in the spaceship they were absent for a total of 40 years. The question is, how much time has elapsed on Earth between the departure and return of the astronaut twin?

Let us consider the first 10 years of the journey. When the astronaut is travelling at speed u relative to an observer on Earth we imagine that they are instantaneously at rest in an inertial frame S' moving at speed u relative to Earth. We know that the acceleration is constant and that $v' = 0$ in S'. Hence, using Eq. (14.2) gives

$$g = \gamma(u)^3 a = \frac{d(\gamma(u)u)}{dt} \tag{14.7}$$

and we have dropped the subscripts since the motion is all taking place in one dimension. We want to compute the time elapsed in the Earth frame given that 10 years have elapsed on the spaceship. We can exploit the fact that the astronaut is instantaneously at rest in S' to give

$$dt' = dt \sqrt{1 - \frac{1}{c^2}\left[\left(\frac{dx}{dt}\right)^2 + \left(\frac{dy}{dt}\right)^2 + \left(\frac{dz}{dt}\right)^2\right]}$$

$$\text{i.e. } dt = \gamma(u)\, dt', \tag{14.8}$$

where t' is the time measured by an observer on the spaceship. Integrating this equation will deliver the result provided we know how u changes with time, which we do know from Eq. (14.7), i.e.

$$\gamma(u)u = gt \tag{14.9}$$

which, after squaring both sides and re-arranging, gives that

$$u = \frac{gt}{\sqrt{1 + g^2 t^2 / c^2}}. \tag{14.10}$$

Substituting for $\gamma(u)$ in Eq. (14.8) gives that

$$\int \frac{dt}{\sqrt{1 + g^2 t^2 / c^2}} = t' \tag{14.11}$$

which leads to

$$\sinh^{-1}\left(\frac{gt}{c}\right) = \frac{gt'}{c},$$

$$\text{i.e. } \frac{gt}{c} = \sinh\left(\frac{gt'}{c}\right). \tag{14.12}$$

Putting $t' = 10$ years and $g = 9.81$ ms^{-2} into this equation gives $t \simeq 14700$ years. Since the sign of g is unimportant in Eq. (14.11) it follows that the total time that the spaceship is away from the Earth is $4 \times 14700 \simeq 59000$ years (recall the astronaut twin has only aged 40 years).

Before leaving the twins paradox we should point out that the fact that the twin who travels (and hence undergoes an acceleration) always ages more slowly regardless of the details of their journey. To see this we only need note that the time elapsed according to the travelling twin ($\Delta t'$) is obtained by integrating Eq. (14.8):

$$\Delta t' = \int_{t_1}^{t_2} dt \sqrt{1 - \frac{u(t)^2}{c^2}} < t_2 - t_1 \tag{14.13}$$

and since $u(t)^2$ is always positive, the integrand is always less than unity and the time registered on the clock of the twin who underwent an acceleration is less than the time interval registered on the clock of the twin who did not. Indeed we see that inertial observers present at any pair of events necessarily experience the maximum possible time interval between those events.

Example 14.1.1 *Show that for small enough speeds the accelerating spaceship in our discussion of the twins paradox follows the trajectory $x \approx \frac{1}{2}gt^2$.*

Solution 14.1.1 *Eq. (14.10) tells us that, for $gt/c \ll 1$*

$$\frac{\mathrm{d}x}{\mathrm{d}t} \approx gt \qquad\qquad (14.14)$$

and hence $x \approx \frac{1}{2}gt^2$ for $x = 0$ at $t = 0$.

14.1.2   Accelerating frames of reference

In the last subsection we considered an accelerating observer but from the point of view of an infinite set of inertial frames and we did not invoke the idea of an accelerating frame of reference. Sometimes we may wish to investigate a piece of physics directly in an accelerating frame of reference, just as we did in Chapter 8 when we discussed non-inertial frames in classical physics. Of course we expect that in so doing we should encounter non-inertial forces.

However, the construction of an accelerating reference frame is not as straightforward in Einstein's theory as it is in classical physics. The most natural way to think would be to suppose we erect a rigid system of rulers and clocks and use these to locate the position of events in the accelerating frame. However, as we shall shortly see, the rods will tend to be bent or buckled and it will not be possible to synchronize the clocks so that they always read the same time. The lack of synchronicity of the clocks need not be a problem; we could accept that time might tick at different rates throughout an accelerating frame. However the buckling of the rulers would make life difficult since we'd need to know all about the physical properties of the rulers in order to compare theoretical predictions for the relationships between events in the accelerating frame and the corresponding observations.

To illustrate these points we'll focus our attention on a very special accelerating frame of reference. Namely one in which the acceleration is time independent and the distance between points in the frame do not vary with time. Clearly this is an appealing frame of reference however, as we shall soon see, it is not a very practical frame. Our goal will be to figure out the equivalent of the Lorentz transformation formulae that relate the co-ordinates of an event in the accelerating frame, S', to the co-ordinates of the same event in our typical inertial frame S. To visualize the accelerating frame let us consider Figure 14.1, which shows a fleet of tiny rocket ships located at different points in S'. The little rockets define the locations of events, i.e. the rocket ship located closest to an event can be used to label the position of that event. The rockets are arranged so that they all accelerate in the

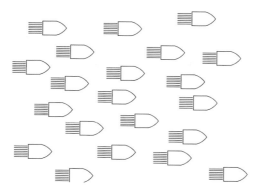

Figure 14.1   A fleet of tiny rocket ships defines a uniformly accelerating frame of reference. The formation of the ships remains the same for all time as measured by an observer on any one of the ships. However, this does not imply that observers on two different ships feel the same acceleration.

x'-direction in such a way that the distance between any two of them does not vary according to an observer on any other rocket in the fleet (i.e. any observer in S'). This is how we define our uniformly accelerating frame of reference.

We already know how the co-ordinates of the rocket that defines the origin in S' relate to the corresponding co-ordinates in the inertial frame since this might be the rocket occupied by the astronaut twin of the previous section:

$$t = \frac{c}{g} \sinh\left(\frac{gt'}{c}\right)$$

$$\text{and } x = \frac{c^2}{g} \cosh\left(\frac{gt'}{c}\right).$$
(14.15)

The second of these equations is obtained by integrating Eq. (14.10), i.e.

$$\int_0^x dx = \int_0^t \frac{gt\,dt}{\sqrt{1 + g^2 t^2/c^2}}.$$
(14.16)

Using Eq. (14.12) we can re-express the right-hand-side as an integral over t'. Thus

$$x = c \int_0^{t'} \frac{\sinh\left(gt'/c\right)\cosh\left(gt'/c\right)dt'}{\cosh\left(gt'/c\right)}$$

$$= \frac{c^2}{g} \cosh\left(\frac{gt'}{c}\right)$$
(14.17)

as claimed. We can say that the origin in S' is located at the space-time point $O = \frac{c^2}{g}\left(\sinh\left(gt'/c\right), \cosh\left(gt'/c\right), 0, 0\right)$ in the basis of an inertial observer in S.

What we'd really like is the location of a general rocket ship. For simplicity, let us confine our attention to a rocket lying at position x' on the x'-axis. This rocket is displaced from the origin in S' by a distance x'. But what are its co-ordinates in S? To obtain this we use a neat trick, we make use of the fact that the four-velocity of the rocket at the origin in S' points in the time direction of space-time according to an observer on the rocket. Hence a four-vector that is orthogonal to this must point along the x' axis and we can therefore locate the four-vector position of the rocket at x'. Let us follow this train of thinking. The four-velocity of the origin in S' is

$$U = c\left(\cosh\left(gt'_c/c\right), \sinh(gt'/c), 0, 0\right) \tag{14.18}$$

and hence the four-vector

$$D = \left(\sinh\left(gt'/c\right), \cosh\left(gt'/c\right), 0, 0\right) \tag{14.19}$$

is a unit four-vector pointing in the x' direction since $U \cdot D = 0$. Thus an event occurring on the little rocket at x', which occurs at a time t', is located at space-time position

$$(ct, x, 0, 0) = O + x'D \tag{14.20}$$

and we now have the general relationship we desire:

$$ct = \left(\frac{c^2}{g} + x'\right)\sinh\left(\frac{gt'}{c}\right)$$

$$\text{and } x = \left(\frac{c^2}{g} + x'\right)\cosh\left(\frac{gt'}{c}\right). \tag{14.21}$$

It is important to realise that the time t' appearing in these two equations is the time according to an observer at the origin in S'. It is called the co-ordinate time, for it is the time co-ordinate we choose to define the location of space-time events in S'. We are not entitled to claim that this is also equal to the time measured on a clock located on the little rocket at x' and indeed we shall soon see that it is not possible for the two to be equal at all times.

The set of co-ordinates defined by the fleet of rockets constitutes a uniformly accelerating frame of reference. We shall now demonstrate that only the little rocket at the origin accelerates at rate g in its own rest frame. All other rockets at $x' \neq 0$ accelerate at a different rate in their respective rest frames. Consider the little rocket located at a particular value of x'. We know that this point must have a four-velocity of magnitude equal to c (see Eq. (12.8)), i.e.

$$c^2 = V \cdot V \tag{14.22}$$

where

$$V = \frac{\mathrm{d}}{\mathrm{d}\tau}\left(\left(\frac{c^2}{g} + x'\right)\sinh\left(\frac{gt'}{c}\right), \left(\frac{c^2}{g} + x'\right)\cosh\left(\frac{gt'}{c}\right), 0, 0\right) \tag{14.23}$$

and τ is the proper time measured by a clock located at x'. Hence, keeping x' fixed so $dx'/d\tau = 0$, gives

$$\left(\frac{c^2}{g} + x'\right)^2 \frac{g^2}{c^2} \left(\frac{dt'}{d\tau}\right)^2 = c^2,$$

i.e. $\dfrac{dt'}{d\tau} = \dfrac{1}{(1 + gx'/c^2)}.$ \qquad (14.24)

This equation relates the co-ordinate time in S', which is the proper time recorded on a clock at the origin in S', to the proper time on a clock at any other x'. Since they are not equal we see that it is impossible to synchronize the clocks in S' for all time. The acceleration of the little rocket at x' can now be determined once we appreciate that the four-acceleration of a particle moving through Minkowski space satisfies Eq. (14.6), i.e.

$$\mathbf{A} \cdot \mathbf{A} = -\alpha^2, \qquad (14.25)$$

where α is the acceleration as determined in the rest frame of the particle. For the rocket at x', its four-acceleration is

$$\mathbf{A} = \left(\frac{dt'}{d\tau}\right) \left(\frac{g}{c}\right) \frac{d}{d\tau} \left(\left(\frac{c^2}{g} + x'\right) \cosh\left(\frac{gt'}{c}\right), \left(\frac{c^2}{g} + x'\right) \sinh\left(\frac{gt'}{c}\right), 0, 0\right)$$

$$(14.26)$$

and hence

$$-g(x')^2 = -\left(\frac{dt'}{d\tau}\right)^4 \left(\frac{g}{c}\right)^4 \left(\frac{c^2}{g} + x'\right)^2,$$

i.e. $g(x') = \left(\dfrac{dt'}{d\tau}\right)^2 g\left(1 + \dfrac{gx'}{c^2}\right).$ \qquad (14.27)

Using Eq. (14.24) then gives our final answer:

$$g(x') = \frac{g}{1 + gx'/c^2}. \qquad (14.28)$$

It is simply not possible to build S' out of a fleet of rockets such that they all accelerate at the same rate in their own rest frame and preserve the distance between each rocket. To do that, as we shall discuss in the next section, means going beyond Minkowski space-time.

We can determine the form of the metric for the accelerating frame. The fleet of rockets is moving through Minkowski space-time hence the interval between any two events satisfies (see Eq. (14.21))

$$(ds)^2 = (c \, dt)^2 - (dx)^2 - (dy)^2 - (dz)^2$$

$$= \left(\left(\frac{c^2}{g} + x' \right) \cosh \left(\frac{gt'}{c} \right) \frac{g}{c} dt' + \sinh \left(\frac{gt'}{c} \right) dx' \right)^2$$

$$- \left(\left(\frac{c^2}{g} + x' \right) \sinh \left(\frac{gt'}{c} \right) \frac{g}{c} dt' + \cosh \left(\frac{gt'}{c} \right) dx' \right)^2$$

$$- (dy')^2 - (dz')^2$$

$$= \left(1 + \frac{gx'}{c^2} \right)^2 (c \, dt')^2 - (dx')^2 - (dy')^2 - (dz')^2. \qquad (14.29)$$

Thus there is some warping of time in the accelerating frame but space is Euclidean (which is not surprising since we constructed it that way).

Let us now turn to another accelerating frame of reference. This time our goal will be to make contact with Chapter 8. Let us consider a non-inertial frame rotating with angular speed ω about the z-axis. It is convenient first to work in cylindrical polar co-ordinates. The space-time interval between two neighbouring events in an inertial frame (i.e. one for which $\omega = 0$) is

$$(ds)^2 = (c \, dt)^2 - (r d\phi)^2 - (dr)^2 - (dz)^2. \qquad (14.30)$$

A most important property of space-time physics is that this interval must be the same in any other system of co-ordinates, even an accelerating system. This is nothing more than the statement that the distance between any two points on a general manifold should be independent of the way we choose to parameterize the manifold. Hence we can choose to work in a non-inertial frame with $\phi' = \phi - \omega t$, $t' = t$, $r' = r$ and $z' = z$ such that

$$(ds)^2 = (cdt')^2 - r'^2(d\phi' + \omega dt')^2 - (dr')^2 - (dz')^2. \qquad (14.31)$$

From now on we will only be interested in the rotating co-ordinates and will subsequently drop the primes. Moreover, we shall find it more convenient to now switch back to Cartesian co-ordinates, i.e.

$$x = r \cos \phi,$$

$$y = r \sin \phi. \qquad (14.32)$$

In which case and after a little algebra it follows that

$$(ds)^2 = (c \, dt)^2 (1 - \omega^2(x^2 + y^2)/c^2) + 2(\omega/c)(y \, dx - x \, dy)(c \, dt) - (dx)^2$$
$$- (dy)^2 - (dz)^2. \qquad (14.33)$$

In the (ct, x, y, z) basis the metric for this system of co-ordinates is

$$
g = \begin{pmatrix}
1 - \omega^2(x^2 + y^2)/c^2 & \omega y/c & -\omega x/c & 0 \\
\omega y/c & -1 & 0 & 0 \\
-\omega x/c & 0 & -1 & 0 \\
0 & 0 & 0 & -1
\end{pmatrix}. \tag{14.34}
$$

Now the motion of a free particle through space-time is some unique trajectory, which cannot depend upon how we choose our co-ordinates, and that means even if we choose a non-inertial co-ordinate system. There must therefore be a genuinely co-ordinate independent way to write the equations of motion of this free particle. Here we quote the answer and defer the proof to Appendix A. The co-ordinate independent way to write the equation of motion of a particle not acted upon by any force is given by

$$
g_{ij} \frac{du_j}{d\tau} + \frac{1}{2} \left(\frac{\partial g_{ik}}{\partial x_l} + \frac{\partial g_{il}}{\partial x_k} - \frac{\partial g_{kl}}{\partial x_i} \right) u_k u_l = 0, \tag{14.35}
$$

where $u_i = dx_i/d\tau$ is the four-velocity of the particle. This is an equation that treats all frames (inertial and non-inertial) on an equal footing. Notice that for inertial co-ordinates all of the derivatives of the metric vanish and we are left with the expected statement that all components of the four-acceleration are constant for a free particle (i.e. $du/d\tau = 0$).

Given Eq. (14.35) we can go ahead and check to see that it gives the expected answer for the motion of a non-relativistic free particle in the rotating frame. We will assume that

$$
\frac{du}{d\tau} = (0, \ddot{x}, \ddot{y}, \ddot{z}), \tag{14.36}
$$

where the dots indicate differentiation with respect to t, which will be fine in the non-relativistic limit. Setting $i = 2$ will then give us the equation of motion in x. We need to evaluate

$$
g_{2j} \frac{du_j}{d\tau} = -\ddot{x}, \tag{14.37}
$$

$$
\frac{\partial g_{2k}}{\partial x_l} u_k u_l = \frac{\partial g_{2l}}{\partial x_k} u_k u_l = \omega \dot{y} \tag{14.38}
$$

and

$$
\frac{\partial g_{kl}}{\partial x_2} u_k u_l = -2\omega^2 x - 2\omega \dot{y}. \tag{14.39}
$$

Hence Eq. (14.35) reduces to

$$
\ddot{x} = 2\omega \dot{y} + \omega^2 x. \tag{14.40}
$$

Similarly with $i = 3$ we obtain

$$
\ddot{y} = -2\omega \dot{x} + \omega^2 y. \tag{14.41}
$$

Eqs. (14.40) and (14.41) are none other than the equations for the Coriolis and centrifugal forces embodied in the equation we derived in Chapter 8, i.e.

$$\ddot{\boldsymbol{x}} = -2\boldsymbol{\omega} \times \dot{\boldsymbol{x}} - \boldsymbol{\omega} \times (\boldsymbol{\omega} \times \boldsymbol{x}). \tag{14.42}$$

Eq. (14.35) has a very interesting interpretation in the mathematics of curved spaces. It is the equation that describes the curved space generalisation of a straight line and it is often referred to as the 'geodesic equation'. The idea of a straight line in a curved space may not be immediately intuitive but the notion is well defined mathematically. One can imagine sliding a tangent vector on the surface along its length. For example, great circles are straight lines on the surface of a sphere. Eq. (14.35) is also our replacement for Newton's First Law. It says that free particles always follow geodesics through space-time and only in inertial frames do these correspond to Euclidean straight lines in space.

This sets the standard: our goal should always be to write all of the laws of physics in a manifestly co-ordinate independent way. It is very important to realise that the space-time about which we have been speaking so far in this book is in all cases Minkowski space-time. Changing co-ordinates to an accelerating frame does not change space-time, it merely makes the geodesic equation more complicated. The mathematics of curved spaces is however also the mathematics of General Relativity: Einstein's theory of gravitation. As we shall shortly discover, in this case the space-time need no longer be Minkowskian.

14.2   A GLIMPSE OF GENERAL RELATIVITY

Newton's Law of Gravitation states that a body of mass m has an acceleration $\boldsymbol{a}$ which is directed towards a body of mass M, i.e.

$$m\boldsymbol{a} = -GMm\frac{\hat{\boldsymbol{r}}}{r^2}. \tag{14.43}$$

At first glance this equation seems fairly unremarkable. However it really is quite astonishing that the mass m on the left hand side is the same as that on the right hand side. It means that all bodies fall with the same acceleration in a gravitational field. This is surprising; what has the mass in Newton's Second Law got to do with the mass appearing in the law of gravitation? There are certainly no other forces in Nature that act upon particles but which induce an acceleration that does not depend upon any intrinsic property of the particle. For example, accelerations in electrodynamics depend upon the ratio q/m. That all bodies fall with the same acceleration in a gravitational field is known as the Equivalence Principle and its consequences are, as we shall very soon see, profound.

Now if at some point in space-time a body experiences a particular acceleration then it is always possible to change co-ordinates so that, at that particular point, the acceleration disappears and, in the infinitesimal neighbourhood of the point, space-time is Minkowskian. If the acceleration is due to gravity then since all bodies experience the same acceleration it follows that we can eliminate the gravitational force at a point by suitably changing co-ordinates.

Generally speaking since the acceleration due to gravity varies over space-time we can't eliminate it everywhere by a single change of co-ordinates but we can eliminate it everywhere if we change co-ordinates differently at different space-time points. This leads to the fascinating possibility that all effects of gravity can be entirely eliminated by a suitable change of co-ordinates. The effect of gravity can therefore be converted entirely into a specification of the geometry of space-time. That the geometry is no longer Minkowskian, but some more general curved space-time follows from the fact that the transformation from a particular co-ordinate basis to a locally inertial (i.e. Minkowski) co-ordinate basis is different at different points in space-time. Put another way, there exists no single co-ordinate transformation that is able to convert the metric tensor into Minkowski form.

Thus gravitation dictates that space-time is locally Minkowskian but globally curved. To illustrate the geometrical ideas involved let us consider the two-dimensional surface of a sphere. We can imagine chopping the surface up into a very large number of small patches. Each patch is approximately flat, with the approximation becoming better the smaller the size of the patch. Physics in the vicinity of any one patch can be described using Euclidean geometry. However, physics that extends over more than one patch is clearly not Euclidean. The curved nature of the sphere is manifest by the fact that it is not possible to represent it by a single Euclidean patch. Free particles will follow straight lines on the surface of the sphere, or more precisely they follow geodesics. Over any patch the path of a free particle is a Euclidean straight line but Eq. (14.35) is needed in order to determine the path of a free particle over a larger portion of the sphere.

The same can be said of gravitation and so Eq. (14.35) tells us how particles move not only in the absence of any external forces but also in the presence of gravity. Conveniently, the geodesic equation is an equation expressed in terms of a single co-ordinate system (i.e. not in terms of one co-ordinate system for each point in space-time). It is the space-time dependence of the metric that tells us how, at any particular space-time point, we can transform co-ordinates so that we are in a locally inertial frame.

To summarize, we have explained how the Equivalence Principle can be used to express the influence of all gravitational fields in terms of the geometry of space-time. Mathematically this information is encoded in the metric tensor g. What we have not yet explained is how one should compute g. Clearly the distribution of matter must play an important role in fixing the space-time geometry. To say more than that takes us beyond the scope of this book but we do hope to have whetted the reader's appetite to study further Einstein's theory of gravitation.

14.2.1 Gravitational fields

As we discussed in the previous section, it is possible to express the invariant distance between a pair of neighbouring events in terms of co-ordinates corresponding to a rigid frame of reference which is accelerating uniformly. The result is given in Eq. (14.29), i.e.

$$(ds)^2 = \left(1 + \frac{gx}{c^2}\right)^2 (c\,dt)^2 - (dx)^2 - (dy)^2 - (dz)^2, \qquad (14.44)$$

where g is the acceleration felt at the origin (recall it is not possible to build a rigidly accelerating frame such that all points within it accelerate at the same rate). Using the Equivalence Principle this must also be the invariant distance in a particular static gravitational field[1]. Armed only with this information we can go ahead and deduce that clocks run faster higher up in a gravitational field. As illustrated in Figure 14.2, let us consider a clock at rest in the gravitational field at a height h above the observer. Then for an observer A adjacent to the clock the space-time interval between two ticks of the clock is given by

$$(\Delta s)^2 = (c\Delta t_A)^2. \tag{14.45}$$

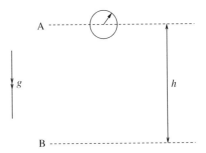

Figure 14.2   A clock in a static gravitational field. The double arrow indicates the direction of the acceleration due to gravity.

Now consider a second observer B for whom the clock is located at $x = h$. For them the same space-time interval is

$$(\Delta s)^2 = (c\Delta t_B)^2 \left(1 + \frac{gh}{c^2}\right)^2, \tag{14.46}$$

where g is the acceleration at B. Equating these two intervals gives

$$\Delta t_A = \left(1 + \frac{gh}{c^2}\right)\Delta t_B \tag{14.47}$$

which means that according to the observer at B the clock runs faster than it does according to the observer at A. One might worry that this is not a very realistic situation because the metric presented in Eq. (14.44) corresponds to a rather artificial field in which the acceleration varies with height according to Eq. (14.28). There is in fact no cause for concern since, provided we assume that $gh/c^2 \ll 1$, it is sufficient to take g as a constant in Eq. (14.47). In any case, we now aim to improve things and describe a truly uniform gravitational field.

---

[1] Although not one in which the field is uniform.

As we have already stated, a uniform field is one in which a particle released anywhere in it feels an acceleration which is the same regardless of when or where the particle was released. The last section revealed that there is no way to arrange this using a single co-ordinate system in Minkowski space. However it can be arranged if we distort space-time accordingly. We start by assuming that the metric yields the following invariant distance:

$$(ds)^2 = f(x)^2(c\,dt)^2 - (dx)^2 - (dy)^2 - (dz)^2. \qquad (14.48)$$

This metric has the virtue that the corresponding co-ordinates are rigid in the sense that the distance between any two points is independent of t. All that we demand is that $f(x)$ be chosen such that any particle released from rest accelerates at the same rate throughout the frame. We can make use of the geodesic equation, Eq. (14.35), to solve the problem for us for it describes the trajectory of a free particle released from rest. The metric is quite simple and, setting $i = 1$ yields the equation

$$-\frac{d^2x}{d\tau^2} - \frac{1}{2}\frac{\partial g_{00}}{\partial x}\left(c\frac{dt}{d\tau}\right)^2 = 0, \qquad (14.49)$$

where $g_{00} = f(x)^2$ and τ is the proper time measured on the particle. Now an analysis identical to that leading up to Eq. (14.24) tells us that

$$\frac{dt}{d\tau} = \frac{1}{f(x)} \qquad (14.50)$$

for a particle at rest (i.e. $dx/d\tau = 0$). Thus the acceleration felt by a particle at rest in a uniform gravitational field is given by

$$\frac{d^2x}{d\tau^2} = -c^2\frac{1}{f}\frac{df}{dx}. \qquad (14.51)$$

We want this to be a constant over the whole space and hence

$$c^2\frac{1}{f}\frac{df}{dx} = g,$$

$$\text{i.e. } f(x) = \exp(gx/c^2) \qquad (14.52)$$

and we have arbitrarily chosen $f(0) = 1$. To recap, we have succeeded in identifying a space-time that is not Minkowskian but which does respresent a uniform gravitational field in which particles released at rest remain equidistant. The invariant distance in this space is

$$(ds)^2 = \exp(2gx/c^2)(c\,dt)^2 - (dx)^2 - (dy)^2 - (dz)^2. \qquad (14.53)$$

The first thing to notice is that this space-time interval is approximately equal to that of the uniformly accelerating frame which we presented in Eq. (14.29) for sufficiently small values of gx/c^2. This is not too surprising on reflection since we would expect that for a sufficiently weak uniform gravitational field there should be an approximation in which the space-time is Minkowski flat.

Example 14.2.1 *The rate at which a clock ticks in the Earth's gravitational field can be approximated by the formula*

$$\Delta t(h) = \left(1 + \frac{\Phi(h)}{c^2}\right) \Delta t(0),$$

where $\Delta t(h)$ is the time interval between two ticks of a clock at height h as determined by an observer at height h and $\Delta t(0)$ is the time interval between the same two events measured by an observer on the ground ($h = 0$). $\Phi(h)$ is the Newtonian gravitational potential, defined such that $\Phi(0) = 0$.

(i) *Show that this expression gives Eq. (14.47) in the case that $h \ll R$ where R is the radius of the Earth.*

(ii) *Now consider a clock on a GPS satellite which orbits the Earth at a speed of 3.9 km/s at an altitude of 20.2×10^3 km. Use the result quoted above to determine by how much the GPS clock speeds up every day compared to an identical clock located on the Earth's surface due to the fact that it is in a weaker gravitational field. Now compute the amount by which the clock slows down as a result of time dilation. Which effect wins? [You may neglect the rotation of the Earth.]*

Solution 14.2.1 (i) *The Newtonian potential at a height h above the Earth's surface is just*

$$\Phi(h) = -\frac{GM}{R+h} + \frac{GM}{R}$$

$$= \frac{GM}{R}\left(1 - \frac{R}{R+h}\right)$$

$$= \frac{GM}{R}\frac{h}{h+R},$$

where M is the mass of the Earth and R is its radius. In the limit $h \ll R$ this reduces to

$$\Phi(h) \approx \frac{GM}{R^2}h,$$

which is equal to gh once we identify $g = GM/R^2$ and hence we have Eq. (14.47).

(ii) *The gravitational speeding of the clock is determined, relative to an observer on the Earth's surface, by the factor*

$$1 + \frac{\Phi(h)}{c^2} = 1 + \frac{gh}{c^2}\frac{R}{h+R}$$

$$= 1 + \frac{9.81 \times 20.2 \times 10^6}{9 \times 10^{16}}\frac{1}{1+20.2/6.4} = 1 + 5.3 \times 10^{-10}$$

and we have put R = 6400 km. Thus the GPS clock gains

$$5.3 \times 10^{-10} \times 24 \times 60^2 = 46 \ \mu s \ per \ day.$$

In contrast, time dilation slows down time on the satellite by a factor

$$\gamma \approx 1 + \frac{1}{2} \left(\frac{3.9 \times 10^3}{3 \times 10^8} \right)^2. \tag{14.54}$$

Thus since $\gamma - 1 \approx 8.5 \times 10^{-11}$ the GPS clock loses about 7 μs every day. The two effects are similar in magnitude with the gravitational effect the larger of the two. The net effect is a 39 μs per day speeding up.

As a final remark, we shall discuss one direct manifestation of the speeding up of time which occurs as one increases altitude in a uniform gravitational field. The time intervals we have been discussing could be the inverse of the frequency of a light wave. Thus Eq. (14.47) becomes

$$\frac{1}{f_A} = \left(1 + \frac{gh}{c^2} \right) \frac{1}{f_B}. \tag{14.55}$$

The upshot is that light emitted from B (which is at the lower altitude) is observed at A to have a lower frequency, i.e. it is red-shifted.

PROBLEMS 14

14.1 The three-force is defined to satisfy

$$\mathbf{f} = \frac{d\mathbf{p}}{dt}.$$

Show that, for the motion of a particle of mass m in one dimension, this equation can be re-written as

$$f = \gamma(u)^3 m \frac{du}{dt}.$$

14.2 A particle of mass m is moving in the laboratory with a speed $u(t)$ and it is subjected to a retarding force of magnitude $\gamma(u)\kappa m$ where $\gamma(u) = (1 - u^2/c^2)^{-1/2}$ and κ is a constant. Given that $u(0) = c/2$ determine the time at which the particle is at rest.

14.3 A particle of mass m moves along the x-axis under an attractive force to the origin of magnitude $mc^2 L/x^2$ where L is constant. Initially it is at rest at $x = L$. Show that its motion is simple harmonic with a period $2\pi L/c$.

14.4 At the CERN Large Electron-Positron Collider (LEP), electrons travelled around a circular particle accelerator of circumference 27 km. Assuming that the electrons had total energy of 45 GeV, determine their proper acceleration as they travel around the accelerator and compare it with non-relativistic expectations.

14.5 In 1959, Pound and Rebka studied 14.4 keV photons (emitted as a result of the radioactive decay of ^{57}Fe) as they travelled the 22.6 m from the roof of their laboratory down to the basement. They observed a frequency in the basement of $(1 + z) f_0$ where $z = (2.57 \pm 0.26) \times 10^{-15}$ and f_0 is the emitted frequency. Confirm that this blue shift is consistent with relativity theory.

14.6 (a) Consider a clock moving (not too fast) at an altitude h in the vicinity of the Earth's surface. If a time $\Delta\tau$ elapses on the clock, convince yourself that

$$\Delta\tau \approx \int \left(1 + \frac{gh}{c^2} - \frac{v(t)^2}{2c^2}\right) dt,$$

where $v(t)$ is the speed of the clock at a time t as measured in an inertial frame at rest relative to the centre of the Earth.

(b) An airplane departs from an airport and travels eastwards above the equator with a constant ground-speed of 1000 km/h and at an altitude of 10 km. After completing one lap of the Earth it returns to the same airport. How does the time registered on a clock on the airplane differ from that registered on a clock in the airport at the end of the journey given that they were initially synchronized? Note that you cannot neglect the rotation of the Earth.

14.7 Two pointlike spacecraft are at rest in an inertial frame S and they are attached by a length of rope which is just taut. Simultaneously in S the spacecraft turn on their identical engines whence they being to accelerate away in the same direction, parallel to the length of rope. What happens to the rope?

Appendix A

Deriving the Geodesic Equation

In this appendix we shall present a derivation of the geodesic equation. We take the approach that a geodesic is the curve through a curved space which corresponds to an extremum of the distance between two points. For example, geodesics on the surface of a sphere are also the curves of shortest length. We shall not prove that these curves correspond to the curves which are generated by sewing together straight lines on locally flat patches, as discussed in the text, but it should not come as too great a surprise that the two are equivalent. The method we shall use is an example of what is called the calculus of variations and it is useful in a number of areas in theoretical physics, perhaps the most notable being the principal of least action from which one can derive the classical equations of motion without resorting to Newton's laws or the notion of force. By eliminating the concept of force, the path to quantum theory is cleared. Here we shall focus only on obtaining the geodesic equation.

Generally the length of a curve between two points A and B is

$$L = \int_A^B ds. \tag{A.1}$$

This is the sum of many line elements each of length ds. Any curve between A and B can be expressed in terms of a set of co-ordinates, i.e. $x_i(s)$ defines a general curve such that

$$(ds)^2 = g_{ij}\, dx_i\, dx_j. \tag{A.2}$$

Consider a general curve between A and B. We can obtain a second curve from this by varying the line elements at each and every point along its path. In this

Dynamics and Relativity   Jeffrey R. Forshaw and A. Gavin Smith
© 2009 John Wiley & Sons, Ltd

way we can express the variation in the length L of a curve:

$$\delta L = \int_A^B \delta(\mathrm{d}s). \tag{A.3}$$

The variation in the element $\mathrm{d}s$ is given by

$$\delta((\mathrm{d}s)^2) = \mathrm{d}x_i \mathrm{d}x_j \delta g_{ij} + 2g_{ij}\mathrm{d}x_i \delta(\mathrm{d}x_j)$$

$$\delta(\mathrm{d}s) = \frac{1}{2}\dot{x}_i \mathrm{d}x_j \delta g_{ij} + g_{ij}\dot{x}_i \mathrm{d}(\delta x_j) \tag{A.4}$$

and we use the dot notation to indicate differentiation with respect to s. Writing $\delta g_{ij} = \delta x_k(\partial g_{ij}/\partial x_k) = \delta x_k g_{ij,k}$ (this comma notation is a common shorthand for differentiation with respect to x) gives

$$\delta(\mathrm{d}s) = \left(\frac{1}{2}\dot{x}_i\dot{x}_j\delta x_k g_{ij,k} + g_{ij}\dot{x}_i\frac{\mathrm{d}(\delta x_j)}{\mathrm{d}s}\right)\mathrm{d}s. \tag{A.5}$$

Under the integration we can re-arrange the second term on the RHS of this equation using integration by parts, i.e.

$$\int_A^B g_{ij}\dot{x}_i\frac{\mathrm{d}(\delta x_j)}{\mathrm{d}s}\mathrm{d}s = \left[\delta x_j g_{ij}\dot{x}_i\right]_A^B - \int_A^B \frac{\mathrm{d}(g_{ij}\dot{x}_i)}{\mathrm{d}s}\delta x_j\mathrm{d}s. \tag{A.6}$$

The first term on the RHS vanishes since we require the variation to vanish at the end points (i.e. the curve must pass through these two points and hence $\delta x_i = 0$ there). Thus the variation in the length is

$$\delta L = \int_A^B \left(\frac{1}{2}\dot{x}_i\dot{x}_j g_{ij,k} - \frac{\mathrm{d}(g_{ik}\dot{x}_i)}{\mathrm{d}s}\right)\delta x_k\mathrm{d}s. \tag{A.7}$$

Now since each element δx_i can be varied independently it follows that the term in parenthesis must vanish identically for the extremal path, i.e.

$$\frac{1}{2}\dot{x}_i\dot{x}_j g_{ij,k} - \frac{\mathrm{d}(g_{ik}\dot{x}_i)}{\mathrm{d}s} = 0. \tag{A.8}$$

We are almost done now. Differentiating the product gives

$$\frac{1}{2}\dot{x}_i\dot{x}_j g_{ij,k} - g_{ik,l}\dot{x}_l\dot{x}_i - g_{ik}\ddot{x}_i = 0 \tag{A.9}$$

which can be re-arranged (utilizing the symmetry of the metric to give the final result a more symmetric appearance and re-naming some of the indices) to give the final answer:

$$g_{ij}\ddot{x}_j + \frac{1}{2}\left(g_{ij,k} + g_{ik,j} - g_{jk,i}\right)\dot{x}_j\dot{x}_k = 0. \tag{A.10}$$

Appendix B

Solutions to Problems

PROBLEMS 1

1.1 (a) $\mathbf{c} = 4\mathbf{i} - \mathbf{k}$; $\mathbf{d} = -2\mathbf{i} + 2\mathbf{j} - 3\mathbf{k}$.

 (b) $|\mathbf{a}| = \sqrt{1^2 + 1^2 + 2^2} = \sqrt{6}$; $|\mathbf{b}| = \sqrt{11}$; $|\mathbf{c}| = \sqrt{17}$.

 (c) $\hat{\mathbf{a}} = \frac{\mathbf{a}}{|\mathbf{a}|} = \frac{1}{\sqrt{6}}\mathbf{i} + \frac{1}{\sqrt{6}}\mathbf{j} - \sqrt{\frac{2}{3}}\mathbf{k}$.

1.2 Choose $x-$axis East and $y-$axis North.

 For the first leg: $\mathbf{l}_1 = 100(\sin 45°, \cos 45°)$ m $\approx (70.7, 70.7)$ m and similar calculations for the second and third legs $\mathbf{l}_2$ and $\mathbf{l}_3$.

 The sum of all three legs is $\mathbf{l} = \mathbf{l}_1 + \mathbf{l}_2 + \mathbf{l}_3 \approx (95.7, -59.2)$ m. Then you should obtain $|\mathbf{l}| = 112$ m and the direction to North $\theta = 122°$.

1.3 Construct displacement vectors $\mathbf{r}_{AB} = \mathbf{r}_B - \mathbf{r}_A$ etc. The sum of these gives the null vector.

1.4 Consider an observer in galaxy G. Using Hubble's Law for galaxy G and galaxy i and subtracting we get

$$\mathbf{v}_i - \mathbf{v}_G = H_0(\mathbf{r}_i - \mathbf{r}_G).$$

 This is the velocity of galaxy i as seen from galaxy G. Since the right-hand-side is just the displacement of i from G we see that the observer in G also "discovers" Hubble's Law.

1.5 $|\mathbf{A}| = \sqrt{3}$; $|\mathbf{B}| = 2\sqrt{3}$; $|\mathbf{C}| = \sqrt{26}$; $|\mathbf{D}| = \sqrt{3}$.

 $\mathbf{A} \cdot \mathbf{B} = -2$; $\mathbf{A} \cdot \mathbf{C} = 0$; $\mathbf{A} \cdot \mathbf{D} = 1$; $\mathbf{B} \cdot \mathbf{D} = -6$.

 $\cos\theta_{AB} = \frac{\mathbf{A} \cdot \mathbf{B}}{|\mathbf{A}||\mathbf{B}|} = -\frac{1}{3}$ so $\theta_{AB} = 1.91$ rad; $\theta_{AC} = \frac{\pi}{2}$ rad; $\theta_{AD} = 1.23$ rad; $\theta_{BD} = \pi$ rad;

 $\mathbf{A} \times \mathbf{B} = 4\mathbf{j} - 4\mathbf{k}$; $\mathbf{A} \times \mathbf{D} = -2\mathbf{j} + 2\mathbf{k}$; $\mathbf{B} \times \mathbf{D} = \mathbf{0}$.

1.6 2×10^{13} ms^{-2}.

Dynamics and Relativity Jeffrey R. Forshaw and A. Gavin Smith
© 2009 John Wiley & Sons, Ltd

1.7 The height of the building is 216 m. The flight time is 6.63 s.

1.8 The velocity is obtained by differentiation, which is straightforward since the basis vectors are constant in time:

$$\mathbf{v} = \frac{d\mathbf{s}}{dt} = 0.3\,\mathbf{i} + 0.5\,\mathbf{j} - 0.01t\,\mathbf{k} \ \ \text{km s}^{-1}.$$

With $t = 0$, $|\mathbf{v}| = \sqrt{0.3^2 + 0.5^2} = 0.58$ km s^{-1}. The velocity at this time lies in the xy plane. At $t = 30$ s, $|\mathbf{v}| = \sqrt{0.3^2 + 0.5^2 + 0.3^2} = 0.66$ km s^{-1}. To work out the angle to the vertical use

$$\cos\theta = \frac{v_z}{|\mathbf{v}|}$$

which will give $\theta = 117°$.

1.9 Running with the travelator, velocities add so $v = 11$ ms^{-1}, against $v = 9$ ms^{-1} so the total time for both legs is

$$t = 50\left(\frac{1}{11} + \frac{1}{9}\right) = 10.1 \text{ s}.$$

Since time is inversly proportional to speed the contribution of the motion of the travelator does not cancel over the two legs in the way that the athelete expected it would.

1.10 To cross the river she must have a net velocity $\mathbf{v}'$ that lies on the line joining the two stations. She must therefore sail with a velocity $\mathbf{v}$ relative to the river, such that

$$\mathbf{v}' = \mathbf{v} + \mathbf{u}$$

where $\mathbf{u}$ is the velocity of the river. Since the three vectors form a right-angled triangle the time taken to cross is

$$t = \frac{d}{\sqrt{v^2 - u^2}}.$$

There is no real solution if $u > v$ making the crossing impossible in this case.

PROBLEMS 2

2.1 (a) The equation for the position vector $\mathbf{r}$ for a general point on the line through $\mathbf{r}_1$ and $\mathbf{r}_2$ can be written

$$\mathbf{r} - \mathbf{r}_1 = \lambda(\mathbf{r}_2 - \mathbf{r}_1)$$

where λ is real. When $0 \le \lambda \le 1$ $\mathbf{r}$ lies between $\mathbf{r}_1$ and $\mathbf{r}_2$. From the definition of the centre-of-mass of two particles you can obtain

$$\mathbf{R} - \mathbf{r}_1 = \frac{m_1}{m_1 + m_2}(\mathbf{r}_2 - \mathbf{r}_1)$$

which proves that the centre-of-mass lies on a line joining the two masses.

(b) The position of the centre-of-mass is given by:

$$\mathbf{R} = \frac{1}{3}(\mathbf{a} + \mathbf{b} + \mathbf{c}).$$

Now the median from $\mathbf{a}$ that intersects the side between $\mathbf{b}$ and $\mathbf{c}$ has the vector equation

$$\mathbf{r} = \mathbf{a} + \lambda\left(\frac{1}{2}(\mathbf{b} + \mathbf{c}) - \mathbf{a}\right) = \mathbf{a} + \frac{\lambda}{2}(\mathbf{b} + \mathbf{c} - 2\mathbf{a}).$$

We can write

$$\mathbf{R} = \mathbf{a} + \frac{1}{3}(\mathbf{b} + \mathbf{c} - 2\mathbf{a})$$

showing that $\mathbf{R}$ lies on this median. Repeat for the other two medians.

2.2 The total mass of the system is $80\,g$ so

$$\mathbf{R} = \frac{1}{80}(130\mathbf{i} + 250\mathbf{j}) = \left(\frac{13}{8}\mathbf{i} + \frac{25}{8}\mathbf{j}\right)\ \text{m}.$$

It doesn't matter to which particle the force is applied, the resulting acceleration of the centre-of-mass is the same

$$\ddot{\mathbf{R}} = \frac{3}{0.08}\,\mathbf{i} = 37.5\,\mathbf{i}\ \text{ms}^2.$$

2.3 Calculate the rate of change of momentum to obtain $\mathbf{F} = 20$ N.

2.4 The friction between the prisoner's hands and the rope acts as a brake providing an upwards force on the prisoner. For the rope to remain in static equilibrium the tension must balance the frictional force. The maximum tension is $T_{\max} = 600$ N. So the acceleration a is obtained from

$$ma = mg - T_{\max}.$$

Once you have the acceleration it is a simple problem to find the speed at impact:

$$v = \sqrt{2h\left(g - \frac{T_{\max}}{m}\right)} = 35\,\text{km hr}^{-1} = 9.6\text{m s}^{-1}.$$

2.5 Draw a free-body diagram and resolve components of the forces parallel and perpendicular to the slope. The force applied by the Egyptians parallel to the slope is

$$F = ma + mg\sin 20° = 18.3\ \text{kN}.$$

The normal force must balance the component of the block's weight normal to the surface, i.e.

$$N = mg \cos 20° = 46.0 \text{ kN}.$$

2.6 Newton's Second Law, and the expression for centripetal acceleration, gives the normal force on the skier to be

$$N = m \left(g - \frac{v^2}{r} \right) = 214 \text{ N}.$$

This is directed upwards, but must be equal in magnitude, and opposite in direction to the force that the skier exerts on the snow, which is 214 N downwards.

2.7 The acceleration of both blocks is given by

$$a = \frac{F}{m_1 + m_2}.$$

The second block accelerates due to the normal reaction force between the blocks. This has the value

$$F = \frac{m_2 F}{m_1 + m_2} = \frac{3}{4} \text{ N}.$$

2.8 (a) The only horizontal force on the upper block (m_1) is friction. This has a maximum value given by $\mu_s m_1 g$ resulting in an acceleration $\mu_s g$. In this case both blocks have the same acceleration and the force on the whole system is 17.7 N.

(b) Both blocks have the same acceleration, $a = 1.47 \text{ ms}^{-2}$. Since the upper block accelerates due to the force of friction we calculate the frictional force to be $2 \text{ kg} \times 1.47 \text{ ms}^{-2} = 2.94 \text{ N}$.

(c) In this case the top blocks slips and we have kinetic friction that causes the top block to accelerate. Thus the acceleration of the top block is $\mu_k g = 1.96 \text{ ms}^{-2}$. The bottom block accelerates due to the resultant of the applied force and the opposing friction due to the top block. Its acceleration is therefore 7.87 ms^{-2}.

2.9 Obtain an equation for the net force **F** on the system as a whole by considering the change in momentum of the "rocket plus exhaust" system. If a mass Δm is ejected you should find that the change in momentum is $\Delta \mathbf{p} = m\Delta \mathbf{v} + (\Delta m)\mathbf{u}$, from which the rocket equation follows after dividing by Δt and taking the limit that Δt goes to zero. Note that there is a sign change because $\Delta m \approx -\frac{dm}{dt} dt$, i.e. the mass of the rocket is decreasing with time. For $\mathbf{F} = \mathbf{0}$ we can write

$$\int_{t_i}^{t_f} \frac{d\mathbf{v}}{dt} dt = \mathbf{u} \int_{t_i}^{t_f} \frac{1}{m} \frac{dm}{dt} dt = -\mathbf{u} \ln \frac{m_i}{m_f}.$$

In a gravitational field the force is no longer zero, instead $\mathbf{F} = m\mathbf{g}$. If the field is uniform this simply introduces a constant $\mathbf{g}$ into the above integration. Rockets burn their fuel quickly so as to minimise t and hence achieve the largest possible final speed.

PROBLEMS 3

3.1 Calculate the change in kinetic energy to be 5×10^5 J. This is equal to the work done. Divide by the time for which the brakes are applied to obtain the average power and you should get 167 kW.

3.2 $\mathbf{F}.\Delta\mathbf{r} = (3\mathbf{i} - 2\mathbf{j}) \cdot (-5\mathbf{i} - 1\mathbf{j}) \times 10^{-2}$ J $= -0.13$ J.

3.3 (a) The centre of mass is at position

$$\frac{mx_1 + mx_2}{2m} = \frac{x_1 + x_2}{2},$$

i.e. midway between the two masses.

(b) The force on the car at x_1 is $-k(x_1 - x_2 - l)$. The force on the car at x_2 is $k(x_1 - x_2 - l)$.

(c) We have

$$ma_1 = -k(x_1 - x_2 - l), \text{ and}$$

$$ma_2 = k(x_2 - x_2 - l).$$

Adding these shows that $m(a_1 + a_2) = 0$. The acceleration of the centre of mass is $(a_1 + a_2)/2$, which is therefore zero. Hence the centre of mass moves with constant velocity as must be the case for a system with no net external force.

(d) Use the state of the system at $t = 0$ to obtain the centre of mass velocity:

$$V_C = \frac{mv_0}{m + m} = \frac{v_0}{2}.$$

(e) Subtract the two equations of motion to show that

$$m(a_1 - a_2) = -2k(x_1 - x_2 - l) = 2ku.$$

Then note that
$$\frac{du}{dt} = a_1 - a_2$$

to get the required equation.

(f) Direct substitution of the trial solution into the differential equation for u proves that $\omega = \sqrt{\frac{2k}{m}}$. We first show that $du/dt = A\omega \cos(\omega t)$. Then at $t = 0$ we get $A\omega = 0 - v_0 = -v_0$.

(g) The motion is a combination of simple harmonic motion and linear motion of the centre of mass at constant speed. It is possible to show from the above solutions that there are times when one of the cars is instantaneously at rest in the lab frame. To an observer in the lab one car moves forward, stops then appears to pull the other car after it.

(h) The kinetic energy due to the simple harmonic motion is $\frac{1}{2}kA^2$, i.e. the potential energy in the spring at maximum extension. This is added to the energy due to motion of the centre of mass: $\frac{1}{2}(m+m)v_0^2/4$ to obtain the mechanical energy of the system. Substitution for A and k using the results from (f) proves that the total mechanical energy is $mv_0^2/2$ as it must be since this is a conservative system.

3.4 The momentum of the rope at any instant is $\lambda y v$. Differentiating with respect to time gives the rate of change of the momentum of the rope to be λv^2. By Newton's Second Law, this must be equal to the sum of the gravitational and applied forces acting on the rope. The gravitational force is $g\lambda y$ from which we deduce that the applied force is $\lambda v^2 + g\lambda y$. We can integrate this to obtain the work done in raising the rope: $\lambda v^2 y + g\lambda y^2/2$. The product of the applied force and the speed gives the power used to raise the rope: $\lambda v^3 + g\lambda y v$. If we compute the rate of change of mechanical energy, directly from the potential energy $(g\lambda y^2/2)$ and the kinetic energy $(\lambda y v^2/2)$ we obtain $\lambda v^3/2 + g\lambda y v$. This is not equal to the power. The discrepancy comes from not including the energy that must be lost through friction in the coil of rope, in order that the coil does not rotate.

3.5 Calculate the reduced mass directly to obtain $(10/3)$ kg. The centre of mass velocity is $(11/3)\mathbf{j}\,\text{ms}^{-1}$. The momentum of particle with mass m_1, relative to the centre of mass, is

$$\mu(\mathbf{v}_1 - \mathbf{v}_2) = (10\mathbf{i} - \frac{10}{3}\mathbf{j})\,\text{kg ms}^{-1}.$$

The momentum of the other particle relative to the centre of mass has equal magnitude but opposite direction to the first, i.e. $-\mu \mathbf{v}_r$.

3.6 Use $F = -dU/dr$ to find the point at which $F = 0$. This gives the solution $r = r_0$, which when substituted into the expression for the potential gives the depth to be $-U(r_0) = \varepsilon$. The range of r is obtained upon realising that the total energy is equal to $K_{\max} - \varepsilon$ and that r must be such that the kinetic energy is always positive, i.e. $U(r) < -\varepsilon + K_{\max}$.

3.7 The proton has initial velocity v and recoils at a speed of $2v/3$. Conservation of momentum gives

$$\frac{5}{3}m_p v = m_u v_u,$$

where m_p the mass of the proton, m_u is the mass of the unknown particle and v_u is its final velocity. Conservation of kinetic energy gives

$$\frac{5}{9}m_p v^2 = m_u v_u^2,$$

from which we get $m_u = 5m_p$.

PROBLEMS 4

4.1 The moment of inertia of the turntable about the rotation axis is $I = \frac{1}{2}MR^2$. Assuming constant angular acceleration we have

$$\tau = \frac{I\omega_0}{\Delta t},$$

where $\omega_0 = 3.49\,\text{s}^{-1}$ is the operating angular speed of the turntable and Δt is the time taken to accelerate from rest. This gives $\tau = 1.48 \times 10^{-2}\,\text{Nm}$. The added mass changes the moment of inertia to

$$I' = I + mr^2,$$

where the mass m is dropped at radius r. Conservation of angular momentum gives the new angular speed

$$\omega' = \frac{I}{I'}\omega_0 = 3.41\,\text{s}^{-1}.$$

4.2 The normal forces on the feet are N_1 and N_2 and the frictional forces are F_1 and F_2. Friction provides centripital acceleration and the normal forces must sum to be equal but opposite to the weight. Thus: $N_1 + N_2 = mg$ and $F_1 + F_2 = \frac{mv^2}{r}$. The centre of mass is accelerating but there is no rotation of the body about it, so sum of all torques must be zero, i.e.

$$\frac{hmv^2}{ar} = N_1 - N_2$$

and, using the weight equation, $N_1 = \frac{1}{2}\left(mg + \frac{mv^2h}{ar}\right)$ and $N_2 = \frac{1}{2}\left(mg - \frac{mv^2h}{ar}\right)$.

4.3 Solve this problem using angular momentum conservation. The ring has mass m radius R so that the moment of inertia about the pivot is $2mR^2$ using the Parallel Axis Theorem. When the bug is opposite the axis, we can write its speed relative to the lab as $v_b = v - 2R\omega$ so that angular momentum conservation gives:

$$0 = 2Rm_b(v - 2R\omega) - 2mR^2\omega,$$

from which

$$\omega = \frac{m_b v}{R(m + 2m_b)}.$$

4.4 The disc has moment of inertia $\frac{1}{2}mr^2$ and rolls without slipping with speed v so that $\omega = v/r$. Hence, the total kinetic energy is

$$K = \frac{1}{2}mv^2 + \frac{1}{2}I\omega^2 = \frac{1}{2}mv^2 + \frac{1}{4}mv^2 = \frac{3}{4}mv^2.$$

4.5 (a) Initially there is no rotation about the centre of mass, so the pivot and the support provide equal normal forces N vertically, which must balance the weight so $mg = 2N$.

   (b) When the support is removed the torque about the pivot due to the weight is $\tau = mgl/2$. Since the moment of inertia about an axis through the pivot is $I = \frac{1}{3}ml^2$ we can write the linear acceleration of the centre of mass $a = mgl^2/(4I) = 3g/4$. The acceleration of the centre of mass is related to the normal force and the weight through Newton's Second Law, so we obtain $N = mg/4$.

   (c) Equating the increase in kinetic energy to the decrease in gravitational potential energy gives us $\frac{1}{2}I\omega^2 = \frac{l}{2}mg\sin\theta$ so $\omega = \sqrt{\frac{3g\sin\theta}{l}}$.

4.6 The forces acting on the disc are friction (F) and gravity (mg). Taking components parallel to the slope we write Newton's Second Law as:

$$mA = mg\sin\theta - F.$$

The torque about the centre of mass of the disc comes only from friction so:

$$I\alpha = \frac{1}{2}mb^2\alpha = Fb.$$

Together with $A = b\alpha$, the above equations give

$$A = \frac{2}{3}g\sin\theta.$$

The total kinetic energy of a rolling disc is $\frac{3}{4}mv^2$, which we can equate to the change in gravitational potential energy after falling through a height $x\sin\theta$, where x is the distance travelled down the slope. Thus,

$$v = \sqrt{\frac{4}{3}gx\sin\theta}.$$

This is exactly the result obtained using the linear acceleration, i.e. $v = \sqrt{2Ax}$.

4.7 The moment of inertia of a solid sphere, calculated from a sum of thin circular discs, is obtained from the integral

$$I = \int_{-R}^{R} \frac{1}{2}\left(R^2 - z^2\right) dm = \frac{\pi\rho}{2}\int_{-R}^{R}\left(R^2 - z^2\right)^2 dz,$$

where $\rho = 3m/4\pi R^3$ is the density of the sphere. Integrating gives

$$I = \frac{2}{5}mR^2.$$

For the cricket ball:

$$\omega = \frac{\tau\Delta t}{I} = \frac{5FR\Delta t}{2mR^2} = 416\,\text{s}^{-1}.$$

4.8 It will help if you draw a diagram. The speed of the ball before the collision with the cushion is v_i, after it is v_f. The collision causes a change in linear momentum

$$\Delta p = m(v_i + v_f),$$

and an angular impulse $\Delta p(h - R)$. If the ball is not to skid or hop, then we must have rolling without slipping both before and after the collision, so

$$\Delta p(h - R) = \frac{I}{R}(v_i + v_f).$$

Now substitute for Δp and $I = \frac{2}{5}mR^2$ and solve to obtain $h = 7R/5$.

PROBLEMS 5

5.1 Need to figure out the speed relative to the ground. In (a) this is $2\,\mathrm{ms}^{-1}$ whilst in (b) it is $0.6\,\mathrm{ms}^{-1}$. Hence the time taken is (a) $20.0/2.0 = 10.0$ s and (b) $20.0/0.6 = 33.3$ s.

5.2 We can prove this by considering the light to move along the x-axis according to $x_1 = ct$ whilst the observer in S' moves according to $x_2 = X_0 - ct$ where X_0 is just the position at $t = 0$. The distance between the two is just $x_2 - x_1 = X_0 - vt - ct$ and the relative speed is the rate of change of this distance, i.e.

$$\frac{\mathrm{d}(x_2 - x_1)}{\mathrm{d}t} = -(v + c).$$

The sign just tells us the distance is decreasing. Notice that no particles of matter are travelling at this speed (in S): it is the relative speed of two different things.

PROBLEMS 6

6.1 $\Delta t = \gamma \Delta t_0$ with $\Delta t_0 = 30$ mins. Since $\gamma = 2.72$ the elapsed time is 81.2 mins.

6.2 $\gamma = 3.20$ and hence in the lab frame the half-life is extended to 5.76×10^{-8} s. Distance travelled is thus 16.4 m.

6.3 $\gamma = 190/12$. Solve for $v = 0.998c$.

6.4 If the spacecraft was not sufficiently high we would need to account for the fact that it moves a significant distance over the time the searchlight is turned on. At ground level, we need to add on the time taken for light to travel the extra distance, i.e. total time is $(0.190 + 0.190 \times 0.998)\mathrm{s} = 0.380$ s. Note that the key word in the previous question is 'see'. It implies an observation using your eyes and hence the need to make this correction. Usually we speak of the intervals between events defined using a network of clocks stationary in some frame: those intervals do not depend upon where the observer is nor whether they have eyes or not.

6.5 $\gamma = 3.20$ and hence length is 31 cm.

6.6 $\gamma = 4.11$ and hence distance is 2.9 km.

6.7 In a spaceship, the 10^5 light years becomes length contracted. Equivalently, a journey time of order 10^5 Earth years can be reduced by time dilation. Suppose we want the journey to take just 20 years. Then we require that the speed relative to Earth, u, should satisfy

$$20u = 10^5 c\sqrt{1 - u^2/c^2}.$$

Solving gives $u = 0.99999998c$. Note you can avoid solving a quadratic by realising that γ is very large and hence u is very close to c so it is a good approximation to solve $\gamma = 10^5/20$ for u giving $u/c \approx 1 - 2 \times 10^{-8}$.

6.8 (a) $\gamma = 1.009$ hence length is 3.57 km. (b) $3600/(4.00 \times 10^7) = 9.00 \times 10^{-5}$ s. (c) $3570/(4.00 \times 10^7) = 8.92 \times 10^{-5}$ s.

6.9 (a) Length measured by friend is $10/\gamma = 6$ m. (b) Since pole is 2 m short of the barn length the time delay is $2\,\text{m}/(0.8c) = 8.33 \times 10^{-9}$ s. (c) Now the barn is contracted to a length of 4.8 m whilst the pole remains at 10 m. (d) From the athlete's viewpoint, the rear of the pole cannot know that the front has struck the wall until at least $10\,\text{m}/c = 3.33 \times 10^{-8}$ s after it has done so (since no signal can travel faster than the speed of light). In this time interval, the barn door can travel "for free" a distance of $10\,\text{m}/c \times 0.8$ $c = 8$ m which is plenty long enough for the pole to fit within the barn. Thus the apparent paradox is resolved.

6.10 Use

$$\lambda = \left(\frac{1 - v/c}{1 + v/c}\right)^{1/2} \lambda_0$$

with $\lambda_0 = 589$ nm and $\lambda = 550$ nm to give $v/c = 0.0684$.

6.11 $t_B - t_A = 1.3$ μs and $x_B - x_A = 1.5$ km. We are also told that $t'_A - t'_B = 0$, where the primes indicate times measured in the observer's rest frame. Need to identify the relevant Lorentz transformation formula and the most useful one is the one involving the given quantities, i.e. we use "$t' = \gamma(t - vx/c^2)$". Subtracting the equation pertaining to event A from that pertaining to event B gives

$$0 = \gamma\left(1.3 \ \mu s - \frac{v}{c^2} \ 1500 \ \text{m}\right)$$

and hence $v = 0.26c$.

6.12 Given $x_B - x_A = 0$, $t_B - t_A = 4$ s and $t'_B - t'_A = 5$ s we are asked to deduce $x'_B - x'_A$. The relevant Lorentz transformation formula states that $\Delta t' = \gamma$ $\left(\Delta t - v\Delta x/c^2\right)$ where $\Delta x = x_B - x_A$ etc. Putting the numbers in gives $\gamma = 5/4$ hence $v/c = 0.6$. Note that you could get this directly from the time dilation formula since the events occur at the same place in one of the frames. The spatial separation is 5 s $\times 0.6c = 9.0 \times 10^8$ m.

6.13 $t_B - t_A = 0$, $x_B - x_A = 1$ km and $x'_B - x'_A = 2$ km we are asked to deduce $t'_B - t'_A$. The relevant Lorentz transformation formula states that $\Delta x' = \gamma$

$(\Delta x - c\Delta t)$ where $\Delta x = x_B - x_A$ etc. Putting the numbers in gives $\gamma = 2$ and hence $v/c = 0.866$. Note that you could get this directly from the length contraction formula since the events occur at the same time in one of the frames. We can make use of "$t' = \gamma(t - vx/c^2)$" to determine the corresponding time interval, i.e.

$$\Delta t' = 2 \left(0 - 0.866 \times \frac{1000}{3 \times 10^8} \right) = -5.77 \times 10^{-6}\text{s}.$$

The minus sign means that the event at larger x is observed first in the primed frame.

6.14 The emission of the two pulses corresponds to two events with $\Delta x = 4$ km and $\Delta t = 5$ μs. Given $\Delta t' = 0$ we need to use $\Delta t' = \gamma(\Delta t - v\Delta x/c^2)$ to determine the speed, i.e. $v/c = c\Delta t/\Delta x = 0.375$.

6.15 Key here is to identify the relevant events. Let event A be the impact of the comet and event B be the event in the party that is simultaneous with event A according to an observer whizzing past the Earth. Then $\Delta t' = 0$ by definition (primes indicate the arbitrary inertial frame) and we want to know the corresponding time interval in the Earth frame (Δt). Since we know $\Delta t'$ and Δx the relevant equation is $\Delta t' = \gamma(\Delta t - v\Delta x/c^2)$. The extreme values of Δt occur when $v = \pm c$ whence $\Delta t = \pm \Delta x/c$. Thus the party must last a time $2 \times 8 \times 10^{11}/(3 \times 10^8)$ s $= 89$ minutes. The impact of the comet occurs (in the Earth frame) at the midpoint of the party whilst the party ends with the students observing the impact using their telescopes.

6.16 A diagram will help to clarify the way in which you must use the velocity addition formula. The speed is $(0.70 + 0.85)c/(1 + 0.7 \times 0.85) = 0.97c$.

6.17 As in the last question, a diagram will help. Required speed is $(0.5 - 0.8)c/(1 - 0.5 \times 0.8) = -0.5c$. The minus sign implies the rocket moves towards the Earth.

6.18 Velocity addition gives $V = (v + c/n)/(1 + v/(nc))$. Note if $n = 1$ then $V = c$. Fizeau would have worked with $v \ll nc$ hence the denominator can be simplified using $(1 + v/(nc))^{-1} \approx 1 - v/(nc)$. This yields the result after neglecting terms suppressed in v/c.

6.19 Work in A's rest frame and determine the velocity of B in that frame, then use $\tan 30° = v_{By}/v_{Bx}$ to determine u. In more detail: use velocity addition to obtain $v_{Bx} = u$ and $v_{By} = u/\gamma(u)$ and hence $\gamma(u) = \sqrt{3}$ which implies $u/c = \sqrt{2/3}$.

PROBLEMS 7

7.1 Use $E = 1.673 \times 10^{-27} \times (2.9979 \times 10^8)^2/(1.6022 \times 10^{-13})$ MeV $= 938$ MeV $= mc^2$. Hence $m = 938$ MeV$/c^2$.

7.2 (a) $\gamma - 1 = 1$ hence $v = 0.866c$. (b) $\gamma - 1 = 5$ hence $v = 0.986c$.

7.3 Power output $= c^2 dm/dt$. Hence rate of mass loss is 4.2 million tonnes per second.

7.4 Classically use $\frac{1}{2}mv^2 = 0.1$ MeV hence $v/c = 0.63$. Relativistically we need $(\gamma - 1)mc^2 = 0.1$ MeV which gives $v/c = 0.55$.

7.5 Mass is given by $\sqrt{11.2^2 - 6^2} = 9.5$ GeV/c^2. The speed can be found from $v/c = cp/E = 6/11.2 = 0.54$.

7.6 Total energy is $\sqrt{65^2 + 80^2} = 103$ GeV. Kinetic energy is 23 GeV. The total energy is not dominated by the rest mass energy and so the particle is relativistic.

7.7 (a) The energy liberated in each fusion event is $(2 \times 2.0136u - 4.0015u)c^2 = 3.84 \times 10^{-12}$ J. 1 kg of deuterium can therefore produce $1/(2 \times 2.0136u) \times 3.84 \times 10^{-12} = 5.74 \times 10^{14}$ J. (b) 17 tonnes.

7.8 (a) $(\gamma - 1)mc^2 = 2$ MeV gives $v/c = 0.943$. (b) 3 MeV. (c) Total momentum before is $\gamma mv = 2.83$ MeV/c. (d) It is not 3 MeV/c^2! Instead use $E^2 = c^2p^2 + m^2c^4$ where $E = 3$ MeV $+ 2$ MeV is the total energy, i.e. $mc^2 = \sqrt{5^2 - 2.83^2} = 4.12$ MeV. (e) Total kinetic energy is $(5 - 4.12) = 0.88$ MeV.

7.9 (a) Need to use energy and momentum conservation. If p_1 and p_2 are the photon momenta then we have $\gamma mc^2 + mc^2 = cp_1 + cp_2$ and $\gamma mv = p_1 - p_2$. Can solve these for p_1 and p_2 once we know the speed of the antiproton, v. Given $(\gamma - 1)mc^2 = 0.667$ GeV we deduce that $\gamma = 1.711$ and $v = 0.811c$. Thus $p_1 = 1.92$ GeV/c and $p_2 = 0.62$ GeV/c. (b) The 0.62 GeV photon travels in the opposite direction to the incoming antiproton. (c) By symmetry, the momenta are of the same magnitude as before but the signs are reversed.

7.10 Conservation of energy: $(\gamma(u_1) + \gamma(u_2))m = \gamma(u)M$. Conservation of momentum gives two equations: $\gamma(u_2)mu_2 = \gamma(u)Mu \sin\alpha$ and $\gamma(u_1)mu_1 = \gamma(u)Mu \cos\alpha$ where α determines the direction of the outgoing particle of speed u. The two momentum equations imply that $(\gamma(u_1)^2u_1^2 + \gamma(u_2)^2u_2^2)m^2 = \gamma(u)^2M^2u^2$. We would like to use this last equation together with the energy conservation equation to obtain an expression for the mass M which depends only upon the γ factors. To do that we need to use the fact that $(u/c)^2 = (\gamma^2 - 1)/\gamma^2$ (and similarly for u_1 and u_2). Hence $(\gamma(u_1)^2 + \gamma(u_2)^2 - 2)m^2 = (\gamma(u)^2 - 1)M^2$. Subtracting this from the square of the energy conservation equation gives the desired result.

PROBLEMS 8

8.1 The length of $\boldsymbol{\omega} \times \mathbf{r}$ is the perpendicular distance from the axis which lies parallel to $\boldsymbol{\omega}$ and runs through the origin.

8.2 In its rest frame, a small mass m of water experiences a horizontal force of $m\omega^2r$ and a downward force of mg. The net force is an angle α below the horizontal, where $\tan\alpha = g/(\omega^2r)$. If $h(r)$ is the water height a distance r from the rotation axis then $dh/dr = \tan(\pi/2 - \alpha) = 1/\tan\alpha$. Integrating gives $h = (\omega r)^2/(2g) + h(0)$ which is a parabola.

8.3 Coriolis force induces an easterly displacement x such that $\ddot{x} \approx 2\omega g t \cos\lambda$. Integrate twice to obtain $x = \omega g(t^3/3)\cos\lambda$. Time to hit ground is $\approx (2h/g)^{1/2}$ and so the deflection is $\approx \omega g \cos\lambda(2h/g)^{3/2}/3$. Put numbers in to find a deflection of 2.4 cm.

8.4 Work in a frame that rotates with the hoop. First obtain the equation of motion for the bead, i.e. consider the torque about the centre of the hoop. Coriolis force is cancelled by reaction from hoop (fixed ω) so net torque is just $-mgR\sin\theta + m\omega^2 R^2 \sin\theta\cos\theta = mR^2\ddot\theta$ where $\theta = 0$ corresponds to the bead at the bottom of the hoop. The bottom of the hoop corresponds to $\ddot\theta = 0$ but to be a stable equilibrium we require $\ddot\theta < 0$ for small $(\theta > 0)$ displacements. Thus $\Omega^2 R^2 = gR$ hence $\Omega = \sqrt{g/R}$. If $\omega > \Omega$ the bead will rise until $\ddot\theta = 0$, i.e. $g/R = \omega^2\cos\theta$ and so $\cos\theta = \Omega^2/\omega^2$.

PROBLEMS 9

9.1 Consider an element of the shell of mass $dM = \frac{M}{4\pi}\sin\theta\,d\theta\,d\phi$. After integrating over azimuth, the potential at a point a distance x from the centre of the shell is

$$\Phi = -\frac{GM}{2}\int_0^\pi \frac{\sin\theta}{y}d\theta,$$

where $y^2 = R^2 + x^2 - 2xR\cos\theta$. Now change variables to obtain an integral over y subject to $R - x < y < R + x$. After integration $\Phi = -GM/R$, which is constant.

9.2 When the body is at a distance x from the centre of the Earth, it feels a force due only to the mass at smaller radii (the previous question proves that there is no force from the mass at larger radii). Hence $\ddot x = -G(x/R)^3 M/x^2$ which corresponds to simple harmonic motion with angular frequency equal to $(GM/R^3)^{1/2}$.

9.3 Consider building up the sphere by adding successive shells brought in from infinity. The total work done in adding a shell of thickness dr to a pre-existing sphere of radius r and mass Mr^3/R^3 is $-GM(r^3/R^3)\,dm/r$ where $dm = M(4\pi r^2 dr)/(4\pi R^3/3)$ is the mass of the shell. Integrating over $0 < r < R$ gives the result. Equating to $\frac{1}{2}I\omega^2$ with $I = \frac{2}{5}MR^2$ gives $\omega = (3GM/R^3)^{1/2}$.

9.4 (i) $L/m = v_0 R = 4.46 \times 10^{15}$ m²s⁻¹ since velocity is tangential at perihelion. (ii) Kinetic energy divided by the mass is $v_0^2/2 = 1.78 \times 10^9$ J kg⁻¹ and the gravitational potential energy divided by the mass is $-GM_\odot/R = -1.78 \times 10^9$ J kg⁻¹. These are equal within errors and so we cannot tell if the orbit is bound or unbound. (iii) Total energy is zero, i.e.

$$0 = \frac{1}{2}v_r^2 + \left(\frac{L}{m}\right)^2\frac{1}{2r^2} - \frac{GM_\odot}{r},$$

where v_r is the radial component of the velocity when the comet is a distance r from the Sun. Putting the numbers in gives $v_r = 29.8$ km s⁻¹. To get the speed we need also the tangential component which is, by the conservation of angular momentum, $(7.48/15)v_0 = 29.7$ km s⁻¹. Speed is therefore 42.1 km s⁻¹.

9.5 First part is obtained by equating the gravitational attraction to the centripetal force. After the dust cloud has been driven away, the total energy of the planet is

$$\frac{1}{2}\frac{G(M + \frac{4}{3}\pi\rho r^3)m}{r} - \frac{GMm}{r}$$

and the orbit is bound if this is negative, hence the result. Determine semi-major axis by equating the total energy to $-GMm/(2a)$. The answer can be re-arranged to read $a = r(1 - M_{\text{dust}}/M)^{-1}$ where $M_{\text{dust}} = \frac{4}{3}\pi\rho r^3$. At distance r the velocity is tangential and so the planet is at pericentre ($r < a$ so cannot be apocentre). Thus $a(1 - \varepsilon) = r$ and so $\varepsilon = M_{\text{dust}}/M$.

9.6 Use $u^2 = (2a/r - 1)GM/a$ to obtain the speed immediately after the firing of the rockets (apocentre of the Eagle's orbit). We have that $r = 1850$ km but need a. Solve for a using $a(1 - \varepsilon) = 1755$ km and $a(1 + \varepsilon) = 1850$ km. Hence speed is 1.61 km s^{-1} which implies a slowing down of around 20 m s^{-1}.

9.7 The difference is due to the fact that the Sun is offset from the centre of the Earth's elliptical orbit by a distance εR. The summertime orbit therefore sweeps out an area $\approx 4\varepsilon R^2$ more than the wintertime orbit. Since the area swept out over the whole year is $\approx \pi R^2$ and since area is swept out at a constant rate this translates into a time difference of $4\varepsilon/\pi$ years ≈ 8 days. This compares well with the 7 days difference between the period 21 March to 22 September and the period 23 September to 20 March.

PROBLEMS 10

10.1 The hoop receives an impulse $\Delta\mathbf{p}$ on the rim which implies an angular impulse $\Delta\mathbf{L} = \mathbf{r} \times \Delta\mathbf{p}$ about the centre of the hoop. The angular momentum before impact is

$$\mathbf{L} = \mathbf{I}\boldsymbol{\omega} = mr^2\omega\hat{\mathbf{n}},$$

where $\hat{\mathbf{n}}$ is horizontal. To only deflect the hoop $\Delta\mathbf{L}$ must also lie in the horizontal plane, i.e. $\mathbf{r}$ must point up and the blow should be applied to the highest point of the hoop. A blow anywhere else will induce a wobble. The change in direction θ is

$$\theta \approx \tan\theta = \frac{r\Delta p}{mr^2\omega} = \frac{\Delta p}{mv},$$

where we have used $v = r\omega$.

10.2 Compute the torque about the centre of mass because the coin accelerates (otherwise we would have to include the effect of fictitious forces). The torque is a result of the normal force and friction at the base of the coin. The normal force is equal in magnitude to the weight and friction provides the centripetal acceleration of the centre of mass. Therefore, the torque has magnitude

$$mgr \sin \phi - \frac{mv^2 r}{R} \cos \phi$$

and it causes the angular momentum to precess about the vertical. Thus the component of $\mathbf{L}$ in the horizontal plane, L_h, rotates with angular speed $\Omega = v/R$. Using

$$L_h = \frac{1}{2} mr^2 \frac{v \cos \phi}{r},$$

gives

$$\left| \frac{d\mathbf{L}}{dt} \right| = L_h \Omega = \frac{mrv^2 \cos \phi}{2R}.$$

Equating this to the torque gives the required result.

10.3 Similar to the example with the pencil but with different moments of inertia. Remember that the sphere is pivoted on the finger. For the solid sphere $I_3 = \frac{2}{5}mr^2$ and $I = \frac{2}{5}mr^2 + mr^2 = \frac{7}{5}mr^2$ (by the Parallel Axis Theorem). This will give a minimum spin of 48 rad s^{-1}. For the spherical shell we have $I_3 = \frac{2}{3}mr^2$ and $I = \frac{5}{3}mr^2$ giving a minimum spin of 31 rad s^{-1}.

10.4 With $x_3 = 0$ for the whole plate we calculate:

$$I_{11} = \frac{M}{2a^2} \int_0^a \int_0^{2a} x_2^2 \, dx_1 \, dx_2 = \frac{1}{3} Ma^2,$$

$$I_{22} = \frac{M}{2a^2} \int_0^a \int_0^{2a} x_1^2 \, dx_1 \, dx_2 = \frac{4}{3} Ma^2,$$

$$I_{33} = \frac{M}{2a^2} \int_0^a \int_0^{2a} (x_1^2 + x_2^2) \, dx_1 \, dx_2 = \frac{5}{3} Ma^2,$$

$$I_{12} = \frac{M}{2a^2} \int_0^a \int_0^{2a} (-x_1 x_2) \, dx_1 \, dx_2 = -\frac{1}{2} Ma^2.$$

So

$$\mathbf{I} = \frac{Ma^2}{6} \begin{pmatrix} 2 & -3 & 0 \\ -3 & 8 & 0 \\ 0 & 0 & 10 \end{pmatrix}.$$

The angular momentum, $\mathbf{L} = \mathbf{I}\boldsymbol{\omega} = \frac{Ma^2 \omega}{6\sqrt{2}}(-1, 5, 0)$. We have constant $\boldsymbol{\omega}$ so we can compute the torque: $\boldsymbol{\tau} = \boldsymbol{\omega} \times \mathbf{L} = \frac{Ma^2 \omega^2}{2}(0, 0, 1)$. The kinetic energy is

$$T = \frac{1}{2} \boldsymbol{\omega} \cdot \mathbf{L} = \frac{1}{6} Ma^2 \omega^2.$$

10.5 The moment of inertia tensor is already diagonal in x_3 so we have the first eigenvector $\boldsymbol{\gamma} = \mathbf{e}_3$ with eigenvalue $I_\gamma = \frac{5}{3}Ma^2$. The other two eigenvectors satisfy

$$\begin{pmatrix} 2-I & -3 \\ -3 & 8-I \end{pmatrix} \begin{pmatrix} a_1 \\ a_2 \end{pmatrix} = 0.$$

Hence

$$\boldsymbol{\alpha} = (1, \sqrt{2}-1, 0) \quad I_\alpha = \frac{Ma^2}{6}(5-3\sqrt{2})$$

and

$$\boldsymbol{\beta} = (1, -\sqrt{2}-1, 0) \quad I_\beta = \frac{Ma^2}{6}(5+3\sqrt{2}).$$

10.6 Use cylindrical co-ordinates (r, θ, z) with the z-axis as the symmetry axis. Then any pair of axes x and y perpendicular to the z axis will serve as principal axes. Calculate

$$I_z = \rho \int_{-h/2}^{h/2} \int_0^{2\pi} \int_0^R r^3 \mathrm{d}r\, \mathrm{d}\theta\, \mathrm{d}z = \frac{\pi \rho R^4 h}{2}$$

and

$$I_x = I_y = \rho \int_{-h/2}^{h/2} \int_0^{2\pi} \int_0^R (r^2 \sin^2\theta + z^2)r\, \mathrm{d}r\, \mathrm{d}\theta\, \mathrm{d}z = \frac{\pi \rho R^4 h}{4}\left(1+\frac{h^2}{3R^2}\right).$$

The barrel of the gun ensures that the angle between the symmetry axis and the angular momentum is small. From the geometry, $I_3/I = 6/19$ so the symmetry axis precesses about 15.8 times per second.

10.7 As with Feynman's plate we can take the $\mathbf{e}_3$ principal axis to be normal to the plate, and $I_1 = I_2 = I$, so that $I_3 = 2I$. Euler's equations then give us

$$I\dot{\omega}_1 + I\omega_2\omega_3 = -k\omega_1,$$
$$I\dot{\omega}_2 + I\omega_3\omega_1 = -k\omega_2,$$
$$\dot{\omega}_3 = 0.$$

As with the free top, ω_3 is constant. Multiply the second equation by i and add to the first equation, then write in terms of η to obtain:

$$I\dot{\eta} - iI\omega_3\eta + k\eta = 0,$$

which has a general solution:

$$\eta = A\exp[-kt/I + i(\omega_3 t + \phi)].$$

Thus $\eta\eta^* = \omega_1^2 + \omega_2^2 = A^2 \exp(-2kt/I)$ decays exponentially. In the lab any initial wobble dies away and the disc then spins only about the symmetry axis.

10.8 To show this use

$$T = \frac{1}{2}\boldsymbol{\omega}\cdot\mathbf{L} = \frac{1}{2}\boldsymbol{\omega}\cdot\mathbf{I}\boldsymbol{\omega}$$

and differentiate the product with respect to time in the principal-axis, body-fixed frame where $\mathbf{I}$ is constant and diagonal and hence in which

$$\dot{\boldsymbol{\omega}}\cdot\mathbf{I}\boldsymbol{\omega} = \boldsymbol{\omega}\cdot\mathbf{I}\dot{\boldsymbol{\omega}} = \boldsymbol{\omega}\cdot\boldsymbol{\tau}.$$

10.9 Write the angular momentum relative to the origin as

$$\mathbf{L} = \sum_i m_i(\mathbf{r}_i + \mathbf{R}) \times (\dot{\mathbf{r}}_i + \dot{\mathbf{R}}_i),$$

expand the brackets and use the definition of $\mathbf{R}$

$$\mathbf{R} = \frac{\sum_i m_i \mathbf{r}_i}{\sum_i m_i}.$$

10.10 (a) Use cylindrical polar co-ordinates (r, θ, z) for the integrals with z along the symmetry axis:

$$I_z = \rho \int_0^h dz \int_0^{zR/h} dr \int_0^{2\pi} d\theta\, r^3,$$

and the density of the cone $\rho = \frac{3m}{\pi R^2 h}$. For the other two principal moments of inertia evaluate

$$I_x = I_y = \rho \int_0^h dz \int_0^{zR/h} dr \int_0^{2\pi} d\theta\, (r^3\cos^2\theta + rz^2).$$

(b) We shall compute the components of $\boldsymbol{\omega}$ in the direction of the principal axes and then use

$$T = \frac{1}{2}I_x(\omega_x^2 + \omega_y^2) + \frac{1}{2}I_z\omega_z^2.$$

There are two sources of rotation contributing to $\boldsymbol{\omega}$: the precessional rotation of the cone about its apex (which points vertically) and the spin of the cone about its symmetry axis. If $\alpha = \tan^{-1}(R/h)$ is the half-angle of the cone, then the precession gives rise to a contribution to ω_z of $\omega\sin\alpha$ and $\omega_x^2 + \omega_y^2 = \omega^2\cos^2\alpha$. The spin contributes only to ω_z and is equal to $-\omega/\sin\alpha$. To see this note that the base of the cone must travel a distance $l = 2\pi R/\sin\alpha$ for each complete precession of the cone, i.e. the cone must spin about its symmetry axis at a rate of $\omega l/(2\pi R)$. Hence $\omega_z = \omega\sin\alpha - \omega/\sin\alpha$. Express $\sin\alpha$ and $\cos\alpha$ in terms of R and h to get the result.

PROBLEMS 11

11.1 We can write $E = mc^2 \cosh \eta$ and $cp/E = \tanh \eta$, i.e. $cp = mc^2 \sinh \eta$. Thus

$$\frac{E + cp}{E - cp} = \exp(2\eta)$$

and the result follows.

11.2

$$\mathbf{G}(u) = \begin{pmatrix} 1 & 0 \\ -u & 1 \end{pmatrix}.$$

Matrix multiplication gives $\mathbf{G}(u)\mathbf{G}(v) = \mathbf{G}(u + v)$. Two Lorentz transformations lead to the quoted result upon using the identities: $\cosh \eta_1 \cosh \eta_2 + \sinh \eta_1 \sinh \eta_2 = \cosh(\eta_1 + \eta_2)$ and $\cosh \eta_1 \sinh \eta_2 + \sinh \eta_1 \cosh \eta_2 = \sinh(\eta_1 + \eta_2)$.

11.3 The left hand side can be written as $\varepsilon_{ijk} R_{j\alpha} V_\alpha R_{k\beta} W_\beta$ which can be further written as $\varepsilon_{abc} R_{ia} R_{jb} R_{kc} R_{j\alpha} R_{k\beta} V_\alpha W_\beta$. Now $\mathbf{R}$ is an orthogonal matrix which means that $\mathbf{R}^{\mathrm{T}} = \mathbf{R}^{-1}$ and so $R_{ij} = (R^{-1})_{ji}$. We can therefore write $R_{jb} R_{j\alpha} = \delta_{b\alpha}$ and $R_{kc} R_{k\beta} = \delta_{c\beta}$ which allows us to simplify the left hand side to $\varepsilon_{a\alpha\beta} R_{ia} V_\alpha W_\beta$ and this is equal to the right hand side.

PROBLEMS 12

12.1 We write the initial four-momentum of the rocket as $m_i(c, 0)$, the final four-momentum as $\gamma m_f(c, u)$ and the total four-momentum of the photons is $\gamma m_f(u, -u)$. The latter holds by virtue of the conservation of momentum and the fact that each photon is massless. Equating the energy components gives $m_i c = \gamma m_f(c + u)$, as required. Solving for u gives $u/c = (m_i^2 - m_f^2)/(m_i^2 + m_f^2)$.

12.2 (a) In the zero momentum frame, the electron and positron must carry the same energy, E_e. Similarly the proton and anti-proton carry the same energy, E_p. Energy conservation dictates that $E_e = E_p$. The minimum of this energy occurs when the proton and anti-proton are produced at rest, i.e. the minimum kinetic energy is $m_p c^2 - m_e c^2 = 937.8$ MeV. (b) In the stationary electron frame we exploit the fact that $(\mathbf{P}_{e^+} + \mathbf{P}_{e^-})^2 = (\mathbf{P}_p + \mathbf{P}_{\bar{p}})^2$ is Lorentz invariant. Thus we can compute it in the zero momentum frame where, at threshold, the proton and anti-proton are produced at rest, i.e. $(\mathbf{P}_{e^+} + \mathbf{P}_{e^-})^2 = 4m_p^2 c^2$. We now work out the same quantity in the electron rest frame where $\mathbf{P}_{e^+} = (E/c, \mathbf{p})$ and $\mathbf{P}_{e^-} = (m_e c, \mathbf{0})$, i.e. $(E/c + m_e c)^2 - p^2 = 4m_p^2 c^2$. This can be re-arranged to give $E = (2m_p^2 - m_e^2)c^2/m_e = 3.45$ TeV.

12.3 Invariant mass satisfies $M^2 c^4 = \left(\sum_i E_i\right)^2 - \left(\sum_i c\mathbf{p}_i\right)^2$. We can compute it in a frame of reference of our choice and, as so often, the zero momentum frame is convenient. In that case $Mc^2 = \sum_i E_i^{CM}$ and this is minimized when the particles are at rest, i.e. $M = \sum_i m_i$. Note that this proof assumes the existence of the zero momentum frame, i.e. it assumes that the particles are not all massless and moving in the same direction. In that case $E_i = c|\mathbf{p}_i|$ and since all the momenta are parallel $M \geq 0$ so the result still holds.

12.4 It is not possible simultaneously to conserve both energy and momentum. A quick way to see it is to note that the momentum four-vector of the photon has zero length whilst that of the electron-positron pair cannot be less than $2mc$ (see previous question) where mc^2 is the mass of the electron/positron.

12.5 Substitute $y = \gamma(y' + Vt')$, $x = x'$, $t = \gamma(t' + Vy'/c^2)$ and re-arrange to obtain the answer provided $k'_x = k$, $k'_y = -\omega\gamma V/c^2 = -uV\gamma k/c^2$ and $\omega' = \gamma\omega$. For **K** to be a four-vector it must transform accordingly, i.e. we require that $k'_x = k_x$, $k'_y = \gamma(k_y - V\omega/c^2)$ and $\omega' = \gamma(\omega - Vk_y/c^2)$. These equations are satisfied since $k_x = k$ and $k_y = 0$. Speed of propagation in S' is

$$u' = \frac{\omega'}{k'} = \frac{\gamma u}{(1 + \gamma^2 u^2 V^2/c^4)^{1/2}}.$$

For $u = c$ we obtain $u' = c$ and for $u \to 0$ we obtain $u' \to u$.

12.6 Conservation of four-momentum states that $\mathbf{P}_\pi = \mathbf{P}_\mu + \mathbf{P}_\nu$. This can be written as $\mathbf{P}_\nu = \mathbf{P}_\pi - \mathbf{P}_\mu$ and squaring both sides gives $0 = \mathbf{P}_\pi^2 + \mathbf{P}_\mu^2 - 2\mathbf{P}_\pi \cdot \mathbf{P}_\mu$. This leads to the result since $\mathbf{P}_\pi^2 = m_\pi^2 c^2$ and $\mathbf{P}_\mu^2 = m_\mu^2 c^2$. Since the muon travels at right angles to the pion we can write $\mathbf{P}_\pi \cdot \mathbf{P}_\mu = E_\pi E_\mu/c^2$. Hence $E_\mu = (m_\pi^2 + m_\mu^2)c^4/(2E_\pi)$. Substituting for $E_\pi = 143.6$ MeV gives a muon kinetic energy equal to 1.4 MeV. The neutrino emerges at an angle $\theta = \arctan(p_\mu/p_\pi)$. Substituting for $p_\mu = 17.0$ MeV/c gives $\theta = 29.6°$.

12.7 Use $p'_x = \gamma(p_x - vE/c^2)$ and $p'_y = p_y$ with $cp_x = E\cos\theta$ and $cp_y = E\sin\theta$. Thus

$$\tan\theta' = \frac{p'_y}{p'_x} = \frac{\sin\theta}{\gamma(\cos\theta - v/c)}$$

which gives $\theta = 68°$. The frequency is obtained from $E' = \gamma(E - vp_x)$ and since $E = hf$, $f' = \gamma f(1 - (v/c)\cos\theta)$, i.e. $f' = 7.4 \times 10^{14}$ Hz.

12.8 (a) In the phi rest frame, momentum conservation dictates that the kaons must be emitted in opposite directions with equal energies. Energy conservation fixes the value, i.e. $E_K = m_\phi c^2/2 = 510$ MeV. The momenta are obtained using $(cp_K)^2 = E_K^2 - (m_K c^2)^2$ and are ± 127 MeV/c. (b) **v** must point in the opposite direction to $\hat{\mathbf{n}}$. Its magnitude can be determined using $E_\phi^2 = c^2 p_\phi^2 + m_\phi^2 c^4 = (\gamma m_\phi c^2)^2$ which can be solved to give $|\mathbf{v}| = 0.947c$. (c) Use $p_K^{\text{lab}} = \gamma(p_K + E_K/c)$ for each kaon in turn and with $\gamma = 3.107$ to obtain momenta of $+1106$ MeV/c and $+1895$ MeV/c.

12.9 At threshold, the particles are produced at rest in the zero momentum frame, i.e. the invariant mass is $(\mathbf{P}_\pi + \mathbf{P}_n)^2 = (m_K + m_\Lambda)^2 c^2$. Since this is invariant we can also compute it in the neutron's rest frame where $\mathbf{P}_\pi + \mathbf{P}_n = (E_\pi/c + m_n c, \mathbf{p}_\pi)$, i.e. $(E_\pi/c + m_n c)^2 - p_\pi^2 = (m_K + m_\Lambda)^2 c^2$. Substituting for $p_\pi^2 = E_\pi^2/c^2 - m_\pi^2 c^2$ gives $E_\pi = ((m_K + m_\Lambda)^2 c^2 - (m_\pi^2 + m_n^2)c^2)/(2m_n) = 897$ MeV.

12.10 In the kaon rest frame the pions are produced back-to-back. Suppose the pion that moves in the lab frame has a speed u in the kaon rest frame. Then it follows that the kaon rest frame must also move with a speed u relative to the lab (in the same direction as the pion) in order that the other pion should be at rest in the lab frame. Drawing a diagram might help you see this. Thus the speed of the moving pion in the lab frame is $2u/(1 + u^2/c^2)$ and u can be deduced using energy conservation in the kaon rest frame, i.e. $E_\pi = m_K c^2/2$ so $\gamma = m_K/(2m_\pi)$ and $u = 0.982c$ which corresponds to an energy of 746 MeV. Alternatively you could determine the energy and momentum in the kaon rest frame and then make a Lorentz transformation to the lab.

12.11 Isotropic decay means that $dP/d\Omega' = 1/(4\pi)$ where $d\Omega' = d(\cos\theta')d\phi'$ is an element of solid angle in the pion rest frame. The normalization is fixed since integration over all solid angles must give a unit probability to find the photon. We need to understand the transformation from (θ', ϕ') to (θ, ϕ). Consider a photon with four-momentum equal to $(E, E\cos\theta, E\sin\theta\cos\phi, E\sin\theta\sin\phi)$. The Lorentz transformation equations tell us that

$$E' = \gamma(E - \beta E\cos\theta),$$

$$E'\cos\theta' = \gamma(E\cos\theta - \beta E),$$

$$E'\sin\theta'\cos\phi' = E\sin\theta\cos\phi,$$

$$E'\sin\theta'\sin\phi' = E\sin\theta\sin\phi,$$

where $\beta = u/c$. The last two equations tell us $\phi = \phi'$ whilst the first two can be used to eliminate the energies, i.e. dividing them gives

$$\cos\theta' = \frac{\cos\theta - \beta}{1 - \beta\cos\theta}.$$

Thus $d(\cos\theta') = d(\cos\theta) \times (1 - \beta^2)/(1 - \beta\cos\theta)^2$ and the result follows.

PROBLEMS 14

14.1 Write $p = \gamma(u)mu$ and differentiate.

14.2 Equation of motion is $\gamma^3 m\dot{u} = -\gamma\kappa m$, i.e. the required time t satisfies

$$\int_{\frac{c}{2}}^{0} \frac{du}{1 - u^2/c^2} = -\kappa t$$

which can be integrated to give $t = (c\ln 3)/(2\kappa)$.

14.3 Equation of motion is $\gamma^3 m\dot{u} = -mc^2 L/x^2$ which is solved by $x = A\cos(\omega t + \phi)$ provided $A\omega/c = 1$ (in which case $\gamma = A/x$). The angular frequency is $\omega^2 = c^2 L/A^3$ and the required solution is obtained upon setting $A = L$.

14.4 Eq. (14.6) relates the proper acceleration to the magnitude of the acceleration four-vector, i.e. $\mathbf{A} \cdot \mathbf{A} = -a^2$. Working in CERN's rest frame, the position four-vector of an orbiting electron is $\mathbf{X} = (ct, R\cos\omega t, R\sin\omega t, 0)$ where $2\pi R = 27$ km. Differentiating twice with respect to the proper time and using $dt/d\tau = \gamma$ gives an acceleration four-vector equal to $\mathbf{A} = -R\omega^2\gamma^2(0, \cos\omega t, \sin\omega t, 0)$. Thus the proper acceleration has a magnitude of $\gamma^2\omega^2 R$ (as one might have anticipated on the grounds of time dilation). Finally we need $\omega = v/R$ where v is the electron's speed. Can deduce v from $\gamma = E/(mc^2) = 88 \times 10^3$, i.e. $v = c$ to a good approximation and $a = 1.6 \times 10^{23}$ ms^{-2}.

14.5 Expect $z = gh/c^2 = 2.46 \times 10^{-15}$.

14.6 (a) $\Delta\tau$ is a proper time interval, i.e. $c\Delta\tau$ is the invariant spacetime distance between the events A and B where A is specified by providing the position of the clock and its time initially and B is specified by providing the position of the clock and its time at the end of the interval. In the vicinity of the Earth's surface we might therefore venture to make use of Eq. (14.53) for a uniform gravitational field, i.e.

$$\Delta\tau \approx \int \left((1 + \frac{gh}{c^2})^2(dt)^2 - \frac{1}{c^2}(d\mathbf{x})^2\right)^{1/2}$$

$$\approx \int \left(1 + \frac{2gh}{c^2} - \frac{1}{c^2}\left(\frac{d\mathbf{x}}{dt}\right)^2\right)^{1/2} dt,$$

where h is the height above the Earth's surface (it is in general a function of t). It is to be understood that the time t refers to the time in an inertial frame that is approximately at rest relative to the centre of the Earth (approximately since we only need to be able to neglect length contraction effects in the specification of h), e.g. when $g = 0$ we regain the Minkowski interval expressed in inertial co-ordinates. The above equation leads directly to the quoted result upon expanding the square root. (b) Let v be the speed of a clock relative to the ground, and let V be the speed of a point fixed on the Earth's surface relative to the inertial frame from part (a), i.e. $V = 2\pi R_{\text{earth}}/(1 \text{ day})$. Now work out the proper time elapsed on (i) the clock at rest in the airplane; (ii) the clock at rest in the airport. For the clock on the airplane we have

$$\Delta\tau_{\text{plane}} \approx \Delta t\left(1 + \frac{gh}{c^2} - \frac{(V+v)^2}{2c^2}\right)$$

(ignoring the variation of h as the airplane ascends and descends) and for the clock on the ground

$$\Delta\tau_{\text{ground}} \approx \Delta t\left(1 - \frac{V^2}{2c^2}\right).$$

Strictly speaking we should account for the fact that the clock on the plane must actually travel slightly faster than $V + v$ (by an amount

$\delta v = 2\pi h/(1 \text{ day}))$ but that leads to a negligible correction. We can eliminate the dependence upon the time measured in the inertial frame by constructing the ratio:

$$\frac{\Delta \tau_{\text{plane}}}{\Delta \tau_{\text{ground}}} \approx 1 + \frac{gh}{c^2} - \frac{1}{2}\frac{(V+v)^2}{c^2} + \frac{1}{2}\frac{V^2}{c^2} \approx 1 + \frac{gh}{c^2} - \frac{1}{2}\frac{v^2}{c^2} - \frac{Vv}{c^2}.$$

Thus the clock on the airplane speeds up by a fractional amount gh/c^2 due to the fact it is in a weaker gravitational field but it slows down in part due to time dilation (by a factor $v^2/(2c^2)$) and in part due to the fact that the Earth is rotating (the Vv/c^2 factor). Putting in the numbers: $gh/c^2 \approx 1.1 \times 10^{-12}$ which corresponds to a speeding up of 3.9 ns/h; $-v^2/(2c^2) \approx -4.3 \times 10^{-13}$ which corresponds to a slowing down of -1.5 ns/h; $-Vv/c^2 \approx -1.4 \times 10^{-12}$ which corresponds to a slowing down of -5.2 ns/h. The total journey time for one circuit of the equator is just over 40 hours and thus the net effect is that the clock on the airplane slows down by $0.11\mu s$ which is an effect large enough to have been measured.

14.7 The tension in the rope would increase until it snaps. This is easiest to see in the inertial frame in which the rockets are initially at rest. If the rockets are initially a distance L_0 apart then they must remain a distance L_0 apart in this frame since they undergo identical accelerations (they are identical rockets). However the rope should suffer a length contraction (it is of length L_0 only in its rest frame) but it is prevented from so doing since it is attached to the rockets. Hence it snaps. Alternatively, we can looking at things from the point of view of an observer on the rocket at the rear. Eq.(14.28) is useful for it informs us that in order to keep the distance between the rockets fixed, the rocket at the front must experience a reduced proper acceleration, but since both rockets are identical this is not the case (observers on each rocket experience the same acceleration) and so the rocket at the front has too great an acceleration and the distance between the rockets increases, again eventually causing the rope to snap.

Index

References to figures are given in italic type; references to tables are given in bold type.

Dynamics and Relativity   Jeffrey R. Forshaw and A. Gavin Smith
© 2009 John Wiley & Sons, Ltd